Trading Fish, Saving Fish

Numerous international legal regimes now seek to address the global depletion of fish stocks, and increasingly their activities overlap. The relevant laws were developed at different times by different groups of states. They are motivated by divergent economic approaches, influenced by disparate non-state actors, and implemented by separate institutions such as the World Trade Organization and the United Nations Food and Agriculture Organization. Margaret A. Young shows how these and other factors affect the interaction between regimes. Her empirical and doctrinal analysis moves beyond the discussion of conflicting norms that has dominated the fragmentation debate. Case studies include the negotiation of new rules on fisheries subsidies, the restriction of trade in endangered marine species and the adjudication of fisheries import bans. She explores how regimes should interact, in fisheries governance and beyond, to offer insights into the practice and legitimacy of regime interaction in international law.

MARGARET A. YOUNG is Senior Lecturer at Melbourne Law School, Australia, and has previous professional experience at the World Trade Organization and the United Nations. She was the inaugural Research Fellow of Public International Law at the Lauterpacht Centre for International Law and Pembroke College, Cambridge.

CAMBRIDGE STUDIES IN INTERNATIONAL AND COMPARATIVE LAW

Established in 1946, this series produces high quality scholarship in the fields of public and private international law and comparative law. Although these are distinct legal sub-disciplines, developments since 1946 confirm their interrelation.

Comparative law is increasingly used as a tool in the making of law at national, regional and international levels. Private international law is now often affected by international conventions, and the issues faced by classical conflicts rules are frequently dealt with by substantive harmonisation of law under international auspices. Mixed international arbitrations, especially those involving state economic activity, raise mixed questions of public and private international law, while in many fields (such as the protection of human rights and democratic standards, investment guarantees and international criminal law) international and national systems interact. National constitutional arrangements relating to 'foreign affairs', and to the implementation of international norms, are a focus of attention.

The Board welcomes works of a theoretical or interdisciplinary character, and those focusing on the new approaches to international or comparative law or conflicts of law. Studies of particular institutions or problems are equally welcome, as are translations of the best work published in other languages.

General Editors	James Crawford SC FBA *Whewell Professor of International Law, Faculty of Law, University of Cambridge*
	John S. Bell FBA *Professor of Law, Faculty of Law, University of Cambridge*
Editorial Board	Professor Hilary Charlesworth *Australian National University*
	Professor Lori Damrosch *Columbia University Law School*
	Professor John Dugard *Universiteit Leiden*
	Professor Mary-Ann Glendon *Harvard Law School*
	Professor Christopher Greenwood *London School of Economics*
	Professor David Johnston *University of Edinburgh*
	Professor Hein Kötz *Max-Planck-Institut, Hamburg*
	Professor Donald McRae *University of Ottawa*
	Professor Onuma Yasuaki *University of Tokyo*
	Professor Reinhard Zimmermann *Universität Regensburg*
Advisory Committee	Professor D. W. Bowett QC
	Judge Rosalyn Higgins QC
	Professor J. A. Jolowicz QC
	Professor Sir Elihu Lauterpacht CBE QC
	Judge Stephen Schwebel

A list of books in the series can be found at the end of this volume.

Trading Fish, Saving Fish: The Interaction Between Regimes in International Law

Margaret A. Young

CAMBRIDGE UNIVERSITY PRESS
Cambridge, New York, Melbourne, Madrid, Cape Town,
Singapore, São Paulo, Delhi, Tokyo, Mexico City

Cambridge University Press
The Edinburgh Building, Cambridge CB2 8RU, UK

Published in the United States of America by Cambridge University Press,
New York

www.cambridge.org
Information on this title: www.cambridge.org/9780521765725

© Margaret Young 2011

This publication is in copyright. Subject to statutory exception
and to the provisions of relevant collective licensing agreements,
no reproduction of any part may take place without the written
permission of Cambridge University Press.

First published 2011
Reprinted 2012

Printed at Print on Demand, World Wide, UK

A catalogue record for this publication is available from the British Library

Library of Congress Cataloguing in Publication data
Young, Margaret A., 1975–
Trading fish, saving fish : the interaction between regimes in international
law / Margaret A. Young.
   p. cm. – (Cambridge studies in international and comparative law)
ISBN 978-0-521-76572-5
1. Fishery law and legislation.   2. Fishery management,
International.   3. International law.   I. Title.
K3895.Y68   2011
343′.07692–dc22
                                                             2010039403

ISBN 978-0-521-76572-5 Hardback

Cambridge University Press has no responsibility for the persistence or
accuracy of URLs for external or third-party internet websites referred to
in this book, and does not guarantee that any content on such
websites is, or will remain, accurate or appropriate.

Contents

Foreword	*page* xi
Acknowledgements	xiv
Table of cases	xvii
Table of Conventions, Declarations and procedures	xxiii
List of abbreviations	xxxi

Part I: Trading Fish, Saving Fish   1

Chapter 1   Introduction   3
- A   Trade law and fisheries sustainability   5
- B   Fragmentation of international law   8
- C   Scope, methodology and use of terms   16
  - 1   A case-study approach   16
  - 2   The vocabulary of regimes   19
  - 3   The legal framework for regime interaction   22
  - 4   The stakeholders   24
- D   Outline of the book   29

Chapter 2   Relevant laws and institutions: an overview   32
- A   The law of the sea   33
  - 1   UNCLOS and the EEZ regime   34
  - 2   UNCLOS's high seas, fish stocks and regional regimes   38
  - 3   FAO fisheries management regime   46
  - 4   Interdependencies with other regimes   50
  - 5   Further links: the UN Consultative Process   53
- B   International environmental law   56
  - 1   Convention on the International Trade in Endangered Species   56
  - 2   Convention on Migratory Species   61

		3	Convention on Biological Diversity	63
		4	Interdependencies with other regimes	65
		5	Further links: UNEP	67
	C	International trade law		68
		1	The WTO Agreements	70
		2	Interdependencies with other regimes	75
		3	Further links: the General Council and the CTE	79
	D	Conclusions		81

Part II: Selected Case Studies   83

Chapter 3	The negotiation of WTO rules on fisheries subsidies		85
	A	Subsidies and overfishing	87
	B	The forum shop: regimes of relevance	89
		1 The WTO's SCM Agreement	91
		2 The Doha negotiations	94
		3 Opposition to the role of the WTO	96
		4 Responses	99
	C	Inter-regime learning	105
		1 WTO members as conduits	105
		2 Participation and observership by others	108
		3 Other forums including UNEP	111
	D	Entrenching interaction	113
		1 Notifications	114
		2 Classifying subsidies using fisheries standards	115
		3 Conditionality through benchmarking and peer review	119
	E	Settling disputes	124
	F	Conclusions	129
Chapter 4	The restriction of trade in endangered marine species		134
	A	Endangered marine species	135
	B	The forum shop: regimes of relevance	138
		1 Listing marine species in the CITES Appendices	139
		2 Opposition to the role of CITES	141
		3 Responses	146

	C	The Memorandum of Understanding between CITES and the FAO		154
		1	Legal status	155
		2	Evolution of the MOU	158
		3	Substantive and procedural constraints	169
		4	National policy coordination	174
	D	Entrenching interaction through the MOU		176
		1	Information-sharing and observership	176
		2	Capacity building	179
		3	Involvement in CITES listing criteria	179
		4	Consultation and review of listing proposals	181
		5	Reporting and resource allocation	184
	E	Settling disputes		184
	F	Conclusions		186
Chapter 5	Adjudicating a fisheries import ban at the WTO			189
	A	Shrimp fisheries and marine by-catch		190
	B	The forum shop: regimes of relevance		190
		1	The US ban on shrimp products	191
		2	Complaint at the WTO	193
	C	Settling the dispute: scope for regime interaction		195
		1	Applicable law	195
		2	Treaty interpretation	197
		3	Relevant facts	204
	D	Settling the dispute: methods of regime interaction		206
		1	Panellists, AB members and the Secretariat	206
		2	The parties' submissions	209
		3	Consultation with scientific experts	211
		4	Consultation with IGO secretariats	215
		5	*Amicus curiae* briefs from NGOs and others	220
	E	Problems and challenges		224
		1	Selection of the adjudicators	226
		2	Framing by the parties	227
		3	Parallel membership of treaties and organisations	229
		4	Legitimacy and the need for guidance in the use of exogenous sources	235
	F	Conclusions		239

Part III: Towards Regime Interaction   241

Chapter 6   From fragmentation to regime interaction   243
- A   From forum shopping to interaction   244
- B   The promotion of regime interaction   249
  1. National policy coordination   249
  2. Learning and information-sharing   253
  3. Allocation of resources   256
- C   Impediments to regime interaction   258
  1. Exclusivity of forum   258
  2. Lack of transparency and openness   261
  3. Need for parallel membership   262
- D   Conclusions   266

Chapter 7   A legal framework for regime interaction   267
- A   The multiple bases of regime interaction   267
  1. Parallel membership   268
  2. Mutual agreement   269
  3. Institutional arrangements   270
- B   Legitimacy of regime interaction   271
  1. Consent and sovereignty   271
  2. Express and implied powers   272
  3. The risk of managerialism   276
- C   Accountable regime interaction   278
- D   Duties to take others into account   284
- E   Conclusions   287

Chapter 8   Implications for international law   288
- A   Appropriate regime interaction in practice   288
  1. Law-making   289
  2. Implementation   292
  3. Dispute settlement   295
- B   Further implications   298
  1. The ILC fragmentation study   298
  2. Beyond fisheries governance   299
  3. Confronting managerialism   302
  4. Situating the participants   304

Appendices	307
A: *Draft consolidated text of the proposed fisheries subsidies disciplines*	307
B: *Final text of the FAO-CITES Memorandum of Understanding*	316
Bibliography	318
Books and edited collections	318
Articles	325
Policy papers, news sources and unpublished works	333
Papers submitted to or produced by international organisations	334
Selected websites	342
Index	344

Foreword

A list of the achievements of international environmental law would undoubtedly include the survival and recovery of the great whales (a process of recovery that has taken place, ironically, under a treaty designed to guarantee the continued exploitability of whales). A list of the failures of international environmental law would likely include the increasingly fragile state of most other pelagic stocks (a process of decline and mismanagement that has taken place under the auspices of regional treaties designed to maintain the sustainability of covered species). As with northern cod in 1991, the road to stock collapse has been paved with good projections.[1]

This record may suggest to the unconverted that the only applicable law in regard to fisheries is the law of unintended consequences. But that would be a mistake. In the more than fifty years since the adoption of the Third Geneva Convention of 1958, international law and international institutions have been a significant force, for good or ill – and this even though the number of contentious cases concerning the merits of fishery conservation measures can be counted on the fingers of one hand.

Instead, the world's pelagic fisheries are managed through some thirty regional fisheries organizations (RFOs) under the general auspices of the Fish Stocks Agreement and Articles 116–120 of UNCLOS. Margaret Young's splendid study is based on the premiss – surely correct – that

---

[1] Cf. *Southern Bluefin Tuna Cases (New Zealand v. Japan; Australia v. Japan), Provisional Measures*, oral argument, ITLOS/PV.99/21/Rev.2, 18 August 1999, p.13; and see M Kurlansky, *Cod: A Biography of the Fish that Changed the World* (Vintage, 1999).

there is 'greater scope for international lawyers to contribute understanding and ideas about collaboration and cohesion within these [fishery] regimes, rather than focussing on *ex post* rules determining priority in later disputes'. This shift enables her to take into account other factors – such as 'soft-law instruments' like the FAO Code of Conduct – and to be soft *ratione personae* as well, given the involvement of 'a multitude of stakeholders on whom fisheries enforcement and monitoring depends, including non-state actors'.

A key notion in her work is that of 'regime', a term used to describe 'a set of laws, processes and institutions that have evolved by addressing a particular problem or function'. In that sense some, but by no means all, current RFOs constitute regimes. Treaties can be concluded with more or less preparation, whereas regimes need time to grow.

But the use of the term 'regimes' does not require exclusivity, or the 'self-contained regime' posited by international law theory. As Dr Young remarks, 'the more pressing concern is to ascertain how CITES, UNCLOS, the Fish Stocks Agreement and the FAO instruments coexist in an effective way'. This story of the struggle for regime interaction she tells clearly and well, ranging across the various fields of practice and the different legal instruments with assurance. Particular highlights include her account of the Memorandum of Understanding between the CITES Secretariat and the FAO, and the work of the WTO, both on fisheries subsidies and dispute settlement (concluding, as to the latter, on the need for greater transparency).

Dr Young draws both practical and theoretical lessons from her case studies. In practical terms she stresses 'the need for national policy coordination in international trade and environmental governance', the need in that context for inter-agency collaboration, the desirability of avoiding a priori determinations of competence of international organisations (typified by the decision of the ICJ in *Use of Nuclear Weapons*), and the clear need to assess the credibility of NGOs.

At the level of theory there are conclusions both particular and general. Among the former is her conclusion that regime interaction 'does not depend on the agreement of all participating states, whether express or implied', and (as a corollary) a rejection of the 'perceived requirement of parallel membership for regime interaction'. Among the latter is the idea that 'the need for representation in governance' is less important than 'the practical need for diverse perspectives'.

Overall this is a major contribution to our understanding of 'the progressive development of international law in the context of fragmentation' as well as 'an attempt to improve the way fisheries governance adapts to complexity and pluralism'. May it be as successful in its particular as in its general aims!

James Crawford
Lauterpacht Centre for International Law
University of Cambridge
January 2011

Acknowledgements

Many generous people have helped since this book began as a PhD thesis at the University of Cambridge. I am particularly indebted to James Crawford and Joanne Scott, inspirational supervisors whose guidance and support were essential. Other scholars who have influenced me (and forced me to better defend my ideas) during my time at Cambridge, New York University and Columbia Law School include Philip Allott, José Alvarez, Jagdish Bhagwati, Gary Horlick, Martti Koskenniemi, Susan Marks, Ricky Revesz and Eleanor Sharpston. My thesis examiners, Ellen Hey and Petros Mavroidis, aided the development of the thesis into a book with encouraging and insightful comments, as did the two anonymous reviewers from Cambridge University Press. I have also profited immeasurably from long conversations and critical comments from family members and friends, especially Stanislav Roudavski and Katharine Young.

Useful professional experience in international organisations has included research assistance at the United Nations International Law Commission in 2003 (for which I thank Václav Mikulka), attendance at the UN Food and Agriculture Organization's Sub-Committee on Fish Trade in 2004 and an internship at the Appellate Body Secretariat of the World Trade Organization in 2005. Consultations with WTO Secretariat staff and trade delegates in Geneva – particularly Doaa Abdel Motaal, Clarisse Morgan, Ana Novik and Werner Zdouc – were invariably useful, as were discussions at the Food and Agriculture Organization, Rome, for which I thank Kevern Cochrane and William Emerson in particular. I also acknowledge the support of King's College, Cambridge, the Cambridge Gates Trust, the Cambridge Faculty of Law, the *Modern Law Review*, the Commonwealth Scholarship and Columbia Law School.

In 2007, I was appointed inaugural research fellow in public international law at Pembroke College and the Lauterpacht Centre for International Law, Cambridge, and I am grateful for the stimulating and collegial academic environment afforded by these institutions. In 2009, I commenced my present role as Senior Lecturer at Melbourne Law School, University of Melbourne, and the thought-provoking discussions and exacting questions from my current colleagues and students continue to shape my ideas.

Parts of the book use materials published earlier: Chapter 5 develops some sections that were first published as 'The WTO's Use of Relevant Rules of International Law: An Analysis of the Biotech Case' (2007) 56:4 *International and Comparative Law Quarterly* 907. Chapter 3 contains research that was first published as 'Fragmentation or Interaction: The WTO, Fisheries Subsidies, and International Law' (2009) 8:4 *World Trade Review* 477. Chapter 4 contains material published as 'Protecting Endangered Marine Species: Collaboration between the Food and Agriculture Organization and the CITES Regime' (2010) 11:2 *Melbourne Journal of International Law* 441.

Margaret Young
Melbourne, 2010

Table of cases

WTO and GATT decisions

Short Title	Full Case Title and Citation
Argentina – Footwear	Appellate Body Report, *Argentina – Safeguard Measures on Imports of Footwear* WT/DS121/AB/R (circulated 14 December 1999) (DSR 2000:I, 515)
Argentina – Textiles and Apparel	Appellate Body Report, *Argentina – Measures Affecting Imports of Footwear, Textiles, Apparel and other Items* WT/DS56/AB/R (circulated 27 March 1998) (DSR 1998:III, 1003); Panel Report WT/DS56/R (circulated 25 November 1997) (DSR 1998:III, 1033)
Australia – Salmon	Appellate Body Report, *Australia – Measures Affecting Importation of Salmon* WT/DS18/AB/R (circulated 20 October 1998) (DSR 1998:VIII, 3327)
Australia – Salmon (Art. 21.5)	Article 21.5 Panel Report WT/DS18/RW (circulated 18 February 2000) (DSR 2000:IV, 2031)
Brazil – Aircraft	Panel Report, *Brazil – Export Financing Programme for Aircraft* WT/DS46/R (circulated 14 April 1999) (DSR 1999:III, 1221)
Brazil – Aircraft (2nd Art. 21.5)	Panel Report, *Brazil – Export Financing Programme for Aircraft* WT/DS46/RW/2 (Second Recourse by Canada to Article 21.5 of the DSU) (circulated 26 July 2001) (DSR 2001:XI, 5481)
Brazil – Retreaded Tyres	Appellate Body Report, *Brazil – Measures Affecting Imports of Retreaded Tyres* WT/DS332/AB/R (circulated 3 December 2007) (DSR 2007:IV, 1527); Panel Report WT/DS332/R (circulated 12 June 2007) (DSR 2007:V, 1649)

Short Title	Full Case Title and Citation
Canada – Aircraft	Appellate Body Report, *Canada – Measures Affecting the Export of Civilian Aircraft* WT/DS70/AB/R (circulated 2 August 1999) (DSR 1999:III, 1377); Panel Report (circulated 14 April 1999) (DSR 1999:IV, 1443)
Canada – Aircraft Credits and Guarantees	Panel Report, *Canada – Export Credits and Loan Guarantees for Regional Aircraft* WT/DS222/R (circulated 28 January 2002) (DSR 2002:III, 849)
Canada – Continued Suspension	Panel Report, *Canada – Continued Suspension of Obligations in the EC – Hormones Dispute* WT/DS321/R (circulated 21 March 2008)
Canada – Herring and Salmon	GATT Panel Report, *Canada – Measures Affecting Exports of Unprocessed Herring and Salmon*, adopted 22 March 1988, BISD 35S/98
EC – Asbestos	Appellate Body Report, *European Communities – Measures Affecting Asbestos and Products Containing Asbestos* WT/DS135/AB/R (circulated 12 March 2001) (DSR 2001:VII, 3243)
EC – Bananas III	Appellate Body Report, *European Communities – Regime for the Importation, Sale and Distribution of Bananas* WT/DS27/AB/R (circulated 9 September 1997) (DSR 1997:II, 589)
EC – Biotech	Panel Report, *EC – Measures Affecting the Approval and Marketing of Biotech Products* WT/DS291/R, WT/DS292/R, WT/DS293/R, (circulated 29 September 2006) (DSR 2006:III, 847)
EC – Chicken Cuts	Appellate Body Report, *European Communities – Customs Classification of Frozen Boneless Chicken Cuts* WT/DS269/AB/R, WT/DS286/AB/R (Complaint by Brazil and Thailand, circulated 12 September 2005) (DSR 2005:XIX, 9157); Panel Report, Complaint by Brazil, WT/DS269/R (DSR 2005:XIX, 9295); Panel Report, Complaint by Thailand WT/DS286/R (DSR 2005:XX 9721)
EC – Geographical Indications	Panel Report, *European Communities – Protection of Trademarks and Geographical Indications for Agricultural Products and Foodstuffs* WT/DS290/R and WT/DS174/R (circulated 15 March 2005) (DSR 2005:X–XI, 4603, 5121 and DSR 2005:VIII–IX, 3499, 4083)
EC – Hormones	Appellate Body Report, *European Communities – Measures Concerning Meat and Meat Products (Hormones)* WT/DS26/AB/R (Complaint by US),

Short Title	Full Case Title and Citation
	WTDS48/AB/R (Complaint by Canada) (circulated 16 January 1998) (DSR 1998:I, 135)
EC – Salmon	Panel Report, *European Communities – Anti-Dumping Measure on Farmed Salmon from Norway* WT/DS337/R (circulated 16 November 2007) (DSR 2008:I, 3)
EC – Sardines	Appellate Body Report, *European Communities Trade Description of Sardines* WT/DS231/AB/R (circulated 26 September 2002) (DSR 2002:VIII, 3359)
EC – Selected Customs Matters	Appellate Body Report, *European Communities – Selected Customs Matters* WT/DS315/R (circulated 13 November 2006) (DSR 2006:IX, 3791)
EC – Tariff Preferences	Appellate Body Report, *European Communities – Conditions for the Granting of Tariff Preferences to Developing Countries* WT/DS246/AB/R (circulated 7 April 2004) (DSR 2004:III, 951); Panel Report WT/DS246/R (circulated 1 December 2003) (DSR 2004:III, 1037)
India – Automotive	Panel Report, *India – Measures Affecting the Automotive Sector* WT/DS146/R, WT/DS175/R (circulated 21 December 2001) (DSR 2002:V, 1827)
India – Quantitative Restrictions	Appellate Body Report, *India – Quantitative Restrictions on Imports of Agricultural, Textile and Industrial Products* WT/DS90/AB/R (circulated 23 August 1999) (DSR 1999:IV, 1763); Panel Report WT/DS90/R (circulated 6 April 1999) (DSR 1999;V, 1799)
Japan – Agricultural Products II	Appellate Body Report *Japan – Measures Affecting Agricultural Products* WT/DS/76/AB/R (circulated 22 February 1999) (DSR 1999:I, 277); Panel Report WT/DS76/R (circulated 27 October 1998) (DSR 1999:I, 315)
Japan – Alcohol II	Appellate Body Report, *Japan – Taxes on Alcoholic Beverages* WT/DS8/AB/R, WT/DS10/AB/R, WT/DS11/AB/R (circulated 4 October 1996) (DSR 1996:I, 97)
Korea – Procurement	Panel Report, *Korea – Measures Affecting Government Procurement* WT/DS/163/R (circulated 1 May 2000) (DSR 2000:VII, 3541)
Spain – Unroasted Coffee	GATT Panel Report, *Spain – Tariff Treatment of Unroasted Coffee*, adopted 11 June 1981 (L/5135 – 28S/102)
Thai-Cigarettes	GATT Panel Report, *Thailand – Restrictions on Importation of and Internal Taxes on Cigarettes*, DS10/R – 37S/200, adopted 7 November 1990

Short Title	Full Case Title and Citation
US – Continued Suspension	Panel Report, *United States – Continued Suspension of Obligations in the EC – Hormones Dispute* WT/DS320/R (circulated 31 March 2008)
US – Gasoline	Appellate Body Report, *United States – Standards for Reformulated and Conventional Gasoline (US – Gasoline)* WT/DS2/AB/R (circulated 20 May 1996) (DSR 1996: I, 3)
US – Lead and Bismuth II	Appellate Body Report, *United States – Imposition of Countervailing Duties on Certain Hot-Rolled Lead and Bismuth Carbon Steel Products Originating in the United Kingdom* WT/DS138/AB/R (circulated 10 May 2000) (DSR 2000:V, 2595)
US – Shrimp	Appellate Body Report, *United States – Import Prohibition of Certain Shrimp and Shrimp Products* WT/DS58/AB/R (circulated 12 October 1998) (DSR 1998: VII, 2755); Panel Report WT/DS58/R (circulated 15 May 1998) (DSR 1998:VII, 2821)
US – Shrimp (Art. 21.5)	Appellate Body Report, *United States – Import Prohibition of Certain Shrimp and Shrimp Products – Recourse to Article 21.5 by Malaysia* WT/DS58/AB/RW (circulated 22 October 2001) (DSR 2001:XIII, 6481); Panel Report WT/DS58/RW circulated 15 June 2001) (DSR 2001:XIII, 6529)
US – Shrimp (Ecuador)	Panel Report, *United States – Anti-Dumping Measure on Shrimp from Ecuador* WT/DS335/R (circulated 30 January 2007) (DSR 2007:II, 423)
US – Shrimp (Thailand)	Appellate Body Report, *United States – Measures Relating to Shrimp from Thailand* WT/DS343/AB/R (circulated 16 July 2008); Panel Report WT/DS343/R (circulated 29 February 2008)
US – Tuna I	GATT Panel Report, *United States – Restrictions on Imports of Tuna* (1991) GATT Doc. DS21/R (Mexico)
US – Tuna II	GATT Panel Report, *United States – Restrictions on Imports of Tuna* (1994) GATT Doc. DS29/R (Europe)
US – Tuna (Canada)	GATT Panel Report, *United States – Prohibition of Imports of Tuna and Tuna Products from Canada*, adopted 22 February 1982, BISD 29S/91

Other cases

Short Title	Full Case Title and Citation
Anglo-Norwegian Fisheries Case [1951] ICJ Rep 116 and 132	*United Kingdom v Norway* [1951] ICJ Reports 116 and 132; 18 ILR 86
Commission v Belgium Case C-2/90 (9 July 1992)	*Commission v Belgium*, Case C-2/90 (9 July 1992)
Fisheries Jurisdiction Cases [1973] ICJ Rep 3; [1974] ICJ Rep 3	*United Kingdom v Iceland (Jurisdiction)* [1973] ICJ Reports 3; 55 ILR 149; *United Kingdom v Iceland (Merits)* [1974] ICJ Reports 3; 55 ILR 238
Gabčíkovo-Nagymaros case [1997] ICJ Rep 7	*Case concerning the Gabčíkovo-Nagymaros Project (Hungary v Slovakia)* [1997] ICJ Reports 7
Ireland v UK (OSPAR) (2003) 42 ILM 1118	Permanent Court of Arbitration: *Dispute Concerning Access to Information Under Article 9 of the OSPAR Convention: Ireland v United Kingdom* – Final Award (2 July 2003) (2003) 42 ILM 1118
Military and Paramilitary Activities in Nicaragua (Merits) [1986] ICJ Rep 16	*Military and Paramilitary Activities in Nicaragua (Merits)* [1986] ICJ Reports 16
Mox Plant Case (2001) 41 ILM 405	Request for Provisional Measures, *Ireland v United Kingdom* (2001) (Order of 3 December 2001) (2002) 41 ILM 405
UN expenses [1962] ICJ Rep 151	Advisory Opinion, *Certain Expenses of the United Nations* [1962] ICJ Reports 151
UN Reparations case [1949] ICJ Rep 174	Advisory Opinion, *Reparation for Injuries Suffered in the Service of the United Nations* [1949] ICJ Reports 174
Use or Threat of Nuclear Weapons [1996] ICJ Rep 226	Advisory Opinion, *Legality of the Threat or Use of Nuclear Weapons* [1996] ICJ Reports 226 (8 July 1996, General List No. 95) (Request for Advisory Opinion by GA)
Use of Nuclear Weapons in Armed Conflict [1996] ICJ Rep 66	Advisory Opinion, *Legality of the Use by a State of Nuclear Weapons in Armed Conflict* [1996] ICJ Reports 66 (8 July 1996 General List No. 93) (Request for Advisory Opinion by WHO)
Southern Bluefin Tuna Case (Australia and New Zealand v Japan) (Award) (2000) (Annex VII Tribunal)	*Australia and New Zealand v Japan (Award on Jurisdiction and Admissibility)* (2000) 119 ILR 508 (Arbitral Award

Short Title	Full Case Title and Citation
Southern Bluefin Tuna Provisional Measures (Australia and New Zealand v Japan) (1999) (ITLOS)	constituted under Annex VII of UNCLOS) (4 August 2000) *Australia and New Zealand v Japan (Provisional Measures)* (1999) 117 ILR 148 (International Tribunal for the Law of the Sea)

Table of Conventions, Declarations and procedures

Short Title	Full Title and Citation
AB Working Procedures	WTO, Working Procedures for Appellate Review WT/AB/WP/5 (4 January 2005)
Aarhus Convention	*Convention on Access to Information, Public Participation in Decision-Making and Access to Justice in Environmental Matters* (1999) 38 ILM 517 (in force 30 October 2001)
Agenda 21	*Agenda 21*, adopted by the Plenary of UNCED on 14 June 1992 (A/CONF. 151/26) (Vols. I–III) (1992)
Agreement on Agriculture	*Agreement on Agriculture* (signed 15 April 1994) in WTO, The Legal Texts (Cambridge University Press) 33
Biosafety Protocol	*Cartagena Protocol on Biosafety to the Convention on Biological Diversity* (2000) 39 ILM 1027 (in force 11 September 2003)
Anti-Dumping Agreement	*Agreement on Implementation of Article VI of the General Agreement on Tariffs and Trade 1994* (signed 15 April 1994) in WTO, The Legal Texts (Cambridge University Press) 147
CBD	*United Nations Convention on Biological Diversity* (1992) 31 ILM 818 (in force 29 December 1993)
CBD Decision on Marine and Coastal Ecosystems (1998)	CBD Conference Decision IV/5 on Marine and Coastal Ecosystems, the Fourth Conference of the Parties to

Short Title	Full Title and Citation
	the Convention on Biological Diversity, Bratislava, 4–15 May 1998, UNEP/CBD/COP/4/27, 84
CCSBT	*Convention for the Conservation of Southern Bluefin Tuna*, 1819 UNTS 360 (10 May 1993)
CITES	*Convention on the International Trade in Endangered Species of Wild Fauna and Flora*, 993 UNTS 243 (in force 1 July 1975)
CITES CoP-12 Decision 12.7	Decision of the CITES, 'Establishment of a Memorandum of Understanding between CITES and the Food and Agriculture Organization of the United Nations (FAO)'
CITES Resolution Conf. 2.8 (1979)	Resolution of the CITES COP, Resolution Conf. 2.8 (1979) (now included in Conf. 11.4, 'Conservation of cetaceans, trade in cetacean specimens and the relationship with the International Whaling Commission' (Rev. CoP12)
CITES Resolution Conf. 8.4 (Rev. CoP14)	Resolution of the CITES COP, Resolution Conf 8.4, 'National Laws for Implementation of the Convention' (revised 3–15 June 2007)
CITES Resolution Conf 9.17	Resolution of the CITES COP, Resolution Conf. 9.17, 'Sharks' (Replaced by CITES Resolution Conf. 12.6)
CITES Resolution Conf. 9.24 (Rev. CoP12) (CITES listing criteria (2007))	Resolution of the CITES COP, Resolution Conf. 9.24 'Criteria for amendment of Appendices I and II' (Last revised at fourteenth COP in 2007)
CITES Resolution Conf. 10.4 (Rev.14)	Resolution of the CITES COP, Resolution Conf. 10.4, 'Cooperation and Synergy with the Convention on Biological Diversity' (Last revised at fourteenth COP in 2007)
CITES Resolution Conf. 11.1 (Rev. CoP14).	Resolution of the CITES COP, Resolution Conf. 11.1, 'Establishment of committees' (Last revised at fourteenth COP in 2007)

TABLE OF CONVENTIONS, DECLARATIONS AND PROCEDURES   XXV

Short Title	Full Title and Citation
CITES Resolution Conf. 12.4	Resolution of the CITES COP, Resolution Conf. 12.4 'Cooperation between CITES and the Commission for the Conservation of Antarctic Marine Living Resources regarding trade in toothfish'
CITES Resolution Conf. 12.5	CITES Resolution Conf. 12.5 'Conservation and management of sharks'
CITES Resolution Conf. 12.8 (Rev. CoP13)	Resolution of the CITES COP, Resolution Conf. 12.8 'Review of Significant Trade in specimens of Appendix-II species' (Last revised at thirteenth COP in 2004)
CITES Resolution Conf. 13.8	Resolution of the CITES COP, Resolution Conf. 13.8, 'Participation of observers at meetings of the Conference of the Parties' (adopted at thirteenth COP in 2004)
CITES Resolution Conf. 14.6 (2007)	Resolution of the CITES COP, Resolution 14.6, 'Introduction from the sea' (adopted at fourteenth COP in 2007)
CITES Standing Committee Rules of Procedure (2009)	Rules of Procedure of the CITES Standing Committee, as amended at fifty-eighth meeting, Geneva, July 2009
CMS	*Convention on the Conservation of Migratory Species of Wild Animals* signed in Bonn, Germany, on 23 June 1979 (in force 1 November 1983)
COFI Rules of Procedure	Rules of Procedure of the Committee on Fisheries, in FAO (2006) II *Basic Texts* Part N
Compliance Agreement	*Agreement to Promote Compliance with International Conservation and Management Measures by Fishing Vessels on the High Seas* (1994) 33 ILM 968 (in force 24 April 2003)
Enabling Clause (1979)	*Decision on Differential and More Favourable Treatment, Reciprocity, and Fuller Participation of Developing Countries*, L/4903, BISD 26S/203 (28 November 1979)

Short Title	Full Title and Citation
FAO Code of Conduct	FAO, *Code of Conduct for Responsible Fisheries* (FAO, Rome, 1995) (adopted 31 October 1995); reprinted in (1995) 11 *International Organizations and the Law of the Sea Documentary Yearbook* 700.
FAO Constitution	FAO Constitution in FAO (2006) I *Basic Texts* Part A
FAO, Cooperation with IGOs	FAO, 'Cooperation with International Governmental Organizations' in FAO (2006), Vol II *Basic Texts*, Part M (See paragraphs 488 to 490 of the Report of the ninth session of the Conference)
FAO, Cooperation with NGOs	FAO, 'Cooperation with International Non-governmental Organizations' (Resolution No. 39/57) in FAO (2006) II *Basic Texts* Part O
FAO General Rules	FAO, General Rules of the Organization, in FAO (2006) I *Basic Texts* Part B
FAO, 'Granting of Observer Status'	FAO, 'Granting of Observer Status (in respect of international governmental and non-governmental organizations)' (Resolution No. 44/57) in FAO (2006) II *Basic Texts* Part Q
FAO Guiding Lines Regarding Relationship Agreements Between FAO and IGOs (Resolution of 10th session of FAO Conference)	Resolution No. 69/59 of the tenth session of the FAO Conference (Guiding Lines Regarding Relationship Agreements Between FAO and Intergovernmental Organizations) Vol II, *Basic Texts* (2006 edition) Part N.
FAO, IPOA-Capacity (1999)	FAO, 'International Plan of Action for the Management of Fishing Capacity' adopted by the twenty-third session of COFI in February 1999 and endorsed by the FAO Council in June 1999 (FAO, Rome, 1999)
FAO, IPOA-IUU Fishing (2001)	FAO, 'International Plan of Action to Prevent, Deter and Eliminate Illegal, Unreported and Unregulated Fishing' (Rome, 2001). The

Short Title	Full Title and Citation
	IPOA-IUU was adopted at the twenty-fourth session of COFI in March 2001 and endorsed by the FAO Council in June 2001.
FAO, IPOA-Seabirds (1999)	FAO, 'International Plan of Action for Reducing Incidental Catch of Seabirds in Longline Fisheries' adopted by the twenty-third session of COFI in February 1999 and endorsed by the FAO Council in June 1999 (FAO, Rome, 1999)
FAO, IPOA-Sharks (1999)	FAO, 'International Plan of Action for the Conservation and Management of Sharks' adopted by the twenty-third session of COFI in February 1999 and endorsed by the FAO Council in June 1999 (FAO, Rome, 1999)
FAO, 'Policy concerning NGO relations'	FAO, 'Policy Concerning Relations with International Non-governmental Organizations' in FAO (2006) II *Basic Texts* Part P
FAO Reykjavik Declaration (2001)	'Reykjavik Declaration on Responsible Fisheries in the Marine Ecosystem' (2001) (FAO Doc. C200/INF/25, appendix I)
Fish Stocks Agreement	*Agreement for the Implementation of the Provisions of the United Nations Convention on the Law of the Sea of 10 December 1982 relating to the Conservation and Management of Straddling Fish Stocks and Highly Migratory Fish Stocks* (1995) 34 ILM 1547 (in force 11 December 2001)
GATT	*General Agreement on Tariffs and Trade 1947* in WTO, The Legal Texts (Cambridge University Press) 423
General Council Rules of Procedure (1996)	General Council, 'Rules of Procedure for Meetings of the General Council' WT/L/161 (25 July 1996)
Import Licensing Procedures Agreement	*Agreement on Import Licensing Procedures* (signed 15 April 1994) in WTO, *The Legal Texts* (Cambridge University Press) 223

Short Title	Full Title and Citation
International Convention on the Conservation of Atlantic Tunas (ICCAT)	*International Convention on the Conservation of Atlantic Tunas (ICCAT)* (in force 21 March 1969)
International Whaling Convention (1946)	*International Convention for the Regulation of Whaling* (1946) 161 UNTS 72 (in force 10 November 1948)
Jakarta Mandate on Marine and Coastal Biodiversity (1995)	CBD Decision on the Conservation and Sustainable Use of Marine and Coastal Biological Diversity of the Convention on Biological Diversity, A/51/312, annex II, decision II/10
Marrakesh Agreement	*Agreement Establishing the World Trade Organization* (signed 15 April 1994) in WTO, *The Legal Texts* (Cambridge University Press) 3
OECD Exports Credit Arrangement (2008)	OECD, 'Arrangement on Officially Supported Export Credits' (2008 revision): See OECD Trade and Agriculture Directorate TAD/PG (2007)28/FINAL. http://www.oecd.org/
OSPAR Convention	*Convention for the Protection of the Marine Environment of the North-East Atlantic* (1992) 32 ILM 1069
Rio Declaration	'Rio Declaration on Environment and Development', adopted by UNCED 1992 (A/CONF.151/26) (Vol. I) (1992)
Rules of Origin Agreement	*Agreement on Rules of Origin* (signed 15 April 1994) in WTO, *The Legal Texts* (Cambridge University Press) 211
SBT Convention	*Convention for the Conservation of Southern Bluefin Tuna* (1993) (in force 20 May 1994)
SCM Agreement	*Agreement on Subsidies and Countervailing Measures* (signed 15 April 1994) in WTO, *The Legal Texts* (Cambridge University Press) 231
SPAW Protocol (2000)	SPAW Protocol (Protocol on Specially Protected Areas and Wildlife) to the *Convention for the Protection and Development of the Marine Environment of the Wider Caribbean Region* (in force 25 April 2000)

Short Title	Full Title and Citation
TBT Agreement	*Agreement on Technical Barriers to Trade* (signed 15 April 1994) in WTO, *The Legal Texts* (Cambridge University Press) 121
UN Convention against Corruption (2003)	*UN Convention against Corruption* General Assembly resolution 58/4 of 31 October 2003 (in force 14 December 2005) (2004) 43 ILM 37
UNCLOS	*United Nations Convention on the Law of the Sea* (1982) 21 ILM 1261 (1982) (in force 16 November 1994).
UNEP Rules of Procedure	The Rules of Procedure of the Governing Council of UNEP (last revised 1988); published online at http://www.unep.org/resources/gov/Rules.asp
United Nations Charter	*Charter of the United Nations* (in force 24 October 1945)
Universal Declaration of Human Rights (1948)	*Universal Declaration of Human Rights* (1948)
VCLT	*Vienna Convention on the Law of Treaties* (1969) 1155 UNTS 331 (in force 27 January 1980)
Vienna Convention on the Law of Treaties between States and International Organizations or between International Organizations (1986) (not yet in force)	*Vienna Convention on the Law of Treaties between States and International Organizations or between International Organizations*, Vienna, 21 March 1986 (not yet in force): A/CONF.129/15
Wellington Driftnet Convention (1990)	*Convention for the Prohibition of Fishing with Long Driftnets in the South Pacific* (1990) 29 ILM 1449 (in force 17 May 1991)
World Summit Plan of Implementation (2002)	'Plan of Implementation of the World Summit on Sustainable Development'(A/CONF.199/20) (4 September 2002)
WTO Decision on Trade and Environment (1994)	Ministerial Decision, 'Decision on Trade and Environment' adopted on 14 April 1994 in WTO, *The Legal Texts* (Cambridge University Press) 411.
WTO Declaration on Coherence in Global Economic Policy Making (1993)	Ministerial Declaration 'Declaration on the Contribution of the World Trade Organization to Achieving Greater Coherence in Global Economic

Short Title	Full Title and Citation
	Policymaking' adopted by the Trade Negotiations Committee on 15 December 1993 in WTO, *The Legal Texts* (Cambridge University Press) 386
WTO Doha Declaration (2001)	WTO Ministerial Declaration adopted on 14 November 2001 (WT/MIN(01)/DEC/1, 20 November 2001) (2002) 41 ILM 746
WTO Guidelines on NGOs (1996)	WTO General Council, 'Guidelines for Arrangements on Relations with Non-Governmental Organizations' adopted on 18 July 1996 (WT/L/162).
WTO Report of the CTE (1996)	WTO Members Report of the Committee on Trade and Environment, WT/CTE/1, 12 November 1996
WTO–UNEP Cooperation Agreement (1999)	Cooperation between the WTO and UNEP Secretariats, Press Release Press/154 (29 November 1999), as established by exchange of letters by the Director-General of the WTO and the Secretary-General of the UN; also published in WTO Doc. TN/TE/S/2/Rev.2

Abbreviations

AB	Appellate Body of the World Trade Organization
ACP	African, Caribbean and Pacific Group of States
AJIL	American Journal of International Law
APEC	Asia-Pacific Economic Cooperation
BYIL	British Year Book of International Law
CBD	Convention on Biological Diversity
CCAMLR	Commission for the Conservation of Antarctic Marine Living Resources
CCSBT	Convention for the Conservation of Southern Bluefin Tuna
CIEL	Center for International Environmental Law
CITES	Convention on the International Trade in Endangered Species of Wild Fauna and Flora
CMC	Center for Marine Conservation
CMS	Convention on Migratory Species
Codex	Codex Alimentarius Commission, a subsidiary body of the FAO and the WHO.
COFI	FAO Committee on Fisheries
COFI-FT	Subcommittee on Fish Trade, COFI, FAO
COP	Conference of the Parties to CITES
CTE	WTO Committee on Trade and Environment
DOALOS	UN Division of Ocean Affairs and the Law of the Sea
DSB	Dispute Settlement Body, WTO
DSU	Understanding on Rules and Procedures Governing the Settlement of Disputes
EC	European Community
ECJ	European Court of Justice
EEC	European Economic Community
EEZ	exclusive economic zone
EJIL	European Journal of International Law
EU	European Union

FAO	United Nations Food and Agriculture Organization
FIELD	Foundation for International Environmental Law and Development
GA	General Assembly of the United Nations
GATT	General Agreement on Tariffs and Trade
GEF	Global Environment Facility
GM	genetically modified
GSP	General System of Preferences
HSI	Humane Society International
IARC	International Agency for Research on Cancer
IATTC	Inter-American Tropical Tuna Commission
ICCAT	International Commission for the Conservation of Atlantic Tunas
ICJ	International Court of Justice
ICLQ	International and Comparative Law Quarterly
ICTSD	International Centre for Trade and Sustainable Development
ICTY	International Criminal Tribunal for the former Yugoslavia
IFAW	International Fund for Animal Welfare
IGO	intergovernmental organization
ILC	International Law Commission
ILM	International Legal Materials
ILO	International Labour Organization
IMF	International Monetary Fund
IMO	International Maritime Organization
IOC	Intergovernmental Oceanographic Commission of UNESCO
IPCS	International Programme on Chemical Safety
IPOA	International Plan of Action
IPPC	International Plant Protection Convention
IR	international relations
ISEAL	International Social and Environmental Accreditation Alliance
ISO	International Organization for Standardization
ITLOS	International Tribunal for the Law of the Sea
IUCN	International Union for Conservation of Nature
IUU	illegal, unreported and unregulated (fishing)

IWMC	An NGO formerly known as the International Wildlife Management Consortium
JECFA	Joint FAO/WHO Expert Committee on Food Additives
JIEL	Journal of International Economic Law
JWT	Journal of World Trade
MEA	multilateral environmental agreement
MFN	most-favoured-nation treatment
MOU	Memorandum of Understanding
MSC	Maritime Stewardship Council
NAFO	Northwest Atlantic Fisheries Organization
NAFTA	North American Free Trade Agreement
NGO	non-governmental organisation
NMRF	US National Marine Fisheries Service
NSG	Nuclear Supplier Group
Oceana	An international ocean environmental advocacy group based in North America, South America and Europe
OECD	Organisation for Economic Co-operation and Development
OIE	World Organisation for Animal Health (formerly International Office of Epizootics)
OSPAR	Oslo and Paris Commissions (see OSPAR Convention)
PCA	Permanent Court of Arbitration
PCIJ	Permanent Court of International Justice
PGE	Permanent Group of Experts
RECIEL	Review of European Community & International Environmental Law
RFMO	Regional Fisheries Management Organisation
SCM	Subsidies and Countervailing Measures
SOCA	Subcommittee on Oceans and Coastal Areas
SOFIA	FAO, *The State of World Fisheries and Aquaculture* (biennial publication)
SPS	sanitary and phytosanitary
TBT	Technical Barriers to Trade
TED	turtle excluder device
TNC	Trade Negotiations Committee
TPRM	Trade Policy Review Mechanism

TRIPS	Trade-Related Aspects of Intellectual Property Rights
UK	United Kingdom
UKTS	United Kingdom Treaty Series
UN	United Nations
UNCED	United Nations Conference on Environment and Development
UNCLOS	United Nations Convention on the Law of the Sea
UNCTAD	United Nations Conference on Trade and Development
UN Consultative Process	United Nations Open-ended Informal Consultative Process on Oceans and the Law of the Sea
UNGA	General Assembly of the United Nations
UN-Oceans	Oceans and Coastal Areas Network, headed by the Executive Secretary of the Intergovernmental Oceanographic Commission of UNESCO
UNEP	United Nations Environment Programme
UNESCO	United Nations Educational, Scientific and Cultural Organization
UNFCCC	United Nations Framework Convention on Climate Change
US	United States of America
USTR	United States Trade Representative
UNTS	United Nations Treaty Series
VCLT	Vienna Convention on the Law of Treaties
WCO	World Customs Organization
WHO	World Health Organization
WIPO	World Intellectual Property Organization
WTO	World Trade Organization
WWF	World Wide Fund for Nature

PART I

Trading Fish, Saving Fish

1 Introduction

International law is a legal system.[1]

International trade law, international environmental law and the law of the sea have been conceived and developed, for the most part, independently. States have agreed to the progressive multilateral liberalisation of trade through the auspices of the GATT, later the World Trade Organization (WTO). They have addressed environmental issues such as the protection of biological diversity through a range of multilateral environmental agreements known collectively as MEAs. The use of ocean resources has been negotiated in various instruments grouped under the 'law of the sea', culminating in the United Nations Convention on the Law of the Sea (UNCLOS) and related agreements. Meanwhile, collective action to ensure freedom from hunger has focused on the utilisation of fisheries and marine products as major goals of the UN Food and Agriculture Organization (FAO). Thus have arisen separate 'regimes' of laws and institutions.[2]

But global problems do not fall neatly within a single regime. The emerging worldwide crisis in fish stocks calls for diverse international legal and political responses. Scientific studies have emphasised that global fisheries are at real risk of collapse. An adequate rate of replenishment of fish, as was commonly achieved pre-industrialisation,[3] is now

---

[1] ILC, Report of the Study Group, Fragmentation of International Law: Difficulties arising from the Diversification and Expansion of International Law: Conclusions (A/CN.4/L.702) (18 July 2006), Conclusion (1) at 7. See also associated Analytical Study finalized by the Chairman (A/CN.4/L.682 and Corr.1) (13 April 2006).
[2] For a more detailed discussion of the definition of regimes, see below n. 83 and accompanying text.
[3] Simon Jennings, Michel Kaiser and John Reynolds, *Marine Fisheries Ecology* (2006) 10.

usually exceeded by catch capacity. Scientists have projected the collapse of seafood-producing species stocks by 2048,[4] and have noted that 63 per cent of assessed fish stocks worldwide require rebuilding, and even lower exploitation rates are needed to reverse the collapse of vulnerable species.[5] The collapse of a fishery leads to massive and lasting social, economic and ecological ramifications.[6]

For the stock groups monitored by the FAO, the estimates are equally alarming:[7]

> 20 per cent are under-exploited or moderately exploited;
> 52 per cent are fully exploited; and
> 28 per cent are either over-exploited, depleted or recovering from depletion.

Within the latter category, there is a high concentration of high seas fishery resources. For example, 50 per cent of the stocks of highly migratory oceanic sharks are thought to be over-exploited or depleted.[8] These figures may be much higher given the possibility of under-reported catch data provided to the FAO.[9] Moreover, although high seas fishery resources are only a fraction of the global fishery resources, they are considered to be 'key indicators of the state of an overwhelming part of the ocean ecosystem'.[10] As such, it is not species but entire ecosystems that are under threat. Overall, it may be an understatement that the 'maximum wild capture fisheries potential from the world's oceans has probably been reached'.[11]

---

[4] Boris Worm et al., 'Impacts of Biodiversity Loss on Ocean Ecosystem Services', (2006) 314 *Science* 787 (3 November 2006) (defining collapse as 10 per cent of unfished biomass). For an alternative definition of collapse, which relies on economic indicators, see Jennings et al., above n. 3, 11 ('Following collapse, the fishery will no longer be profitable'). The latter definition will presumably be subject to the variable conditions affecting economic viability.

[5] Boris Worm et al., 'Rebuilding Global Fisheries' (2009) 325 *Science* 578.

[6] See e.g. the situation of the Newfoundland cod fishery, which was fished under management at low but increasing levels from 1977 until it collapsed in 1992, leaving an entire industry without employment. Jennings et al. state that the size of the spawning stock had fallen from an estimated 1.6 million tonnes in 1962 to 22,000 tonnes in 1992: above n. 3.

[7] FAO, *The State of World Fisheries and Aquaculture* (2008) 7 (SOFIA).

[8] FAO, *FAO Fisheries Technical Paper No. 495*, 'The State of World Highly Migratory, Straddling and Other High Seas Fishery Resources and Associated Species' (2006).

[9] *FAO Fisheries Technical Paper No. 389*, Stefania Vannuccini, *Shark Utilization, Marketing and Trade*, (1999), para 3.4.

[10] FAO, SOFIA (2008) 35.

[11] Ibid. 7. The Report calls for 'a more closely controlled approach to fisheries management'.

Meanwhile, the international trade in fish and fish products is increasing rapidly, and fish is one of the most highly traded food and feed commodities.[12] The yearly value of exports of fish and fish products was last recorded as US$85.9 billion, representing a 32.1 per cent increase in the period 2000 to 2006.[13] Apart from value increases, the volume of exported fisheries product now represents 37 per cent of the total estimated yearly production of 144 million tonnes.[14] Globally, the per capita average fish consumption is almost double that of fifty years ago.[15]

This growing international fish trade has a serious negative effect on fish stocks and marine ecosystems.[16] For example, although causation is difficult to establish, the dramatic increase in the international trade in sharks corresponds to a massive decline in shark species, of which several are now critically endangered.[17] Depletion of marine resources has multiple impacts, including on ecology, global food security, culture and economic prosperity. This book explores the ways in which international law, fragmented into regimes and applied by diverse institutions, governs the trading and 'saving' of fish.

A Trade law and fisheries sustainability

The growing awareness of the ecological crisis and the increased level of trade in fish products has led to a proliferation of legal and institutional responses. The sustainability of global fish stocks is impacted by, if not a direct objective of, more than one international regime. Three examples are illustrative of the trend: the growing multilateral effort to discipline fisheries subsidies that have harmful environmental effects; the decision by participating states to restrict the trade in certain endangered marine

---

[12] FAO, *The State of World Fisheries and Aquaculture* (SOFIA) (2006) 7.
[13] FAO, SOFIA (2008) 8 (figures are adjusted for inflation).   [14] Ibid., 8.
[15] The FAO records world apparent per capita fish consumption as an average of 9.9 kg in the 1960s, 11.5 kg in the 1970s, 12.5 kg in the 1980s, 14.4 kg in the 1990s, and 16.4 kg in 2005: ibid., 8.
[16] Caroline Dommen, 'Fish for Thought: Fisheries, International Trade and Sustainable Development' (1999) *Natural Resources, International Trade and Sustainable Development Series, No. 1* (ICTSD and the IUCN) 2–3 (pointing to perceptions of a negative effect of international trade on fish stocks, but noting that empirical evidence is lacking in this debate). See also Marc Allain, *Trading Away Our Oceans* (2007). Other effects on fish species include climate change and ocean acidification: see e.g. Duncan E J Currie and Kateryna Wowk, 'Climate Change and CO_2 in the Oceans and Global Oceans Governance' (2009) 4 *Carbon and Climate Law Review* 387.
[17] See further Chapter 4, p. 136.

species through the Convention on the International Trade in Endangered Species of Wild Fauna and Flora (CITES); and the use of unilateral trade measures for the preservation of marine species.

The first example of the increasing relevance of trade principles to fisheries relates to subsidies.[18] A major contributing factor to the global fisheries crisis is overfishing. That so many fish are being caught is a function of the sheer size, concentration and technological capacity of the global fishing fleet. In the development of policies aimed at reducing overfishing and overcapacity, attention has focused on the current economic structure and behaviour of the fishing industry. The sector is heavily supported by national governments. This support takes the form of a range of subsidies, including financial transfers, infrastructure development, income support and provision of access rights. Although the levels are difficult to quantify, a range of studies indicate that subsidies could amount to a quarter of the value of revenues in the fisheries sector or even higher.[19]

The potential for these subsidies to have trade-distorting and environmentally harmful effects has led to proposals for new international policies within the international trade regime. The 153 members of the WTO are currently considering reforms to the Subsidies and Countervailing Measures Agreement (SCM Agreement).[20] This Agreement disciplines the use by WTO members of subsidies and regulates the actions other WTO members can take to counter the effects of subsidies. As currently framed, the Agreement is inadequate to deal with fisheries issues. In negotiations to amend the Agreement, WTO negotiators have been forced to differentiate between subsidies that are environmentally harmful, such as state aid to increase vessel capacity, and subsidies that are beneficial to fisheries sustainability, such as the provision of management or research resources. This attempt by the WTO to incorporate environmental objectives gives rise to questions about its mandate, its expertise and the existence of an institutional 'trade' bias. These questions, in turn, give rise to issues about the degree of deference shown to other regimes such as UNCLOS and FAO fisheries instruments in the framing and implementation of the subsidies disciplines.

The second example of the use of economic policies in fisheries is through controlling trade in selected species.[21] CITES is the international

---

[18] See generally Chapter 3. [19] See further Chapter 3 n. 7.
[20] WTO Doha Declaration (2001) para 28; see also para 31.
[21] See generally Chapter 4.

regime that restricts trade in identified endangered species. The regime allows for the listing of species that are being depleted at unsustainable levels. As popularly conceived, CITES is aimed at curbing the destruction of prominent species such as elephants and tigers, through making the trade in ivory and game-trophies unlawful. However, it is open for any party to propose the listing of any wildlife species, including marine species.

Increasingly, some of the Convention's parties have sought to safeguard threatened marine species such as sharks, seahorses, queen conch and bluefin tuna through proposals for listing. This intervention is not supported by major fishing nations, which perceive CITES to be an inappropriate tool to conserve and manage marine species. In providing Secretariat support for listing proposals of marine species, the CITES Secretariat is reliant on the expertise of the FAO, although this collaboration is treated suspiciously by some states, as manifest in difficulties surrounding a Memorandum of Understanding (MOU) between the two intergovernmental organizations (IGOs).[22]

Apart from restricting trade through CITES, some countries are increasingly using unilateral trade measures to achieve environmental goals, and these provide a third example of the growing influence of trade law.[23] These trade restrictions are not necessarily targeted at the trade in endangered species. Instead, they may address the trade in certain products, which has observable impacts on the sustainability of other species. For example, the United States has sought to conserve sea turtles by banning the import of shrimp harvested using nets known to cause sea turtle mortality.[24] It has also sanctioned the use of labels that bear information about the ecological impact of certain products, such as dolphin logos on canned tuna to signify that dolphins were not harmed by the tuna harvesting methods.[25] The WTO-consistency of

---

[22] The Memorandum of Understanding between the FAO and the CITES Secretariat (2006) is considered in detail in Chapter 4.
[23] See generally Chapter 5.
[24] Apellate Body Report, *US-Shrimp*. The United States' restrictions on the import of tuna caught using techniques harmful to dolphins were the subject of challenge in GATT Panel Report, *US – Tuna I* and GATT Panel Report, *US-Tuna II*.
[25] Such 'eco-labels' are often supplied by NGOs such as the Maritime Stewardship Council (MSC), which was originally founded as a partnership between Unilever and the WWF, and which now operates independently to certify fisheries and to 'harness consumer preference': see www.msc.org. Consumer guides to seafood, which list species that are fished sustainably and species that are not, are also increasingly available around the world: see e.g. www.panda.org/what_we_do/how_we_work/conservation/marine/sustainable_fishing/sustainable_seafood/seafood_guides/.

these labels is uncertain,[26] and the issue is currently the subject of discussions between WTO members.[27]

The trade measures relating to fish products are an observable response to the environmental sustainability of the fishing sector and give rise to many questions about consistency with the WTO agreements and cohesiveness with fisheries regimes. This has led to disputes at the WTO, where panels and the Appellate Body have been asked to rule on questions relating to fisheries and fisheries sustainability.

B Fragmentation of international law

The largely autonomous and uncoordinated growth of regimes such as trade law, law of the sea, and species protection is part of a more general proliferation of laws and institutions. Regimes such as international investment,[28] international human rights[29] and international criminal law[30] have developed rapidly and independently.

These regimes represent parts of the 'system' of international law.[31] Yet, like the problem of fisheries sustainability, current issues do not implicate just one regime. For example, the problem of state corruption, which is addressed by a specialised UN Convention,[32] demands to be tackled by other institutions, including international arbitration tribunals called on to enforce suspect investment contracts. Restraints on the use of nuclear weapons under the laws of armed conduct are accompanied by

---

[26] But note that the US labelling programme for dolphin-safe tuna survived the GATT challenge in *US – Tuna I*, at 204. See generally Arthur Appleton, *Environmental Labelling Programmes: International Trade Law Implications* (1997). The issue has re-emerged in a current dispute at the WTO: see *United States – Measures Concerning the Importation, Marketing and Sale of Tuna and Tuna Products* (DS 381) (panel composed 14 December 2009).
[27] See WTO Doha Declaration (2001) para 32 (iii).
[28] This amorphous body of law is most holistically discussed when disputes arise: see the International Centre for Settlement of Investment Disputes: www.icsid.worldbank.org.
[29] As contained in a number of international and regional instruments: see e.g. *Universal Declaration of Human Rights* (1948).
[30] The most significant development of international criminal law has been the establishment of the International Criminal Court and its associated cases: see www.icc-cpi.int.
[31] See ILC Study Group, above n. 1. Cf H. L. A. Hart, *The Concept of Law* (1961) 230–1 ('it is submitted that there is no basic rule providing general criteria of validity for the rules of international law, and that the rules which are in fact operative constitute not a system but a set of rules, among which are the rules providing for the binding force of treaties').
[32] UN Convention against Corruption (2003).

trade restrictions and principles of environmental law.[33] The protection of individuals in times of armed conflict varies according to international human rights law and international humanitarian law.

The diversification and specialisation of international law gives rise to challenges of 'fragmentation', where there is potential for

> conflicts between rules or rule-systems, deviating institutional practices and, possibly, the loss of an overall perspective on the law.[34]

The proliferation of the regimes that seek to address fisheries sustainability gives rise to the possibility that the governance of fisheries is fragmented. It is unclear how international laws and institutions cohere, and indeed whether they could or *should* do so.[35]

The possibility that fisheries governance is fragmented gives rise to specific legal questions. For example, what is the legal basis for the WTO to cooperate with the FAO in the trade-related aspects of fisheries policies? Is it relevant that only some of the WTO members have joined the FAO? Do FAO Secretariat staff have observer-status at the WTO and if not, why not? Moreover, when fishery subsidy laws are negotiated at the WTO, under what conditions should states and secretariats collaborate with and scrutinise other regimes? Is such institutional interaction appropriate in the implementation of existing laws, such as recommendations to list endangered species on the appendices of CITES? What is the role of Memoranda of Understanding between secretariats? What about such collaboration in the resolution of disputes? Can adjudicating bodies within one regime, such as the panels of the WTO, apply laws, interpret treaties or establish relevant facts using the principles and evidence of another regime? What happens if the law of the sea and international environmental law advance conflicting principles, such as the precautionary approach to species protection versus a strict quantitative test? Which has priority? And which body should decide?

Uncertainty about the interaction between these regimes presents practical and conceptual problems for those responsible for maintaining,

---

[33] See *Use or Threat of Nuclear Weapons* (1996) ICJ Rep 226; See also Nuclear Supplier Group (NSG) *Guidelines for Transfers of Nuclear-Related Dual-Use Equipment, Materials, Software and Related Technology* (INFCIRC/254, Part 2), which govern the export of nuclear related dual-use items and technologies.

[34] ILC Analytical Study, above n. 1, para 8.

[35] A timely warning not to address the challenges of fragmentation with unreflective calls for coherence was provided by Martti Koskenniemi and Päivi Leino, 'Fragmentation of International Law? Postmodern Anxieties' (2002) 15 *Leiden Journal of International Law* 553.

understanding and complying with international law, and for international law's beneficiaries. To address this uncertainty, it is important to investigate the relevant laws and institutions in order to ascertain whether there is indeed fragmented global fisheries governance. The next step is to determine strategies to address problems associated with such fragmentation, not to contribute to coherence in international law (which may not be a useful goal in itself, given possible advantages of diversity),[36] but to promote effective efforts, for both institutions and the law, in achieving fisheries sustainability.

In domestic systems of law, overlapping or conflicting laws are resolvable by the 'sovereign' or even by the constitution itself,[37] and the courts act as a check on the domestic government's choices for public policies and the interaction between such policies. International law has a radically different structure, with scant notions of hierarchy.[38] Beyond pleading with states to have regard to the 'effect on the international statute book as a whole'[39] before entering into new laws, international lawyers gain little from domestic analogies in promoting solutions to the problems of fragmentation. Indeed such analogies may simply mislead.

In response to the international phenomenon of fragmentation, recent literature has concentrated on one particular aspect, namely the problem of conflicting norms.[40] This problem is usually manifest at the stage of dispute settlement in international law, when states have already negotiated and implemented relevant laws. For example, if the International Court of Justice (ICJ) is called upon to adjudicate the retaliation by one state against an alleged abuse of another state's diplomatic mission, it will permit the retaliation only if it is within the accepted law of diplomatic immunities and, as the Court did in 1980, may even go so far as to suggest that such law constitutes a 'self-contained regime'.[41] Disputing states may disagree over the applicable law relevant to their disputes, which may extend to the filing of disputes within different international

---

[36] Ibid.
[37] See e.g. section 109 of the Constitution of Australia, which provides that the laws of the Commonwealth shall prevail over those of a state to the extent of any inconsistency.
[38] Notions of hierarchy are implicit in a limited set of norms: see further below nn. 50 and 52.
[39] Wilfred Jenks, 'Conflict of Law-Making Treaties' (1953) 30 *BYIL* 401, 452.
[40] Joost Pauwelyn, *Conflict of Norms in Public International Law: How WTO Law Relates to Other Rules of International Law* (2003).
[41] *Consular Staff in Tehran (USA v. Iran)* [1979] ICJ Rep 7.

tribunals.[42] Even if the dispute is filed in only one tribunal, there may be disputes about whether the norms of a particular regime, such as WTO law, should have priority over the norms of another.[43]

The question of conflicting norms was considered by the International Law Commission (ILC) as part of its mandate to contribute to the progressive development and codification of international law. In 2002, it convened a Study Group to consider 'The Fragmentation and Diversification of International Law'.[44] The Study Group, chaired by Martti Koskenniemi, concentrated on situations where international norms operated in a relationship of interpretation or conflict.[45]

The report of the ILC Study Group in 2006 contained a series of recommendations to determine the primacy of existing international norms. For example, the Study Group discussed the rule of *lex specialis derogat legi generali*, which is based on the primacy of the specific over the general, and observed that a more specific treaty will usually trump the general treaty.[46] According to the same principle, a more specific regime will usually have priority over general international law.[47] The Study Group also discussed the principle of *lex posterior derogat legi priori*, which gives primacy to a more recent treaty over an earlier one,[48] and the harmonising effect of treaty interpretation under the Vienna Convention on the Law of Treaties (VCLT) Article 31(3)(c).[49] The Study Group pointed also to peremptory norms, norms 'accepted and recognized by the international

---

[42] See e.g. the dispute between the EU and Chile over swordfish: Marcos Orellana, 'The Swordfish Dispute between the EU and Chile at the ITLOS and the WTO' (2002) 71 *Nordic Journal of International Law* 55; see also the dispute between Ireland and the UK filed at an OSPAR arbitral tribunal, the ECJ and an UNCLOS Annex VII Tribunal: Robin Churchill and Joanne Scott, 'The Mox Plant Litigation: The First Half-Life' (2004) 53 *ICLQ* 643.
[43] In *EC – Biotech*, the United States, Argentina and Canada claimed that the relevant laws were the WTO agreements. It was arguably open to the EC, as respondent, to claim that various environmental Conventions had precedence: see further Chapter 5.
[44] See above n. 1.
[45] ILC Study Group Conclusions, above n. 1, Conclusion (2), 7–8 (distinguishing between situations where one norm assists in the interpretation of another and where the application of two norms would lead to incompatible decisions).
[46] Ibid., Conclusion (5), 8–9, citing, in particular *Mavrommatis Palestine Concessions* case, PCIJ Series A, No. 2 (1924) 31.
[47] Ibid., Conclusions (11)–(16), 12–13.
[48] VCLT, Art. 30. See ibid., Conclusions (24)–(30), 17–19.
[49] The Study Group called this a principle of 'systemic integration': see ILC Study Group Conclusions, Conclusions (17)–(23), 13–17. See further Campbell McLachlan, 'The Principle of Systemic Integration and Article 31(3)(c) of the Vienna Convention' (2005) 54 *ICLQ* 279.

community of States as a whole from which no derogation is permitted',[50] such as the prohibition of slavery or genocide. It emphasised that these norms will trump all others in the event of conflict.[51] Finally, the Study Group considered special treaty clauses within regimes, which set out the priority of conflicting norms. The Study Group referred in particular to Article 103 of the United Nations Charter, which states that members' obligations under the Charter will prevail over their obligations under any other international agreement.[52]

The ILC summarised a 'tool-box' of principles to assist in a wide range of situations where norms are found to conflict in international law. However, anyone faced with the fragmented response to fisheries sustainability may find only limited assistance from these principles. For example, species protection treaties are regularly updated by conferences of the parties.[53] A different approach exists for the FAO and its constitutive instruments, which are not regularly updated by members but which do encompass a changing work agenda. An application of the *lex posterior* principle might give primacy to the species protection regimes regardless of these differences.[54] Similarly, the coexistence between a number of soft-law principles, such as those encompassed in the FAO's Code of Conduct for Responsible Fisheries, and conventions such as CITES is difficult to assess because the 'tool-box' only applies to concrete legal norms.[55] The ILC approach may not sufficiently accommodate the managerial approach to treaty compliance that relies on cooperative rather than coercive approaches[56] or more general notions of governance in international law. [57]

---

[50] VCLT, Art. 53.   [51] ILC Study Group Conclusions, above n. 1, Conclusion (32), 20.
[52] Such priority was based on Art. 103 and the special character of the UN: see ibid., Conclusion (36), 22.
[53] See Robin Churchill and Geir Ulfstein, 'Autonomous Institutional Arrangements in Multilateral Environmental Agreements: A Little-Noticed Phenomenon in International Law' (2000) 94 *AJIL* 623.
[54] Indeed, the Study Group acknowledged the difficulty of applying this principle to norms from differing regimes: see ILC Study Group Conclusions, above n. 1, Conclusion (25).
[55] See the ILC Study Group's qualification that it would not consider issues of soft law: ILC Analytical Study, above n. 1, para 490.
[56] For an influential theory of compliance, see Abram Chayes and Antonia Handler Chayes, *The New Sovereignty: Compliance with International Regulatory Agreements* (1995).
[57] On the use of the term 'governance', see e.g. Joseph Weiler, 'The Geology of International Law – Governance, Democracy and Legitimacy' (2004) 64 *ZaöRV* 547, 550 ('Couple the regulatory layer of treaties with the international practice of management and a new form of international legal command may justifiably be conceptualized as governance').

Another example of the inadequacy of the ILC's 'tool-box' for fisheries governance is the trumping effect of peremptory norms. Rather than involving such norms, the sustainable utilisation of fisheries is an issue of justifiable (but often conflicting) economic, social and environmental interests.[58] In such a context, no single norm is peremptory, and the relationship between conflicting goals and interests cannot be resolved by giving a categorical priority to just one of them.[59]

The ILC Study Group's focus on conflicting norms also accounts for its rather limited relevance to the problem of fragmented fisheries governance. New laws affecting fisheries sustainability are being developed and implemented. These laws are *in progress* and yet there is considerable uncertainty about how regimes should interact in both their formulation and future application. In the context of the varying membership and objectives of different regimes, such uncertainty seems sometimes to extend to mutual suspicion. There is greater scope for international lawyers to contribute understanding and ideas about collaboration and cohesion within these regimes, rather than focusing on *ex post* rules determining priority in later disputes. Yet the ILC Study Group focused on the latter issue, the resolution of conflicts. As such, its 'tool-box' does not assist in providing principles of institutional cooperation and collaboration when states within one regime attempt to formulate new rules that impact upon another regime, or when states within one regime attempt to implement existing rules that have inter-regime consequences.

A further limitation of the ILC study is in its refusal to incorporate an institutional perspective in its analysis of fragmentation.[60] The Study Group's mandate expressly excluded an institutional analysis of normative

---

[58] Cf arguments that the Fish Stocks Agreement gives rise to obligations *erga omnes*: Jost Delbrück, '"Laws in the Public Interest" – Some Observations on the Foundations and Identification of *Erga Omnes* Norms in International Law' in Volkmar Götz, Peter Selmer and Rüdiger Wolfrum (eds.) *Liber Amicorum Günther Jaenicke: Zum 85. Geburtstag* (1998) 17, 26–7 ('the common interest in ... protecting endangered species is so overwhelming that no State may be permitted not to comply ... regardless of whether or not it has consented'). The emerging movement for 'environmental rights' will have a growing influence on the relevance of peremptory norms to fisheries.

[59] See, for a related view, Martti Koskenniemi, 'International Law: Constitutionalism, Managerialism and the Ethos of Legal Education' (2007) 1 *European Journal of Legal Studies* (online) 'there will be no hierarchy between the various legal regimes in any near future... not only because the international realm lacks a *pouvoir constituent* but because if such presented itself, it would be empire' (footnote omitted).

[60] The ILC's mandate is noted at ILC Analytical Study, above n. 1, 13 (para. 13).

conflicts for obvious reasons – as a part of the UN itself, an inquiry by the ILC into the competences of the ICJ vis-à-vis the International Criminal Tribunal for the former Yugoslavia (ICTY) or the WTO vis-à-vis the International Tribunal for the Law of the Sea (ITLOS) would have been controversial. Perhaps in recognition of the limitations caused by this restricted mandate, the ILC emphasised in its conclusion that there was a need for increased attention 'to the collision of norms and regimes and the rules, methods and techniques for dealing with such collisions'.[61] In addition to its 'tool-box' recommendations, the Study Group acknowledged the need for such attention to be focused on 'the notion and operation of "regimes"'.[62]

Legal scholarship emerging from the fragmentation phenomenon has not had to be as circumspect in its analysis of institutions. A body of work that has considered linkages between international trade law and other areas such as human rights and environmental law has closely engaged with institutions such as the WTO.[63] This field of study, known colloquially as 'trade and ...', has turned its attention to the functional strengths of the trade regime and has considered whether and how it can achieve other objectives. Early writers sought to bolster other regimes so as to achieve more balance against the strength of the trade regime, through such institutional reforms as the unsuccessful proposal for the establishment of a World Environment Court.[64] Other writers called instead for the restructuring of the trading regime so as to incorporate the perspectives and voices of other policy areas.[65]

Like the ILC Study, the 'trade and ...' literature focuses largely on dispute settlement.[66] A somewhat different approach is taken by

---

[61] Ibid., 249 (para 493). [62] Ibid., 249.

[63] See, among many contributions, the symposia on 'Linkage as Phenomenon: An Interdisciplinary Approach' (1998) 19:2 *University of Pennsylvania Journal of International Economic Law*, and 'Boundaries of the WTO' (2002) 96:1 *AJIL*.

[64] See e.g. the call by Daniel Esty and others for the establishment of a Global Environment Organisation: *Greening the GATT: Trade, Environment, and the Future* (1994); 'Stepping Up to the Global Environmental Challenge' (1996–97) 8 *Fordham Environmental Law Journal* 103.

[65] Robert Howse, 'From Politics to Technocracy – and Back Again: The Fate of the Multilateral Trading Regime' (2002) 96 *AJIL* 94.

[66] One exception is Kenneth Abbott and Duncan Snidal, 'International Action on Bribery and Corruption: Why the Dog Didn't Bark in the WTO' in Daniel Kennedy and James Southwick (eds.) *The Political Economy of International Trade Law: Essays in Honour of Robert E Hudec* (2002) 177, 177 (emphasising the benefit of considering a policy the WTO has failed to address).

writers in the field of systems theory, and specifically those who focus on the study of 'communicative systems'.[67] Scholars emerging from this tradition have acknowledged that the conflict between regimes in international law reflects wider societal conflicts. Rather than attempting to develop rules to achieve harmony in the event of conflict, these writers focus on the practices within the specific 'discursive networks' of the regimes, demonstrating certain biases and preferences. Empirical work within international trade law, for example, has led to findings that there is a bias against environmental principles within much of its discourse.[68] Although WTO adjudication features heavily in the analysis, other institutional aspects are considered as well.

Studies that address the law-making power of international organisations also provide a useful institutional perspective.[69] Although not focused on the laws produced by the interaction *between* institutions, this literature is invaluable in directing attention to the influence of international organisations and in demonstrating the diversity of their constituent parts; aside from their member states, IGOs are made up of secretariat staff and are advised by expert bodies and, in some cases, non-governmental organisations (NGOs), and these all influence law-making.[70] Similarly, studies carried out under the rubric of 'global administrative law' note the role of 'transnational administrative bodies', including IGOs and informal groups of officials, which are not directly controlled by states, but which shape law through administrative and regulatory practices.[71]

Diverse actors and non-traditional conceptions of law-making are also the focus of a body of work that investigates 'new governance'.[72] This literature, which concentrates in the main on the legal jurisdictions of the European Union and the United States, investigates how

---

[67] Andreas Fischer-Lescano and Gunther Teubner, 'Regime-collisions: The Vain Search for Legal Unity in the Fragmentation of Global Law' (2004) 25 *Michigan Journal of International Law* 999; Oren Perez, *Ecological Sensitivity and Global Legal Pluralism: Rethinking the Trade and Environment Conflict* (2004).
[68] Perez's work applies also to international construction contracts: ibid.
[69] José Alvarez, *International Organizations as Law-Makers* (2005).
[70] Ibid. 607–10. See also Alan Boyle and Christine Chinkin, *The Making of International Law* (2007) 41–93.
[71] See Benedict Kingsbury, Nico Krisch and Richard Stewart, 'The Emergence of Global Administrative Law' (2005) 68 *Law and Contemporary Problems* 15.
[72] Gráinne de Búrca and Joanne Scott, 'Introduction: New Governance, Law and Constitutionalism' in Gráinne de Búrca and Joanne Scott (eds.) *Law and New Governance in the EU and the US* (2006) 1; see also symposium 'Narrowing the Gap? Law and New Approaches to Governance in the European Union' (2007) 13:3 *Columbia Journal of European Law*.

laws are made and shaped by practices and processes that fall outside traditional conceptions of regulation.[73] Empirical studies of different policy areas such as health care, environmental protection and employment practices are brought together to contribute to understanding about the effect of diverse governing structures.[74] By studying the involvement of actors who are 'affected' by these policy areas, rather than simply those actors who have representative functions, this body of work seeks to accommodate 'stakeholders'. By moving beyond traditional conceptions of 'government', it incorporates into its analysis instruments of soft law such as voluntary codes, non-binding action plans and declarations.

The approaches of scholars that are attentive to 'new governance', 'global administrative law' or other institutional perspectives are useful for addressing the challenges of fragmented laws and institutions relevant to fisheries sustainability. The range of soft-law instruments such as the FAO Code of Conduct, the involvement of a multitude of stakeholders on whom fisheries enforcement and monitoring depends, including non-state actors such as the World Wide Fund for Nature (WWF), and the dependence upon IGO secretariats for the administration of several interconnecting policies, calls for a broad and inclusive methodology that moves beyond a focus on states.

C Scope, methodology and use of terms

1 A case-study approach

In contrast to the approach of the ILC Study Group, I have incorporated an institutional analysis. My methodology has been empirically driven, in that I have closely followed the norms and practices of the FAO, CITES, the WTO and other relevant IGOs, and conducted interviews with key participants. My analysis extends to a wide range of actors

---

[73] The term 'regulation' is commonly further narrowed by the preceding terms 'command-and-control'.
[74] See e.g. Louise Trubek, 'New Governance Practices in US Health Care' in Gráinne de Búrca and Joanne Scott (eds.) *Law and New Governance in the EU and the US* (2006) 245; Bradley Karkkainen, 'Information-forcing Regulation and Environmental Governance' in ibid., 293; Susan Sturm, 'The Architecture of Inclusion: Advancing Workplace Equity in Higher Education' (2006) 29 *Harvard Journal of Law & Gender* 247; Joshua Cohen and Charles Sabel, 'Sovereignty and Solidarity: EU and US' in Jonathan Zeitlin and David Trubek (eds.) *Governing Work and Welfare in a New Economy: European and American Experiments* (2003) 345.

affected by fisheries sustainability. I have used the language of 'governance' to refer collectively to both traditional forms of regulation and soft-law approaches to fisheries sustainability. In attempting to move beyond an analysis of conflicting norms, I have assessed problems of fragmentation that arise during the full spectrum of international law-making and implementation.

This methodology acknowledges that international law is a process.[75] One way to understand this process is to separate three main stages of law-making, implementation and adjudication, notwithstanding obvious overlap between them.[76] These 'stages' are not always temporally sequenced, especially as law may be 'made' before it is formally negotiated or implemented.[77] Problems associated with the fragmentation of international law impact upon all three stages. My research is based on the hypothesis that an enhanced understanding about the processes and principles of regime interaction from one stage will enhance an understanding about another stage. Moreover, the examples from fisheries governance will be relevant to the problems raised by the phenomenon of fragmentation in other areas of international law.[78]

In aiming to provide greater understanding about actual regime interaction – about the actors, the processes and the relevant legal frameworks that impact upon the normative and institutional inter-regime relations – the book selects examples from law-making, implementation and dispute settlement in contested areas of global fisheries governance, as follows:

(i) The negotiation of multilateral rules on fisheries subsidies at the WTO. This example is useful because new law at the WTO is being made amongst existing soft-law norms relating to state aid to the fishing

---

[75] On international law as a process, see Rosalyn Higgins, *Problems and Process: International Law and How We Use It* (1994) and associated writings of the New Haven School of policy-oriented jurisprudence. See also Harold Koh, 'Why Do Nations Obey International Law?' (1997) 106 *Yale Law Journal* 2599.

[76] Cf New Haven Policy School's conception of legal process as comprising three communicative streams, 'policy content, authority signal and control intention': W. Michael Reisman, 'International Lawmaking: A Process of Communication' (1981) 75 *American Society of International Law Proceedings* 101.

[77] As implicit in Lauterpacht's judicial function: see Hersch Lauterpacht, *The Function of Law in the International Community* (1933).

[78] One similarity I share with the New Haven Policy School's methodology is my selection of a policy area to discuss wider legal and institutional issues; see e.g. Johnston's selection of fisheries to demonstrate the policy-oriented approach: Douglas Johnston, *The International Law of Fisheries: A Framework for Policy-Oriented Inquiries* (1985; first published 1965).

sector, which are overseen by the FAO. The ensuing challenges include considerable uncertainty within the WTO about whether FAO experts should participate in the negotiations, whether the FAO should play a role of peer reviewer in the implementation of the resulting rules and whether the conduct of WTO members should be regulated according to existing FAO instruments. NGOs such as the WWF have also had a major monitoring role that may be difficult to accommodate by the WTO. Moreover, although most of the states formulating these rules are members of both the WTO and the FAO, the membership of these institutions is not identical. This normative and institutional fragmentation has major effects on efforts to make law.

(ii) The restriction of trade in marine species according to the CITES regime. This example demonstrates the implementation of CITES norms amongst existing FAO rules and procedures. The proposal to add marine species to CITES' remit is part of the ongoing implementation of CITES which takes place through regular conferences to update the listings of endangered species. Some states consider that a greater role for CITES will be an incursion on the role of the FAO and its associated fisheries management laws. The membership of CITES and the FAO is not identical, although the states that are apparently distrustful of CITES are parties to it. Amidst these disagreements, the Secretariats have had to collaborate on scientific issues, and NGOs have sought access to discussions between the two bodies. The development of a Memorandum of Understanding between the CITES Secretariat and the FAO demands analysis.

(iii) The adjudication of a trade dispute involving a fisheries import ban. In *US – Shrimp*, a WTO panel was asked to rule upon the WTO-consistency of an import ban on shrimp that was claimed to have the environmental objective of reducing turtle mortality. A number of international environmental laws were relevant, yet it was unclear how much the WTO panel could take them into account as applicable law or interpretative tools, or in establishing facts. The settlement of disputes such as this provides useful lessons for regime interaction.

These case-studies involve a close analysis of the institutional impact of uncertainties about the legal basis of regime interaction. The uncertainties are not apparent from a purely textual analysis. My methodology draws on interviews with secretariat staff and country delegates as well as my own experience at international organisations.[79] For example, members of the WTO Secretariat have freely used dictionaries to find

---

[79] Empirical engagement with the professional practice of international law has not, of course, been limited to the New Haven Policy School: see further, Martti Koskenniemi, *The Gentle Civilizer of Nations: The Rise and Fall of International Law, 1870–1960* (2002) 474–80.

meanings of particular words, but have been reticent about using manuals produced by other IGOs. FAO staff were keen to enter into a Memorandum of Understanding with their counterparts at the CITES Secretariat, but had to wait until the states that were opposed to CITES listings of marine species formally agreed to the text. NGOs with proven expertise in fisheries were blocked from some meetings in some regimes, while apparently less credible NGOs were given greater scope in others. Such institutional difficulties have led me to address legal and theoretical problems of regime interaction that have too often been ignored. Moreover, by engaging closely with particular subject areas across the FAO, the WTO and CITES, I have confronted the assumption prevalent in much of the literature on 'trade and environment' that the WTO is the only international forum of relevance.[80] The book accounts for the pluralism and institutional diversity of trade-related aspects of fisheries.

2 The vocabulary of regimes

'Regimes' is the term I use to describe a set of laws, processes and institutions that have evolved by addressing a particular problem or function.[81] This definition can be compared to the notion of 'self-contained regimes', which describes unions of rules,[82] or even fields of professional or academic specialisation.[83] My conception of 'regimes'

---

[80] Perez refers to 'Greens' and 'Free-traders' who both assume that 'the WTO constitutes the epitome of the trade and environment conflict': see Perez, above n. 67, 7; such assumptions are also observed by Andrew Lang, 'Reflecting on 'Linkage': Cognitive and Institutional Change in The International Trading System' (2007) 70 *Modern Law Review* 523.

[81] Cf Steven Ratner, 'Regulatory Takings in Institutional Context: Beyond the Fear of Fragmented International Law' (2008) 102 *AJIL* 475, 485 ('[t]o say that there is a regime on X ... is not quite the same as saying that there is a law governing X, as it requires institutions as well').

[82] See B Simma, 'Self-Contained Regimes' (1985) 16 *Netherlands Yearbook of International Law* 111, 115 and 117 (drawing on the *Tehran Hostages* case, above n. 41, and discussing 'sub-systems' of international law that were defined as 'a set of primary rules linked to the specific legal consequences of their breach'; self-contained regimes were a limited sub-category of sub-systems of international law, which embraced 'a full (exhaustive and definite) set of secondary rules').

[83] The ILC Analytical Study compared three notions of 'self-contained regimes'. See ILC Analytical Study, above n. 1 at 68–70 ('In a narrow sense, the term is used to denote a special set of secondary rules under the law of State responsibility that claims primacy to the general rules concerning consequences of a violation. In a broader sense, the term is used to refer to interrelated wholes of primary and secondary rules, sometimes also referred to as 'systems' or 'subsystems' of rules that cover some

also contrasts with the definition proposed in international relations (IR) scholarship, which conceives of regimes as encompassing sets of norms and procedures.[84]

Although international legal scholars had studied the operation of regimes for many years, regimes became the subject of close attention of IR scholars in the 1980s, when political scientists called for greater attention to the operation of discrete groups of norms such as international economic law.[85] The attention of IR scholars to the operation of regimes arguably changed the emphasis of the discipline from the realist political assumption that states rarely cooperate on issues outside their direct national interest towards an acknowledgement of the possibility of regular international cooperation.[86] The IR regime literature usually analyses only one regime, rather than regime interactions. However, there is a growing body of work on the role of 'epistemic communities' in building and bridging regimes,[87] and of the coordination by treaty negotiators of

> particular problem differently from the way it would be covered under general law . . . . Sometimes whole fields of functional specialization, of diplomatic and academic expertise, are described as self-contained (whether or not that word is used) in the sense that special rules and techniques of interpretation and administration are thought to apply' [footnote omitted]). It is this third notion of self-contained regimes that relates most closely to my conception of regimes. See also Fischer-Lescano and Teubner, above n. 67, 1013.

[84] Stephen Krasner, 'Structural Causes and Regime Consequences: Regimes as Intervening Variables' in Stephen Krasner (ed.) *International Regimes* (1983) 2 ('Regimes can be defined as sets of implicit or explicit principles, norms, rules, and decision-making procedures around which actors' expectations converge in a given area of international relations'). On regime theory and international law, see Anne-Marie Slaughter, 'International Law and International Relations Theory: A Dual Agenda' (1993) 87 *AJIL* 205, 206 ('As international lawyers soon realised, [regime theory] was international law by another name').

[85] Ibid.; see also Robert Keohane and Joseph Nye, 'The Club Model of Multilateral Cooperation and Problems of Democratic Legitimacy' in Roger Porter *et al.* (eds.) *Efficiency, Equity, and Legitimacy: The Multilateral Trading System at the Millennium* (2001) 264, 266 (discussing how regimes, 'decomposable' from the rest of the system, have operated as a 'club'). For use of regime theory in the WTO context, see Mary Footer, *An Institutional and Normative Analysis of the World Trade Organization* (2006).

[86] Political realism in the IR tradition usually assumes that states are primarily motivated by the desire for military and economic power or security, rather than by ideals or ethics: see Hans Morgenthau, *Politics Among Nations* (2nd edn, 1954); see further Slaughter, above n. 84. For a description of the 'normative turn' in IR scholarship, notably among constructivists, and advocacy of an alternative 'interactional' theory of international law that emphasises law's internal morality, see Jutta Brunnée and Stephen Toope, 'International Law and Constructivism: Elements of an Interactional Theory of International Law' (2000) *Columbia Journal of Transnational Law* 19, 24ff.

[87] Peter Haas, 'Introduction: Epistemic Communities and International Policy Coordination' (1992) 46 *International Organization* 1.

'regime complexes'.⁸⁸ Scholars have begun to consider the interaction of regimes within the environmental and fisheries fields.⁸⁹

Legal scholars, too, emphasise the role of institutions as 'organizational catalysts' that build bridges between disciplines and actors.⁹⁰ Ethnographic fieldwork within international organisations is particularly useful to analyse how institutional culture affects the ability of regimes to incorporate international norms.⁹¹ Such approaches may be particularly useful in confronting the risk posed by 'the politics of regime definition',⁹² where the reduction of a 'trade' or 'environment' regime to a single policy upon which to ground cooperation forestalls political contestation. In documenting and investigating the interaction between regimes affecting fisheries sustainability, this book draws on empirical accounts and seeks to avoid essentialising characterisations of regimes.

In describing and analysing instances of regime interaction, it is often necessary to refer to particular sets of norms within particular regimes such as 'the WTO covered agreements' or 'multilateral environmental agreements'. The reverse situation of referring to laws by reference to their position vis-à-vis other regimes also gives rise to problems. For example, it is tempting to call the *pacta sunt servanda* principle a 'non-WTO norm' vis-à-vis the WTO covered agreements, but this is misleading. The *pacta sunt servanda* norm is exogenous to the WTO regime in the sense that it was not part of the positive law developed by the WTO negotiators, but it nevertheless applies to the WTO regime. The WTO covered agreements are part of the international legal system but not endogenous to the law of the sea. I have found it necessary to refer at times to 'non-WTO sources'⁹³ to describe treaties and other instruments

---

⁸⁸ Kal Raustiala and David Victor, 'The Regime Complex for Plant Genetic Resources' (2004) 58 *International Organization* 277.

⁸⁹ See e.g. Oran Young, *International Cooperation: Building Regimes for Natural Resources and the Environment* (1989); Olav Schram Stokke (ed.) *Governing High Seas Fisheries: The Interplay of Global and Regional Regimes* (2001); see also Sebastian Oberthür and Thomas Gehring (eds.), *Institutional Interaction in Global Environmental Governance: Synergy and Conflict Among International and EU Policies* (2006).

⁹⁰ See e.g. Sturm, above n. 74.

⁹¹ Galit Sarfaty, 'Why Culture Matters in International Institutions: The Marginality of Human Rights at the World Bank' (2009) 103 *AJIL* 647.

⁹² The phrase is from Martti Koskenniemi, 'The Fate of Public International Law: Between Technique and Politics' (2007) 70 *Modern Law Review* 1, 27.

⁹³ I note that the ILC Study Group uses the phrase 'non-WTO law' to describe international law that is extrinsic to the WTO covered agreements: see e.g. ILC Analytical Study, above n. 1, 88 (para 167). On 'sources' of international law, see Statute of the International Court of Justice, Art. 38.

produced by international bodies outside of the WTO, but such language should be read with this caveat in mind.

I make further distinctions in my analysis of regime interaction. I draw on the distinction made between the language of 'organisations' and 'institutions'.[94] However, the distinction made by IR scholars between 'collaboration' and 'cooperation' within systems is not reproduced in this work.[95] More useful is the differentiation between the mutual support and learning provided between regimes (in essence, a kind of collaboration, most clearly demonstrated by secretariat interaction), and the various kinds of critical review to which one regime subjects another (more aptly described as 'scrutiny'). Related to this latter type of regime interaction, the phrase 'cross-fertilisation' is often used to describe the influence of norms from a different jurisdiction or subject area in international jurisprudence. Others have described the ability of an adjudicatory panel to have regard to exogenous norms as 'cognitive openness'.[96] I treat collaboration, scrutiny and cross-fertilisation as all encompassed by the phrase 'regime interaction'.

Notwithstanding similarities in the use of terms, my research extends beyond the IR literature by focusing on the *legal basis* for regime interaction and the associated problematic theoretical and conceptual issues. In particular, the book tackles the notion and consequences of international law as a legal system.

3 The legal framework for regime interaction

The focus of the present study is on the legal framework for the interaction between different regimes in fisheries governance. The case studies demonstrate how selected regimes interact at different stages. My analysis of these case studies and related examples extends to a consideration of how regimes *should* interact.

On one level, the book is an investigation into the capability of a fragmented international legal system to advance the goal of fisheries

---

[94] Institutions are defined as principles and processes, a subset of which are organisations, agreements and commitments: see, e.g., Young, above n. 89.
[95] See distinctions between coordination games and collaboration games in systems analysis described by David Victor, Kal Raustiala and Eugene Skolnikoff, 'Introduction and Overview' in David Victor, Kal Raustiala and Eugene Skolnikoff (eds.) *The Implementation and Effectiveness of International Environmental Commitments: Theory and Practice* (1998) 1, 10, and sources contained therein.
[96] Perez, above n. 67, 68. Perez describes as 'ecological cognizance' the ability of the WTO to take into account the environmental implications of trade liberalisation: at 58.

sustainability. On another level, it is an inquiry into the co-existence of regime interaction with several major tenets of international law. For example, traditional accounts of 'state sovereignty' in international law emphasise the state as supreme authority. Kelsen preferred to describe these accounts as perspectives emphasising the 'primacy of national law'.[97] The perspectives consider that although states are bound by the general system of international law, they are not subject to specific international laws except by their consent;[98] Crawford implies that this is a rebuttable presumption.[99] Within particular regimes, an emphasis on state sovereignty also seeks to contain the ability of adjudicative organs to 'add to or diminish the rights and obligations' of states as set out in treaty text, as provided in the oft-cited Understanding on Rules and Procedures Governing the Settlement of Disputes (DSU) of the WTO.[100] These perspectives operate ostensibly to safeguard international law's legitimacy.

An emphasis on state sovereignty in international law may well be tested in the meeting of international regimes. For example, a state that has joined the WTO but not the FAO may find itself subject to FAO norms if the FAO has an active role as peer reviewer of WTO disciplines on fisheries subsidies. If that state is a party to CITES, its rights might be diminished by the application of FAO scientific principles to the listing of marine species on the CITES appendices. On the other hand, a state that is not a party to CITES may nonetheless find that its WTO rights are affected by a listing of species if such a listing is found to be determinative in a WTO dispute to which it is a party.

If regime interaction is *not* based on the consent of the formal participants of international law – the states themselves – it is important to

---

[97] Hans Kelsen, 'Sovereignty and International Law' (1960) 48 *Georgetown Law Journal* 627, 634 (this definition related to the monist conception; Kelsen rejected the dualist conception of international and national law on different grounds). Kelsen noted that such perspectives led to false conclusions such as that states were not bound by majority decisions of a collegiate organ or tribunal: at 637.
[98] An extreme formulation of this principle is found in the *Lotus* case, where the Permanent Court stated that '[r]estrictions upon the independence of states cannot therefore be presumed'. This is commonly assessed under the principle of 'sovereignty' (see e.g. Ian Brownlie, *Principles of Public International Law* (2003) 74) or as the right of a state to 'independence' (see e.g. Malcolm Shaw, *International Law* (2001) 150). See also the principle that a treaty does not create either obligations or rights for a third state without its consent: VCLT Art 34 (*pacta tertiis*).
[99] James Crawford, *The Creation of States in International Law* (2nd edn, 2006) 41 ('Derogations from these principles will not be presumed').
[100] DSU, Art. 3.2.

inquire into other bases for regime interaction and to ascertain whether these alternative bases are legitimate. For example, the use by a WTO panel of exogenous sources from a number of IGOs may depend on the ability of that panel to differentiate between the authority and relevance of a large number of instruments. Equally, a WTO panel that admits *amicus* briefs from NGOs may be inundated by a range of actors of varying levels of 'credibility'.[101] An acceptance of unchecked regime interaction risks arbitrary and undisciplined decision-making. Relevant to these questions are ideas for possible alternative theoretical and legal frameworks for accountability and legitimacy in regime interaction in global fisheries governance and in public international law.

4 *The stakeholders*

To give context to the global problem of fisheries sustainability, and to the case studies, it is important to identify the multiple interests involved.[102] A multitude of differing interests are put forward by states in the regimes of the WTO, CITES and the law of the sea. Coastal states and the states that seek to fish in distant waters often have divergent interests.[103] In addition to national interests in fishing opportunities, there is a growing global fish trade.[104]

Developing countries in particular have a significant interest in the trade in fish and fishery products. The net exports of developing countries have grown considerably in recent decades, and the value of the fish trade is now significantly higher than other agricultural commodities such as rice, meat, coffee, rubber and tea combined.[105] There is thus a significant 'development' component to the national interests in

---

[101] The concept of credibility demands analysis in its own right, and may include notions of representativeness, accountability and other factors: see further Chapter 7.
[102] A methodology that is promoted, for example, by such diverse approaches as the New Haven Policy School (see above n. 75) and new governance (see above n. 72).
[103] Major distant-water fishing nations include China, Japan, Korea, Poland, Russia, Spain, the Taiwan province and the United States: see e.g. United Nations Department of Public Information, 'Special Session of the General Assembly to Review and Appraise the Implementation of Agenda 21' (23–27 June 1997) www.un.org/ecosocdev/geninfo/sustdev/fishery.htm.
[104] The countries that export the largest amount of fish and fishery products in value terms are, respectively, China, Norway, Thailand, the United States, Denmark, Canada, Spain, Chile, Netherlands and Vietnam: SOFIA (2006), Table 11. The countries that import the largest amount of fish and fishery products in value terms are Japan, the United States, Spain, France, Italy, China, United Kingdom, Germany, Denmark and Korea.
[105] Ibid., 44.

fisheries. A variant of this interest in trade is the fact that many countries consider fisheries to be integral to their 'food security'.[106] Some countries may also have economic interests apart from the trade in fish products, such as the maintenance of ecology and biodiversity for ecotourism and other commercial activities. The vested interests of these countries in particular species and ecosystems may be quite different from those nations that simply wish to ensure a continuing harvest.

It is also necessary to accommodate interests in fisheries sustainability that are advanced in the forums I study, but *not* by participating states. In general terms, the receptiveness of the modern administrative state to goals of industrial coordination and efficiency has been contrasted to its poor ability to define environmental problems.[107] Many interests in fisheries are promoted by non-state actors rather than states. For example, consumers concerned about biodiversity and ecology have an interest in knowing the source of fish products through eco-labelling, yet states rarely conceive of eco-labels as part of the 'national interest'.[108] Also commonly outside the purview of states is the cultural and religious significance that fish hold for many people.[109] Relatedly, the interests of artisanal and subsistence fishermen are often different from those fishermen employed by the commercial sector, and may not find voice in state representation. Where advocates for these interests exist,[110] they are commonly environmental and consumer NGOs.[111]

---

[106] Fish is said to be the source of at least 20 per cent of the daily protein intake of 2.6 billion people, mostly in developing countries: see (2007) 11:16 *Bridges Digest* (9 May 2007). On food security, see FAO, *Code of Conduct for Responsible Fisheries* (1995), Art. 6.2.

[107] Douglas Torgerson, 'Limits of the Administrative Mind: The Problem of Defining Environmental Problems' in John Dryzek and David Schlosberg (eds.) *Debating the Earth* (1999) 110.

[108] In general terms, states meeting at the WTO have historically advanced *producer*-led rather than *consumer*-led interests: see e.g. Ernst-Ulrich Petersmann, 'Addressing Institutional Challenges to the WTO in the New Millennium: A Longer-Term Perspective' (2005) 8 *JIEL* 647. But note emerging efforts to promote labelling by the EU: see Mar Campins Eritja (ed.) *Sustainability Labelling and Certification* (2004).

[109] Apart from the historic religious significance of fish, see religious fishing practices noted in Jennings *et al.*, above n. 3, 117–18.

[110] On the rise of advocacy networks, see e.g. Margaret E Keck and Kathryn Sikkink, *Activists Beyond Borders: Advocacy Networks in International Politics* (1998); on the growth of networks of bureaucrats and elites with purported representative functions, see Anne-Marie Slaughter, *A New World Order* (2005).

[111] Although there is no generally accepted definition of NGOs, a useful set of characteristics is offered by Anna-Karin Lindblom, *Non-Governmental Organisations in*

These differing interests and perspectives in fisheries are not mutually exclusive or exhaustive.[112] It is useful to articulate them, not only to give context to the problem addressed by this book, but also to give a sense of the number and types of relevant 'stakeholders' that have an interest in fisheries sustainability. These stakeholders are not merely states, and a study into regime interaction must take account of their influence and participation. NGOs, in particular, may play an important role in the interaction between regimes, possibly because they are attempting to fill the perceived gap in one regime (caused by an absence of state representation) by recourse to another.

This conception of the 'stakeholders' in fisheries sustainability may appear controversial. The sovereign theory of international law assumes that states sufficiently represent the interests of all their citizens, without inquiring about the methods and success of that representation. This assumption has been historically linked to early influences on the emerging structure of international law that gave primacy to states over individuals.[113] States have deviated from this assumption to enforce new forms of state sovereignty,[114] but have not directly extended participation in international law to individuals.[115]

The assumption that states have a monopoly of representation of the interests of citizens is uncritically reproduced by many trade lawyers and scholars.[116] By contrast, a body of international literature of various

---

International Law (2005) 36, 46–52 (NGOs are non-governmental, non-profit-making organisations, which do not use or promote violence and which have democratic and representative structures, but not necessarily legal personality).

[112] For a similar attempt to list the perspectives of those interested in reducing overfishing, see Christopher Stone, 'Too Many Fishing Boats, Too Few Fish: Can Trade Laws Trim Subsidies and Restore the Balance in Global Fisheries?' in Kevin P Gallagher and Jacob Werksman (eds.) *International Trade and Sustainable Development* (2002) 286, 289 (listing perspectives as 'consumer satisfaction', 'food security', 'biodiversity' and 'regional development').

[113] See Allott describing the influence of Vattel's idea that 'Nations or sovereign States must be regarded as so many free persons living together in the state of nature': Philip Allott, *The Health of Nations: Society and Law beyond the State* (2002) 58.

[114] See further Simon Chesterman, *Just War or Just Peace? Humanitarian Intervention and International Law* (2002); see also Susan Marks, *The Riddle of All Constitutions: International Law, Democracy, and the Critique of Ideology* (2003).

[115] Exceptions might exist in human rights, which accrue regardless of citizenship, notwithstanding that, in practice, such norms can only be enforced against the state. On the protection of the individual in international law, see, e.g., (2007) 20 *Leiden Journal of International Law* 729ff. In addition, the 'direct effect' of European Union law attests to a vertical relationship between supranational law and the individual.

[116] See, e.g., Brian Hindley, 'What Subjects are Suitable for WTO Agreement?' in Daniel Kennedy and James Southwick (eds.) *The Political Economy of International Trade Law:*

critical[117] and prescriptive[118] bents challenges this assumption. From a variant tradition, public choice literature acknowledges that trade policy has often been pursued on the basis of lobbying by special interest groups who seek protection for their industries; trade delegations clearly do not represent all interests affected by trade policy.[119] This study also recognises the limitations of a model of international law-making that preserves states as sole units of representation for all affected parties, and I attempt to move beyond state structures in recognising the interests of those affected by fisheries governance.

An attempt to categorise all those 'affected by fisheries governance' is of course problematic. The 'affected by' standard of inclusiveness has been criticised as wide, vague and unworkable.[120] In short, constituting the demos has been labelled as one of the most neglected parts of political theory.[121] One response proposes a two-part standard, where an individual is 'affected by' a decision or action of a polity if: (i) the decision or action 'imposes governance norms on her' (the principle of self-governance); or (ii) the decision or action 'could otherwise reasonably be expected to impose external costs on her' (the principle of internalisation).[122] These principles help to make workable an idea that the book develops practically through engagement with case studies.

Many people and groups have an interest in fisheries sustainability and are affected by the interaction of trade, environmental and fisheries regimes. The regimes impose governance norms on some individuals by

---

Essays in Honour of Robert E. Hudec (2002) 157, 168 ('The convention that governments represent the interests of their populations, however hard it may sometimes be to accept it, seems to be the only available foundation for a viable constitution for international trading relations').

[117] See, e.g., Joseph Stiglitz, *Globalization and its Discontents* (2002) 22–3 ('we have a system ... in which a few institutions – the World Bank, the IMF, the WTO – and a few players – the finance, commerce and trade ministries, closely linked to certain financial and commercial interests – dominate the scene, but in which many of those affected by their decisions are left almost voiceless').
[118] Philip Allott, *Eunomia: New Order for a New World* (1990).
[119] See, e.g., studies in relation to US government protection of its steel producers cited in Dan Sarooshi, *International Organizations and their Exercise of Sovereign Powers* (2005) 80.
[120] See e.g. Robert Keohane, 'Global Governance and Democratic Accountability' in David Held and Mathias Koenig-Archibugi (eds.) *Taming Globalization* (2003) 130, 141 (if 'being affected' led to consultation rights, 'virtually nothing could ever be done').
[121] Robert E Goodin, 'Enfranchising All Affected Interests, and Its Alternatives' (2007) 35 *Philosophy & Public Affairs* 1, 41. The issue is also known as the 'boundary problem' or 'the problem of inclusion'.
[122] Eric Cavallero, 'Federative Global Democracy' (2009) 40 *Metaphilosophy* 42, 55.

requiring particular action, such as the denial of entry into ports of vessels suspected of engaging in illegal fishing.[123] In addition, there are many decisions and actions that impose external costs on individuals, such as the subsidising practices of fishing nations.

The following non-exhaustive list includes a number of participants in formal or informal processes that are relevant to the case studies:

(i) Participating states of the WTO, the FAO, CITES and other relevant IGOs such as The United Nations Environment Programme (UNEP). Membership of these regimes is overlapping for many states, but is not uniform. It is clear that the interests of states vary from the major fishing nations – including Japan, Spain, Portugal, Norway, the United States, Canada, Korea, Taiwan and increasingly China[124] – to the nations for whom commercial fishing is less of a priority.

(ii) Secretariats of the IGOs identified in (i).

(iii) NGOs – their members and employees, including where relevant marine scientists and other fisheries experts – such as the WWF, the International Centre for Trade and Sustainable Development (ICTSD), Oceana, and Humane Society International (HSI).

(iv) Standard-setting bodies, including private organisations such as the Maritime Stewardship Council (MSC).

(v) Individuals and companies who work and profit from the fisheries sector and their trade unions and lobby groups.

(vi) Individuals and companies who work in and profit from sectors that compete with the fish sector, such as land-based meat production (which competes as a protein source), or ecotourism (whose interest is in preserving commercially fished species).[125]

(vii) Individual consumers and their representatives, such as the UK consumer association that lobbied in the *EC-Sardines* dispute.[126]

This list is necessarily non-exhaustive, and is compiled according to those who have contributed to current debates in fisheries governance, either through formal or informal avenues. There will be many voices excluded from the list. For example, individuals with extrinsic interests in fisheries conservation who have not expressed their views through membership of any groups or more general participation – either through domestic politics, consumer behaviour or civil society groups such as NGOs – do not appear as participating stakeholders, although they may meet the 'affected by' standard. These gaps must be borne in

---

[123] See e.g. *Agreement on Port State Measures to Prevent, Deter and Eliminate Illegal, Unreported and Unregulated Fishing*, agreed in 2009 but not yet in force. See further Chapter 2, p. 43.
[124] See e.g. the major subsidising nations discussed in Chapter 3.
[125] See further Chapter 4.   [126] See further Chapter 5.

mind in the attempt to monitor the participation of stakeholders in the intersections between the relevant regimes.

D Outline of the book

This book is structured in three Parts. Part I provides the present introduction (Chapter 1), followed by a detailed overview of the relevant international laws and institutions that impact upon the trade and utilisation of fisheries (Chapter 2). Separating my analysis along the rough lines of 'regimes', I first describe the prominent Conventions and institutions of the law of the sea, with particular attention to UNCLOS and the Fish Stocks Agreement. The institutional framework provided by these Conventions is limited to a small UN unit known as the Division for Ocean Affairs and the Law of the Sea. It is the FAO's Fisheries and Aquaculture Department that plays the major international institutional role for fisheries and I describe the work of the FAO's inter-governmental Committee on Fisheries. I point also to the institutional role of regional fisheries management organisations. A number of these instruments and institutions expressly recognise the inter-connectedness with other areas of international law, and the chapter points to relevant conflict clauses and other institutional provisions.

The next part of Chapter 2 describes selected aspects of 'international environmental law' that are relevant to fisheries. I give a summary of the regulation of international trade of certain endangered species, as set out in CITES. I also discuss provisions for the conservation and sustainable use of marine biological diversity included in the Convention for Biological Diversity. These conventions expressly recognise their interdependencies with other international conventions and I introduce relevant provisions and practices that are the basis of my analysis of the case studies.

The final part of Chapter 2 describes how international trade law impacts upon the sustainable trade of fisheries. I give a short overview of relevant WTO agreements, with particular attention to provisions regulating the use by WTO members of subsidies. I also introduce the legal provisions that envisage interdependencies with other areas of international law. In particular, I point to institutional innovations like the WTO Committee on Trade and Environment and ongoing discussions about synergies between trade and environment under the WTO's Doha Development Agenda.

In line with my case-study approach, Part II draws on these provisions to examine selected examples of institutional and normative interdependencies between the law of the sea, trade and environmental law. Three chapters (Chapters 3–5) investigate the legal bases for regime interaction in global fisheries governance by reference to the three examples I listed above.

In assessing regime interaction in the negotiation of international laws relevant to fisheries preservation, Chapter 3 analyses the current negotiation of rules to discipline fisheries subsidies. It describes how the WTO rules on subsidies are currently inadequate to constrain fishery subsidies and explains that WTO members have sought to address this inadequacy by negotiating new rules under the Doha agenda. There is much scope for regime interaction in these negotiations. For example, one of the main problems in framing new rules has been to differentiate between fishery subsidies that improve fisheries sustainability and those that instead lead to overfishing by enhancing vessel capacity. Learning and collaboration between multiple stakeholders including the WTO, FAO, UNEP and NGOs is an obvious way to overcome this problem, yet such concerted action is opposed by many WTO members. A recurring concern is that the mandate to negotiate does not allow for the participation of stakeholders that advocate different interests from those that are formally represented at the WTO. This gives rise to the question of whether regime interaction may be legally justified on bases other than the consent of the state negotiators – which is based on the question of whether consent by states is required for institutional collaboration to occur in international law (the 'consent to regime interaction' issue indicated above).

Moving from interdependencies in the negotiation of international law to interdependencies in implementation, Chapter 4 examines the restriction of trade in endangered marine species. It describes how the FAO and the CITES Secretariat have collaborated in the revision of CITES listing criteria and the inclusion of marine species in the CITES appendices. A major tool for this regime interaction has been a Memorandum of Understanding between the two organisations. The chapter examines the legal basis for this MOU. The chapter assesses the notion of institutional hierarchy in the context of the evolution of the MOU and offers some conclusions about its efficacy in promoting inter-regime collaboration.

Chapter 5 focuses on international adjudication, and assesses interdependencies between regimes in the adjudication of claims based on previously agreed international rules. I take a WTO dispute over fisheries import bans as a case study to demonstrate how WTO panels and the Appellate Body of the WTO (AB) collaborate with other relevant regimes. The chapter focuses on the *US – Shrimp* case involving the United States' prohibition on the import of shrimp harvested by ecologically destructive methods and assesses the scope in this case for the use of international environmental law in interpreting the WTO covered agreements and establishing relevant facts. In exploring legal bases and methods for regime interaction, the chapter compares a number of other WTO cases. Prominent amongst these is the *EC – Biotech* decision, in which a WTO panel suggested that non-WTO treaties could only be relevant to WTO treaty interpretation if all WTO members had ratified them. Conceptually, this is the same problem that arises in Chapters 3 and 4; i.e. whether state consent is needed for regime interaction. The *EC – Biotech* panel also drew on a wide range of sources to interpret the 'ordinary meaning' of treaty terms, and the chapter critiques its blanket acceptance of such sources. The need for regime scrutiny by adjudicating bodies provides a further link to my investigation into the need for critical inquiry and ongoing review by IGOs in the framing of WTO subsidy rules and the CITES-led restriction of trade in Chapters 3 and 4.

Part III analyses the implications of my case studies for international law. In Chapter 6, I describe processes that promote and obstruct inter-regime collaboration. This leads me to challenge, in Chapter 7, the conceptual problem of state consent to regime interaction. The conception of 'stakeholders' is relevant to this issue and I investigate substitutes for state consent such as evolving norms of institutional accountability in fisheries and beyond. Chapter 7 develops a legal framework for appropriate legal interaction that redefines the capacity of IGOs to collaborate and stresses procedural safeguards to ensure openness, transparency and participation. This framework is of crucial importance to the fragmented international legal order and I conclude in Chapter 8 by noting implications for international law and by identifying further possible research to support international law's response to global policy problems.

The book's presentation of empirical and doctrinal research and theoretical and normative engagement is interdependent. In combination, these approaches seek to contribute to international law's response to the challenge of global fisheries depletion.

2 Relevant laws and institutions: an overview

> Ensuring the sustainable development of the oceans requires effective coordination and cooperation, including at the global and regional levels, between relevant bodies ...[1]

This chapter describes the international laws and institutions that have historic and emerging relevance to the trade-related aspects of fisheries. Within the law of the sea, various instruments, including UNCLOS, the Fish Stocks Agreement and voluntary FAO codes, seek to govern fisheries management. These approaches meet the definition of legal regimes offered in Chapter 1, and the first part describes first UNCLOS and the EEZ regime, then high seas, fish stocks and regional regimes and then the FAO fisheries management regime. The chapter then moves to examine international environmental law. The CITES regime is increasingly relevant to the trade in endangered marine species, while the Convention on Migratory Species addresses conservation obligations of range states and the Convention on Biological Diversity manages marine living resources. The final part of the chapter considers the WTO regime. In promoting liberalised international trade, the WTO covered agreements affect the trade in fish and fish products in many ways, and new disciplines on fisheries subsidies promise to focus trade law on fisheries sustainability in the future. In my consideration of these selected regimes within the law of the sea, international environmental law and international trade law, I also examine ways in which these areas of law 'look outwards' to other areas of international law. This discussion of interdependencies is augmented in the following

---

[1] Plan of Implementation of the World Summit on Sustainable Development (A/CONF.199/20), para 30.

chapters by my specific examination of law-making in Chapter 3 (looking especially at trade/law of the sea interdependencies), the implementation of existing law in Chapter 4 (focusing on environmental law/law of the sea interdependencies), and the resolution of disputes in Chapter 5 (where trade law disputes relate to environmental regimes).

A The law of the sea

States have sought to create a 'law of the sea' through a single Convention: the 1982 *United Nations Convention on the Law of the Sea* (UNCLOS), which recognises that 'the problems of ocean space are closely interrelated and need to be considered as a whole'.[2] UNCLOS's greatest impact on fisheries regulation has been in extending the exclusive economic zone ('EEZ') of states to 200 nautical miles from their coast.[3] This regime creates special rights and duties for coastal states and continues to interact with existing regional approaches, as discussed in the next section. The remainder of the oceans outside the EEZs continue to be classified by UNCLOS as 'high seas',[4] and the following section describes the rules applicable to these areas, as implemented by the 1993 Fish Stocks Agreement. UNCLOS also preserves general obligations on states to 'protect and preserve the marine environment'[5] and to cooperate on a global or regional basis to 'formulat[e] or elaborat[e] international rules, standards and recommended practices and procedures' to achieve this end.[6] These latter provisions are focused on ocean pollution, but are arguably applicable to the conservation and management of fisheries resources in areas both within and outside national jurisdiction.

Aside from the UNCLOS regimes, states have met in other intergovernmental forums to devise principles and binding rules. The third section in this part considers the normative and institutional contributions of the FAO to fisheries conservation and management. The final section then considers the ways in which these regimes have acknowledged their interdependencies with each other and with other areas of international law.

---

[2] *United Nations Convention on the Law of the Sea* (UNCLOS), Montego Bay, 10 December 1982; in force 16 November 1994; 21 *ILM* 1245 (1982), Preamble.
[3] See UNCLOS, Arts. 55–75, esp. Art. 57.
[4] Ibid., Arts. 116–120.   [5] Ibid., Art. 192.   [6] Ibid., Art. 197.

1 UNCLOS and the EEZ regime

The framework for the law of the sea is set out in UNCLOS. After three separate rounds of negotiations, the third one spanning nine years, UNCLOS opened for signature in 1982, updating earlier multilateral arrangements which had been agreed in 1958.[7] The multilateral negotiations were attended by large state delegations and by small NGOs.[8] Heralded at the time as a 'constitution for the oceans',[9] it now has 160 parties.[10]

UNCLOS recognises the desirability of conservation and equitable and efficient utilisation of fisheries resources.[11] Underlying these goals is an aspiration for a 'just and equitable international economic order', and due consideration of the special interests and needs of developing countries is pronounced to be particularly vital to this end.[12]

UNCLOS sought to settle the highly controversial question of the jurisdiction of coastal states over the fishing resources located in the seas adjacent to their coasts. This issue had long been debated in international law, including through a series of decisions on territorial delimitation and fisheries rights by the International Court of Justice.[13] Under UNCLOS, coastal states have full jurisdiction over

---

[7] Convention on the Territorial Sea and the Contiguous Zone, Convention on the High Seas, Convention on the Continental Shelf and Convention on the Fishing and Conservation of the Living Resources of the High Seas, Geneva, 1958, agreed after the First Conference on the Law of the Sea, Geneva, Switzerland, 1966 (UNCLOS I). On the third conference (known as UNCLOS III), see generally William Burke, *The New International Law of Fisheries: UNCLOS 1982 and Beyond* (1994). For references to the law of the sea prior to UNCLOS III, as developed through the work of the ILC, see Arthur Watts (ed.) *The International Law Commission 1949–1998* (1999) 23–137.

[8] See Ralph B Levering and Miriam L Levering, *Citizen Action for Global Change: The Neptune Group and the Law of the Sea* (1999).

[9] See 'A Constitution for the Oceans', remarks by Tommy T B Koh of Singapore, President of the Third United Nations Conference on the Law of the Sea, upon the adoption of the text in: Official Text of the United Nations on the Law of the Sea with Annexes and Index (1983) E 83V5, xxxiii, available at www.un.org/Depts/los/convention_agreements/texts/koh_english.pdf.

[10] www.un.org/Depts/los/convention_agreements/convention_agreements.htm (as at 8 January 2010). A notable exception is the United States.

[11] UNCLOS, Preamble.   [12] Ibid.

[13] See *Anglo-Norwegian Fisheries Case* [1951] ICJ Rep 116; *Fisheries Jurisdiction cases* (UK v. Iceland) [1973] ICJ Rep 3; [1974] ICJ Rep 3. The Court recognised the validity in certain circumstances of a zone of preferential fishing rights in high-seas fisheries resources for coastal states; see further Robin Churchill and Vaughan Lowe, *The Law of the Sea* (3rd edn, 1999) 285.

their 'territorial sea', the area adjacent to the coast and extending seawards to twelve nautical miles.[14] Extending further seawards, up to 200 nautical miles, is the EEZ, which is subject to a set of UNCLOS provisions that embody a 'special legal regime'.[15]

The conceptualisation of EEZs was in part a response to the growing dissatisfaction with the historic concept of freedom of fishing and accompanying institutional mechanisms. Significant and growing competition between states to fish in areas outside of national jurisdiction had led in the early twentieth century to the establishment of Regional Fishing Management Organisations (RFMOs).[16] These were considered to be forums that allowed states to determine 'who gets access and how much', rather than forums for cooperation on conservation matters.[17] RFMOs comprised states with well-developed fishing industries and vessel capacity, such as Japan, the United Kingdom, South Korea, Norway, Spain and the United States, which were seeking to expand their fishing opportunities in distant waters.[18] The distant-water fishing fleets are considered to have placed the greatest pressure on fishery stocks.[19] Other coastal states, such as Canada and Iceland, either encouraged the development of their own private coastal fleets[20] or sought assistance against distant water competitors,[21] establishing their own regimes of over-exploitation.[22] Even where regional conservation arrangements existed, the fisheries commissions themselves had little power to enforce the obligations of member states.[23] Thus, in mutually exploiting a global commons, there were disincentives for states and RFMOs to adopt conservation measures, especially where such measures could benefit 'free-riding' states.

---

[14] See UNCLOS Art. 2; see also Arts. 19, 21. The determination of a twelve-mile zone for the territorial sea is set out in UNCLOS Art. 3.

[15] UNCLOS, Art. 55.

[16] The other acronym often used is 'RFBs' – Regional Fishery Bodies.

[17] M J Peterson, 'International Fisheries Management' in Peter Haas, Robert Keohane and Marc Levy (eds.) *Institutions for the Earth: Sources of Effective International Environmental Protection* (1993) 249, 260.

[18] Ibid., 259. [19] Burke, above n. 7, 23.

[20] Including Peru and other developing countries: Peterson, above n. 17, 260.

[21] Including Iceland, Canada, some European nations and the United States: ibid.

[22] See, e.g., Kaye's analysis of pollock stocks in the Bering Sea and the exploitation by EEZ states, particularly the United States: Stuart Kaye, *International Fisheries Management* (2001) 312–13.

[23] Post-war fishing bodies were severely restricted in both enforcement powers and independence: see Burke, above n. 7, 23–4.

Ocean areas outside of national jurisdiction could thus be conceived as the paradigmatic 'tragedy of the commons'.[24] Such a characterisation is often an imprimatur for the allocation of property rights, and indeed the EEZ is a form of ocean enclosure. However, the rights for coastal states within the EEZ are not unfettered within UNCLOS's EEZ regime.

According to the EEZ regime, access to fish resources is largely subject to coastal state jurisdiction: coastal states are granted sovereign rights in their EEZ with respect to economic activities, marine science research and conservation.[25] They are to promote an objective of 'optimum utilisation' and to that end must determine the total allowable catch of the EEZ.[26] Extending the EEZ in this way ensured that 90 per cent of fish resources are now subject to national jurisdiction.[27]

The rights of coastal states in the EEZ are not absolute, however, and the coastal state must have due regard to the rights and duties of other states.[28] After determining the capacity of their own vessels to maximise the catch (ie, vessels registered within the state), coastal states are to grant access to any surplus on an equitable basis. Candidates for access include those states that are land-locked and geographically disadvantaged, states whose nationals 'habitually fished in the zones'[29] and

---

[24] See theories developed from the 'tragedy of the commons' premise, most famously described by Hardin, where multiple users vie for access and face disincentives in attempts to adopt conservation measures that might benefit competitors that remain outside any regime: Garrett Hardin, 'The Tragedy of the Commons' (1968) 162 *Science* 1243. The tragedy is not inevitable: see particularly Elinor Ostrom, *Governing the Commons: The Evolution of Institutions for Collective Action* (1990) (analysing conditions that alleviate commons tragedy such as the ability of individuals affected by relevant rules to participate in modifying the rules and monitoring compliance).

[25] UNCLOS, Art. 56. The acquisitiveness of coastal states in establishing these rights in the face of competing interests has been noted: see Douglas Johnston, *The International Law of Fisheries: A Framework for Policy-Oriented Inquiries* (1985; first published 1965) LXXVII.

[26] Ibid., Arts. 61 and 62.

[27] Burke, above n. 7, 23 (citing FAO Committee on Fisheries, 'UNCED and its Implications for Fisheries', FAO Doc. COFI/93/Inf/8, January 1993). This enclosure of most of the oceans appears to accord with public choice theory, which predicts that property rights will ensure the preservation of a common resource. But cf Newfoundland fishery, which was a managed fishery before collapse. Note also the inability of public choice theory to account for the agreement, through UNCLOS, on this issue: see e.g. Jonathan Wiener, 'On the Political Economy of Global Environmental Regulation' (1999) 97 *Georgetown Law Journal* 749 (describing limits of public choice theory in accounting for origins of international environmental law).

[28] UNCLOS, Art. 56.   [29] Ibid., Art. 62(3).

nearby developing states.[30] Thus, UNCLOS recognises rights and duties of non-coastal states on the basis of the need to minimise economic dislocation.[31] In addition, there is increasing recognition of conservation and management obligations for coastal states within EEZs,[32] with some suggesting that coastal states must act as 'stewards' over these areas.[33]

As part of the EEZ regime, coastal states can exploit their EEZ economically rather than through direct fishing activities. For example, many coastal states have granted fishing access to vessels from distant waters as a means to obtain revenue from other states. The states that procure fishing rights have large vessel capacity (and may even have depleted the fisheries within their own EEZ). The EU, for example, has entered into twenty bilateral fishing agreements, primarily with developing countries in Africa, which allow EU vessels to fish within the EEZs of these countries.[34] The role of coastal states as environmental stewards over these areas may be subject to some doubt given recent critiques of the ecological effects of such arrangements.[35] Indeed, it could be suggested that fishing nations procuring such access might also be subject to stewardship duties.

Conflicts between the interests of coastal and other states within the EEZs are to be resolved on the basis of equity.[36] Notably, such conflicts are not subject to the binding dispute settlement mechanism that is otherwise lauded as an important part of UNCLOS's enforcement armoury.[37] This 'mandatory system of dispute settlement',[38] which is contained in

---

[30] Ibid., Art. 62. Note also that grants of access were be made according to, *inter alia*, 'nutritional needs' of populations within those states seeking access, indicating a concern for poorer nations.
[31] Ibid., Art. 62(3).
[32] See further Richard Barnes, *Property Rights and Natural Resources* (2009) 219ff.
[33] As discussed by Barnes, ibid., 298 (referring to Vaughan Lowe and Philip Allott, among others). These obligations are related to principles outside the EEZ regime, such as the FAO Code of Conduct discussed below at p. 48.
[34] An example is the recently renegotiated agreement between the EU and Mauritania, under which the EU will pay Mauritania EUR 75.25 million a year to catch 250,000 tonnes of fish species, including octopus, crab, crawfish, sardines, anchovies and lobster, from 1 August 2008 to 31 July 2012. Eastern European and Asian vessels, particularly from Russia and China, also fish within EEZ zones of African states, while companies from countries such as China also co-own private vessel fleets: see 'EU, Mauritania Renegotiate Fisheries Access Agreement' (2008) 8:5 *Bridges Trade BioRes*.
[35] See Emma Witbooi, 'Governing Global Fisheries: Commons, Community Law and Third Country Coastal Waters' (2008) 17 *Social Legal Studies* 369 (drawing on EU–Senegal fisheries relations as a case study).
[36] UNCLOS, Art. 59.
[37] Ibid., Art. 297(3). See further Natalie Klein, *Dispute Settlement in the UN Convention on the Law of the Sea* (2005) 185ff.
[38] See Koh, above n. 9, xxxiii.

Part XV of the Convention, establishes the International Tribunal for the Law of the Sea (ITLOS) together with avenues for compulsory dispute settlement under other ad hoc tribunals and the ICJ. Yet the exercise by a coastal state of its discretionary powers within the EEZ regime, including in setting the terms and conditions of its conservation and management laws, is not subject to this system without its consent. This curtails the ability of states (and international law) to remedy the failings of a coastal state in carrying out its responsibilities under the EEZ regime.[39]

There is no direct institutional support for the implementation of the fisheries provisions of UNCLOS. The UN Division of Ocean Affairs and the Law of the Sea (DOALOS) provides advice and assistance on the implementation of UNCLOS but does not have specific functions relating to fisheries.[40] The General Assembly of the UN regularly discusses oceans matters and makes annual resolutions on sustainable fisheries.[41] These resolutions have broad oversight of the EEZ regime and other regimes, but do not deal with specific matters of compliance and implementation. Instead, the EEZ regime relies on states protesting about a lack of treaty compliance by coastal states, without guaranteed recourse to third party procedures. This situation is not effective in ensuring that states comply with the conservation and management responsibilities within the EEZ.[42] The next section examines whether a similar situation exists for fisheries in areas outside national jurisdiction, beyond the EEZs.

2 UNCLOS's high seas, fish stocks and regional regimes

Ocean areas seawards of EEZs continue to be classified in the traditional language of 'high seas'.[43] These areas, although less productive than

---

[39] See Richard Barnes, 'The LOSC: An Effective Framework for Domestic Fisheries Conservation?' in David Freestone, Richard Barnes and David Ong (eds.) *The Law of the Sea: Progress and Prospects* (2006) 233, 246.

[40] The functions of DOALOS include the provision of advice, studies, research and technical assistance on the implementation of UNCLOS and the servicing to the UNGA on the law of the sea and ocean affairs: see excerpts from the Secretary-General's Bulletin ST/SGB/1997/8 available at www.un.org/Depts/los/doalos_activities/about_doalos.htm. For a brief overview, see Louise de La Fayette, 'The Role of the United Nations in International Oceans Governance' in David Freestone, Richard Barnes and David Ong (eds.) *The Law of the Sea: Progress and Prospects* (2006) 63.

[41] See e.g. UNGA Resolution on Sustainable Fisheries of 8 December 2006, Resolution 61/105, at UN Doc A/RES/61/105 (6 March 2007).

[42] Barnes, above n. 39, 258.

[43] UNCLOS, Arts. 116 – 120. For a summary of the historical influences of the idea of oceans that were open to all, including the writings of Hugo Grotius, see Robin Warner,

coastal areas, contain a number of highly migratory species of significant commercial value, in addition to 'straddling stocks' which move back and forth across EEZ boundaries.[44] Thus, fish stocks within these commons are particularly attractive to distant water fishing fleets with high technological capacities.[45]

UNCLOS provides that all states will continue to enjoy the traditional freedoms of fishing in these areas, but this right is qualified by the need to conserve living resources.[46] UNCLOS entrenches a duty to cooperate by coastal states and other fishing states operating in the region.[47] For highly migratory species, this cooperation is to be undertaken through the twin lens of 'optimum utilisation' of the species and conservation.[48]

Under UNCLOS, such cooperation is to be achieved through existing or to-be-established organisations providing institutional support for fisheries management. As mentioned above, the main historical form of cooperation for states has been through RFMOs. As such, RFMOs are supported by UNCLOS and can be conceived as part of its legal regimes. Yet given their separate normative and institutional basis, and the specific issue-areas originally sought to be addressed by RFMOs, it is also useful to consider them as independent regional regimes.[49]

Some 20 RFMOs have been established since 1945.[50] Many, like the Northwest Atlantic Fisheries Organization (NAFO),[51] are formed pursuant to regional agreements. Others, like the Asia-Pacific Fishery Commission established in 1948, have been established under the

---

Protecting the Oceans Beyond National Jurisdiction: Strengthening the International Law Framework (2009) 28–30.

[44] UNCLOS, Art. 64. [45] See tragedy of the commons referred to above n. 24.
[46] UNCLOS, Arts. 116–119. In the Swordfish dispute between Chile and the EC, Chile requested ITLOS to decide whether the EC had complied with these provisions to ensure conservation of swordfish in the high seas adjacent to Chile's EEZ: see further Chapter 1, n. 42.
[47] Ibid., Art. 118.
[48] Ibid. For a list of highly migratory species covered by Art. 64, see UNCLOS Annex I.
[49] See e.g. Kaye, above n. 22, who applies regime theory to fisheries management agreements operating in the polar regions: the Convention on the *Conservation and Management of Pollock Reserves in the Central Bering Sea*, done at Washington DC on 16 June 1994, entered into force 8 December 1995; reprinted in 34 ILM 67 (1995), known as the Doughnut Hole Convention, and the *Convention on the Conservation of Antarctic Marine Living Resources*, entered into force 6 April 1981; UKTS No 48 (1982).
[50] Tore Henriksen, Geir Hønneland and Are Sydnes, *Law and Politics in Ocean Governance: The UN Fish Stocks Agreement and Regional Fisheries Management Regimes* (2006) 3.
[51] The NAFO was formed under the *Convention on Future Multilateral Cooperation in the Northwest Atlantic Fisheries* 1978 and replaced its 1949 predecessor, the ICNAF (International Commission of the Northwest Atlantic Fisheries).

auspices of the FAO.[52] Yet others take a species-specific approach, such as the International Commission for the Conservation of Atlantic Tunas (ICCAT).[53]

Like the early criticisms of RFMOs mentioned above,[54] the effectiveness of RFMOs in meeting conservation and management objectives has been doubted.[55] To demonstrate their deficiencies, use might be made of discussions about institutional effectiveness by Keohane, Haas and Levy. In their studies of environmental institutions, the authors point to 'three Cs' of institutional effectiveness. Bodies are assessed according to their ability to increase levels of *concern* about the exploitation of the resource, provide a hospitable *contractual environment* in which interests can be mediated and action monitored, and improve *capacity* of states to change exploitative behaviour through political or administrative arrangements.[56] Most RFMOs fail to meet these requirements, and instead continue to operate as clubs that are, in the main, focused on access rights to exploit.[57] In particular, there are grave deficiencies in the monitoring role played by RFMOs, which rely on member states to enforce measures by sending vessels to often remote and inhospitable ocean environments. This is especially difficult for the polar regions, where, for example, member states of the Commission for the Conservation of Antarctic Marine Living Resources (CCAMLR)[58] must police vast and difficult areas.[59]

UNCLOS – and the RFMOs – rely on member states to control their fleets, which are identified through the flags they fly. Yet the phenomenon to which those assessing global market failures are now turning

---

[52] See FAO Constitution, Art. XIV. The FAO lists the ten RFMOs that it currently supports on its website at www.fao.org/fishery/rfb.

[53] Formed pursuant to the *Convention for the Conservation of Atlantic Tunas* 1966 and responsible for the conservation of tunas and tuna-like species in the Atlantic Ocean and adjacent seas.

[54] See above n. 17 and surrounding text.

[55] FAO Fisheries Circular No. 940, Rome, April 1999, Gail Lugten, 'A Review of Measures Taken by Regional Marine Fishery Bodies to Address Contemporary Fishery Issues' (concluding that 'very few [RFMOs] have started to implement the conservation and management measures provided for in the post-1982 fishery instruments'); see also FAO, SOFIA (2006) 53–4 and FAO, SOFIA (2008) 9.

[56] Robert Keohane, Peter Haas and Marc Levy, 'The Effectiveness of International Environmental Institutions' in Peter Haas, Robert Keohane and Marc Levy (eds.) *Institutions for the Earth: Sources of Effective International Environmental Protection* (1993) 3.

[57] See further Peterson, above n. 17.

[58] Formed under the *Convention on the Conservation of Antarctic Marine Living Resources*, above n. 49.

[59] See www.ccamlr.org/pu/e/gen-intro.htm.

their attention,[60] that of mobile global capital searching for jurisdictions with weak regulatory and enforcement inclination, has been affecting fisheries for many decades. The flag system allows vessels to register with states of choice, regardless of nationality or port-location requirements. Nationals of a state that has signed up to a RFMO (and attendant conservation regime) can re-flag their vessels with non-member states to escape regulation. For example, within the CCAMLR regime, Norwegian nationals have been identified as operating out of third-party states in order to harvest the imminently endangered Patagonian Toothfish.[61] This phenomenon is part of the problem of fishing that is 'illegal, unreported and unregulated' (IUU),[62] which has been recognised to be of serious and increasing concern for all capture fisheries. Separate attempts to compel members of RFMOs to enforce national compliance have, however, failed.[63] Indeed, some have suggested that lack of effective management of high seas areas has been supported by states that considered themselves disadvantaged by the EEZ regime and wanted to maintain full freedoms in commons areas.[64]

In 1995, attempts to improve the implementation of UNCLOS high seas provisions and the effectiveness of RFMOs led to the *Agreement for the Implementation of the Provisions of the United Nations Convention on the Law of the Sea of 10 December 1982 relating to the Conservation and Management of*

---

[60] See, e.g., Eyal Benvenisti, 'Exit and Voice in the Age of Globalization' (1999) 98 *Michigan Law Review* 167–213.

[61] Richard Herr, 'The International Regulation of Patagonian Toothfish' in Olav Schram Stokke (ed.) *Governing High Seas Fisheries: The Interplay of Global and Regional Regimes* (2001) 303, 317 and sources therein. On the relevance of CITES to the Patagonian toothfish, see Chapter 4, n. 248 and accompanying text.

[62] IUU has been defined to include illegal fishing within EEZs or within high seas governed by RFMOs, and unregulated fishing in areas where there are no applicable conservation or management measures 'and where such fishing activities are conducted in a manner inconsistent with State responsibilities for the conservation of living marine resources under international law': see FAO, IPOA-IUU Fishing, para 3, described below n. 125 and surrounding text.

[63] Herr, above n. 61, 316–17, citing a recent failed attempt by New Zealand to seek greater enforcement by CCAMLR members of the activities of their nationals. At the 17th meeting of the Commission, members instead reasserted their preference for flag-state authority in conformity with UNCLOS Part VII.

[64] For example, it has been suggested that the lack of effective regulation of the high seas actually allowed for agreement by distant-water fishing nations to the extension of the EEZ in UNCLOS: see Rosemary Rayfuse *et al.*, 'Australia and Canada in Regional Fisheries Organizations: Implementing the United Nations Fish Stocks Agreement' (2003) 36 *The Dalhousie Law Journal* 47, 49.

Straddling Fish Stocks and Highly Migratory Fish Stocks ('Fish Stocks Agreement').[65] There are currently 77 parties.[66] Notwithstanding that the Fish Stocks Agreement was implemented as part of UNCLOS,[67] not all of the parties are parties to UNCLOS, and vice versa.[68]

The Fish Stocks Agreement applies to high seas and, in a limited way, to certain fish stocks in EEZs.[69] It aims to improve cooperation between states to achieve the long-term conservation and sustainable use of straddling fish stocks and highly migratory fish stocks.[70] Coastal states and states fishing on the high seas are to adopt necessary conservation and management measures, protect biodiversity in the marine environment, collect and share data on fishing activities and undertake effective monitoring, control and surveillance.[71]

The Fish Stocks Agreement preserves the reliance on RFMOs for enforcement, but seeks to strengthen those bodies. It provides that coastal states and states fishing on the high seas shall pursue cooperation in relation to straddling and highly migratory fish stocks 'either directly or through appropriate subregional or regional fisheries management organizations or arrangements'.[72] Parties who fail to become members of an existing regional regime and refuse to apply conservation and management measures can be denied access to the relevant fishery.[73] The Fish Stocks Agreement also provides for the application of a precautionary approach to the conservation, management and exploitation of straddling and highly migratory fish stocks.[74]

The Fish Stocks Agreement also attempts to address the problem of 'flags of convenience'. It requires parties to 'take measures ... to deter the activities of vessels flying the flag of non-parties which undermine the effective implementation of this Agreement'.[75] In addition, it extends port-state jurisdiction to disallow landings and transhipments 'where it has been established that the catch has been taken in a

---

[65] See UN Doc A/CONF.164/37; entered into force 11 December 2001; see further www.un.org/Depts/los/convention_agreements/convention_agreements.htm.
[66] See www.un.org/Depts/los/convention_agreements/convention_agreements.htm (as at 8 January 2010).
[67] See Fish Stocks Agreement, Art. 4 ('Nothing in this Agreement shall prejudice the rights, jurisdiction and duties of States under [UNCLOS]. This Agreement shall be interpreted and applied in the context of and in the manner consistent with [UNCLOS]').
[68] See e.g., the United States, which has acceded to the Fish Stocks Agreement but is not a party to UNCLOS. By contrast, China is a party to UNCLOS but not the Agreement.
[69] Fish Stocks Agreement, Art. 3.   [70] Ibid., Preamble.   [71] Ibid., Art. 5.
[72] Straddling Fish Stocks, Art. 8(1).   [73] Ibid., Art. 8(3) and (4).
[74] Ibid., Art. 6; see also Art. 5(c).   [75] Ibid., Art. 33.

manner which undermines the effectiveness of subregional, regional or global conservation and management measures on the high seas'.[76]

These provisions appear to extend the applicability of the Fish Stocks Agreement to more than the relatively few signatories. In an innovative manipulation of state sovereignty, the Agreement seems to create duties for states even if they have not consented to the regime.[77] Some consider such legal obligations for third states to flout the *pacta tertiis* rule.[78] Others, however, conclude that states which remain outside the UNCLOS high seas regimes and relevant RFMOs may still be subject to concrete obligations.[79]

The efforts to utilise port state measures to combat IUU fishing, included in the Fish Stocks Agreement, may be enhanced in the future. The *Agreement on Port State Measures to Prevent, Deter and Eliminate Illegal, Unreported and Unregulated Fishing* ('Port State Measures Agreement') was adopted by the FAO in 2009.[80] It provides that port states may deny the entry of vessels into its port if the vessel fails to confirm that it has not engaged in IUU fishing or related activities in support of such fishing.[81]

While promising in creating enforceable procedures against rogue vessels, the resolution of disputes arising from the high seas regime is subject to continuing controversy and uncertainty. As mentioned above, a

---

[76] Ibid., Art. 23(3). The Agreement also allows inspections for non-flag states in relation to marine pollution rules and standards: see further Geir Hønneland, 'Recent Global Agreements on High Seas Fisheries: Potential Effects on Fishermen Compliance' in Olav Schram Stokke (ed.) *Governing High Seas Fisheries: The Interplay of Global and Regional Regimes* (2001) 121. For general provisions for boarding and inspecting vessels, see Douglas Guilfoyle, *Shipping Interdiction and the Law of the Sea* (2009) 105.

[77] See Rosemary Rayfuse, 'The United Nations Agreement on Straddling and Highly Migratory Fish Stocks as an Objective Regime: A Case of Wishful Thinking?' (1999) 20 *Australian Year Book of International Law* 253.

[78] Erik Franckx, '*Pacta Tertiis* and the Agreement for the Implementation of the Straddling and Highly Migratory Fish Stocks Provisions of the UNCLOS' (2000) 8 *Tulane Journal of International and Comparative Law* 49, 73 (concluding that, in strict application of the *pacta tertiis* principle, the Fish Stocks Agreement does not create any legal obligations for third states). See also Joost Pauwelyn, *Conflict of Norms in Public International Law* (2003), 101-4. *Pacta tertiis* is noted in Chapter 1, n. 98.

[79] Tore Henriksen, 'Revisiting the Freedom of Fishing and Legal Obligations on States Not Party to Regional Fisheries Management Organizations' (2009) 40 *Ocean Development and International Law* 80.

[80] Approved at the thirty-sixth session of the FAO Conference (22 November 2009) but not yet in force. The agreement will enter into force 30 days after the twenty-fifth instrument of ratification or accession: see Art. 29. As at 28 September 2010, there were 17 signatories. See further www.fao.org/Legal/treaties/037s-e.htm. The FAO is discussed in more detail below at pp. 46ff.

[81] Port State Measures Agreement, Art. 9.

major innovation of UNCLOS was the inclusion of a series of provisions for binding dispute settlement for disputes concerning the interpretation or application of the Convention or the Agreement.[82] Yet the application of compulsory procedures entailing binding decisions by bodies such as ITLOS is highly constrained for fishery disputes, not just because of the freedom of coastal states to refuse binding dispute settlement within the EEZ regime,[83] but also because the operation of regional fisheries regimes may affect the compulsory nature of the procedures under the high seas regimes, including for straddling and highly migratory fish stocks.

In *Southern Bluefin Tuna*[84] a fisheries dispute arose between Australia and New Zealand on the one hand and Japan on the other over the latter's exploitation of southern bluefin tuna, which was alleged to contravene UNCLOS obligations concerning highly migratory species and the high seas.[85] The dispute was filed pursuant to UNCLOS Part XV and provisional measures were granted by an ITLOS Tribunal.[86] At the jurisdiction phase, an Arbitral Tribunal constituted under UNCLOS Part XV found that the procedures for dispute settlement in UNCLOS fall 'significantly short of establishing a truly comprehensive regime of compulsory jurisdiction'.[87] The majority found that a reference to non-binding dispute settlement mechanisms in a 1993 species-specific regional treaty[88] preserved the choice of Japan and therefore trumped the binding dispute settlement mechanisms of UNCLOS Part XV. As such, the dispute was ruled inadmissible. The majority's approach has led some commentators to favour the characterisation of the procedures of Part XV as 'quasi-compulsory', because the emphasis is on choice of a variety of modes of dispute settlement before compulsory jurisdiction arises.[89] The case demonstrates the challenges faced in

---

[82] UNCLOS, Part XV (see especially Arts. 286-296); see Fish Stocks Agreement Art. 30.
[83] UNCLOS, Art. 297(3), noted above n. 37.
[84] *Australia and New Zealand v. Japan (Award on Jurisdiction and Admissibility)* (2000) 119 ILR 508 (Arbitral Award constituted under Annex VII of UNCLOS) (*Southern Bluefin Tuna Case*).
[85] UNCLOS Art. 64 (highly migratory species) and Arts. 116-119 (high seas).
[86] *Australia and New Zealand v. Japan (Provisional Measures)* (1999) 117 ILR 148 (ITLOS) ('*Southern Bluefin Tuna – Provisional Measures*').
[87] *Southern Bluefin Tuna Case*, above n. 84, 45 [62].
[88] The 1993 *Convention for the Conservation of Southern Bluefin Tuna* (CCSBT).
[89] See e.g. Klein, above n. 37, 30. Even if this characterisation is correct, other commentators note the benefits of the *Southern Bluefin Tuna* litigation for the parties involved: see Tim Stephens, 'The Limits of International Adjudication in International Environmental Law: Another Perspective on the Southern Bluefin Tuna Case' (2004) 19 *International Journal of Marine and Coastal Law* 177, 188.

providing judicial oversight of fisheries governance, in that a party may escape the compulsory jurisdiction of UNCLOS by entering into regional arrangements.[90]

The potential for compulsory third-party dispute settlement in disputes within the high seas, including for straddling highly migratory fish stocks, is not entirely emasculated. The majority in the *Southern Bluefin Tuna* Arbitral Tribunal were careful to point out that the Fish Stocks Agreement expressly refers to the dispute settlement procedure in UNCLOS Part XV.[91] Perhaps more importantly, a persuasive judgment from the dissenting arbitrator rejected the trumping of UNCLOS's procedures by a regional agreement, and referred instead to the shared desires of states for a central, comprehensive and compulsory jurisdiction during the negotiations leading to UNCLOS.[92] More recent cases involving interests other than fishing suggest that this approach may be preferred in subsequent cases, and that if states intend that UNCLOS disputes are to be resolved *outside* of the Part XV procedure, they must make this *express*.[93] The new Port State Measures Agreement's dispute settlement procedures may fall in this category: although it sets out its own framework for resolving disputes, the procedures require the parties to 'continue to consult and cooperate with a view to reaching settlement of the dispute in accordance with the rules of international law relating to the conservation of living marine resources.[94] As such, future fishing disputes relating to the high seas may well be subject to 'compulsory' jurisdiction under UNCLOS.

Even with strong dispute settlement procedures, the high seas regime, including for straddling and highly migratory fish stocks, continues to suffer from a lack of institutional support. RFMOs remain weak and ineffective.[95] There is no global body with oversight on their operation,[96]

---

[90] See Alan Boyle, 'The Southern Bluefin Tuna Arbitration' (2001) 50 *ICLQ* 447.
[91] *Southern Bluefin Tuna Case*, above n. 84, 48 [71].
[92] Ibid., (dissent of Sir Kenneth Keith) 54-6 [23-8]. Sir Kenneth considered that without an express provision, the CCSBT could not purport to exclude UNCLOS's compulsory jurisdiction.
[93] See e.g. *Mox Plant Case*, Request for Provisional Measures (*Ireland* v. *United Kingdom*) (Order of 3 December 2001) (2002) 41 ILM 405, separate Judgment of Judge Wolfrum, at 426 (ITLOS required an 'intention to entrust the settlement of disputes concerning the interpretation and application of the Convention to other institutions [to] be expressed explicitly in respective agreements'). See further Victoria Hallum, 'International Tribunal for the Law of the Sea: The *Mox Nuclear Plant Case*' (2002) 11 RECIEL 372, 374.
[94] Port State Agreement, Art. 22.   [95] FAO, SOFIA (2006) 53-4 and FAO, SOFIA (2008) 9.
[96] As discussed with respect to DOALOS: see above n. 40 and surrounding text.

an absence identified as a key problem in achieving sustainability in marine areas beyond national jurisdiction.[97]

In sum, the norms and institutions of the law of the sea are generally weak in achieving fisheries sustainability. Notwithstanding high hopes for the taming of the behaviour of over-exploiting states,[98] UNCLOS's provisions for fisheries conservation are considered to be the Convention's chief deficiency.[99] The Fish Stocks Agreement has failed to remedy UNCLOS's problems due to poor implementation by states and by RFMOs.[100] Hopes remain that the new Port State Measures Agreement will remedy deficiencies by using national measures to deny port access. These weaknesses have led to new forms of governance under an alternative UN body, the Food and Agriculture Organization. This fisheries management regime targets a wider range of stakeholders and involves policies and strategies of other international regimes, such as international environmental law and international trade law.

3 FAO fisheries management regime

The FAO is a UN specialised agency which was established in 1945.[101] It is not necessary to be a party to UNCLOS to be a member of the FAO, and indeed their membership is not identical.[102] The FAO did, however,

---

[97] See Rosemary Rayfuse and Robin Warner, 'Securing a Sustainable Future for the Oceans Beyond National Jurisdiction: The Legal Basis for an Integrated Cross-Sectoral Regime for High Seas Governance for the 21st Century' (2008) 23(3) *The International Journal of Marine and Coastal Law* 399. The authors call for an international oceans trust to set standards for activities on the high seas, and, at 408–411, use important international law concepts such as 'common heritage of mankind'. See also Warner, above n. 43, 233–4.

[98] The duty to cooperate was seen as tempering 'sovereignty': see Johnston, above n. 25, LCIII.

[99] Patricia Birnie and Alan Boyle, *International Law and the Environment*, (2002) 647; FAO Fisheries Report No. 293 William Burke, '1982 Convention on the Law of the Sea Provisions on Conditions of Access to Fisheries Subject to National Jurisdiction', Report of the Export Consultation on the Conditions of Access to the Fish Resources of the Exclusive Economic Zone (1983) 23 (commenting on the weakness of coastal state responsibilities); Churchill and Lowe, above n. 13, 279–323.

[100] See COFI, *FAO Fisheries Report* No 780 (2005), 2 (para 13) (urging states acting through RFMOs 'to ensure that they took further steps to implement the relevant provisions of the post-UNCED fisheries instruments').

[101] UN Charter, Art. 57; FAO Constitution, Art. XII.

[102] See, e.g., United States is a member of the FAO but not a party to UNCLOS. There are currently 191 Member Nations of the FAO, one 'Member Organization', the European Community and one 'Associate Member', the Faroe Islands: www.fao.org/unfao/govbodies/membernations3_en.asp.

provide 'technical back-up' to the UN conference that led to the Fish Stocks Agreement.[103] The functions of the FAO are to assist member states in the utilisation of their agriculture, including fisheries, by providing information and recommending research and concerted action with respect to, *inter alia*, the conservation of natural resources.[104] The FAO collects fisheries data, monitors fisheries stocks and publishes information such as the biennial *FAO State of the World Fisheries* report (SOFIA). To facilitate information sharing, the FAO has established a Fisheries Global Information System, which is intended to develop into a global inventory of fish resources. As well as acting as an information base, the FAO provides structures for the negotiation and ongoing revision of rules. It is made up of several governing bodies, including sectoral committees in which members may apply to participate.

The FAO Committee on Fisheries (COFI) was established by the FAO Conference in 1965. Membership, which currently numbers 137,[105] is open to FAO members and to any non-member eligible to be an observer of the FAO. Representatives of certain NGOs may participate but do not have the right to vote.[106] Similar access requirements apply to representatives of the UN, UN bodies and specialised agencies and RFMOs. The responsibility of COFI, as the 'only global inter-governmental forum' on fishery issues, is to review work programmes, conduct periodic general reviews of international fishery problems and examine possible solutions 'with a view to concerted action by nations, by FAO, inter-governmental bodies and the civil society'.[107] COFI may also establish sub-committees on certain specific issues; in 1985, it established the Sub-Committee on Fish Trade, and in 2001, the Sub-Committee on Aquaculture.

In 1993, the FAO Conference adopted the *Agreement to Promote Compliance with International Conservation and Management Measures by Fishing Vessels on the High Seas* ('Compliance Agreement').[108] This represented an attempt to

---

[103] FAO Code of Conduct, Preface.   [104] FAO Constitution, Art. I.
[105] See www.fao.org/unfao/govbodies/fishfinal_en.asp.
[106] COFI Rules of Procedure, Rule III; FAO General Rules, Rule XVII.
[107] www.fao.org/fishery/about/cofi. The reference to civil society reflects a 2008 amendment to the website.
[108] Entered into force 24 April 2003 (after twenty-fifth acceptance: see Art. 11). (Contracting parties include the EC, Argentina, Benin, Canada, Cyprus, Georgia, Japan, Madagascar, Myanmar, Namibia, Norway, Seychelles, Sweden, Tanzania, United States, Uruguay and Australia.

target flag-state problems. Each state party is required 'to ensure that fishing vessels entitled to fly its flag do not engage in any activity that undermines the effectiveness of international conservation and management measures'.[109] The Compliance Agreement also establishes mechanisms to ensure the free exchange of information about the authorisation of vessels.[110]

Disputes arising under the Compliance Agreement are not subject to mandatory dispute settlement; parties are instead required to 'consult and cooperate with a view to reaching settlement of the dispute in accordance with the rules of international law relating to the conservation of living marine resources'.[111] Disputes can be heard by the ICJ, ITLOS or an arbitral body but only with the parties' consent.[112]

In 1995, the FAO adopted the Code of Conduct for Responsible Fisheries ('FAO Code of Conduct') of which the Compliance Agreement is now an integral part.[113] The FAO Code of Conduct is a voluntary code addressed at states, IGOs, NGOs and 'all those involved in fisheries'.[114] It provides guidance for responsible fishing and is to be applied in accordance with norms of international environmental law.[115] General principles include sustainable utilisation,[116] a loosely described ecosystem approach[117] and a commitment to the objectives of the WTO.[118] Significantly, the Code affirms a commitment to the precautionary principle,[119] the need to consider the special interests of developing countries[120] and accessibility for state and non-state representatives.[121]

To further the implementation of the FAO Code of Conduct, members have adopted 'International Plans of Action' (IPOAs) on specific species and issues. These urge members to adopt national plans for the conservation of seabirds and sharks,[122] and the management of fishing capacity and IUU fishing.[123] IPOAs are elaborated within the framework of the FAO Code of Conduct[124] and are implemented by

---

[109] Compliance Agreement, Art. III.   [110] Ibid., Art. VI.   [111] Ibid., Art. IX.
[112] Ibid., Art. IX(3).   [113] FAO Code of Conduct, Art. 1.1   [114] Ibid., Annex 2.
[115] Ibid., Art. 3.2(c), see below n. 143 and accompanying text.
[116] FAO Code of Conduct, Art. 6.3.   [117] Ibid., Art. 6.2.   [118] Ibid., Art. 6.14.
[119] Ibid., Art. 6.5.   [120] Ibid., Art. 5.2.   [121] Ibid., Art. 7.1.6.
[122] See FAO, 'International Plan of Action for Reducing Incidental Catch of Seabirds in Longline Fisheries' ('IPOA-Seabirds') (1999) and FAO, 'International Plan of Action for the Conservation and Management of Sharks' ('IPOA-Sharks') (1999).
[123] FAO, 'International Plan of Action for the Management of Fishing Capacity' ('IPOA-Capacity') (1999); FAO, 'International Plan of Action to Prevent, Deter and Eliminate Illegal, Unreported and Unregulated Fishing' ('IPOA-IUU Fishing') (2001).
[124] See FAO Code of Conduct, Art. 2(d).

countries, which notify their 'national plans of action' to the FAO on a voluntary basis.

The FAO IPOA-IUU Fishing aims to prevent, deter and eliminate IUU fishing through the following principles and strategies: participation and coordination; a phased implementation of national plans of action; a comprehensive and integrated approach; conservation; transparency; and non-discrimination.[125] National plans to implement the IPOA on IUU fishing have been notified by Australia, Argentina, Canada, Chile, Japan, Korea, New Zealand and the United States, and regional plans have been notified by the Lake Victoria Fisheries Organization and the European Commission.[126]

The FAO IPOA-Capacity was adopted in 1999.[127] It reinforces the assumption made by UNCLOS that coastal states will continually assess the capacity of their fleets to harvest their EEZ fisheries[128] and links this assessment to the aim for vessel capacity reduction contained in the FAO Code of Conduct.[129] The IPOA urges states and RFMOs to assess the possible impact of subsidies in contributing to overcapacity, and provides that states should reduce and progressively eliminate subsidies that undermine fishing sustainability.[130] This aspiration appears to have been largely ignored by states and RFMOs – only three states and two RFMOs have submitted plans of action to date.[131]

Both the Compliance Agreement and the FAO Code of Conduct are intended to function within the overall framework of the law of the sea, including UNCLOS, the Fish Stocks Agreement and the new Port State Measures Agreement. Various provisions provide for their interrelationship,[132] and it is possible to group the EEZ regime, the high seas, species-specific and regional regimes, and the FAO fisheries management regime under the rubric of the law of the sea. The following section considers these interdependencies in more detail, and considers the capacity of the law of the sea to look outwards to other international regimes.

---

[125] FAO IPOA-IUU Fishing, para. 9.
[126] Ibid. For notifications, see www.fao.org/fishery/publications/iuu/npoa.
[127] FAO, IPOA-Capacity, see above n. 123.   [128] UNCLOS, Art. 62.2.
[129] FAO Code of Conduct, Arts. 6.3, 7.2.2.
[130] IPOA-Capacity, paras. 25–6.   [131] See Chapter 3, n. 22.
[132] Compliance Agreement, Preamble; FAO Code of Conduct, Art. 3.1–2. See further Rosemary Rayfuse, 'The Interrelationship between the Global Instruments of International Fisheries Law' in Ellen Hey (ed.) *Developments in International Fisheries Law* (1999) 107.

4 Interdependencies with other regimes

There are a number of legal and institutional mechanisms through which the law of the sea recognises its interdependence with other areas of international law. As to the former, UNCLOS contains a conflicts clause in Article 311:

> 2. This Convention shall not alter the rights and obligations of States Parties which arise from other agreements compatible with this Convention and which do not affect the enjoyment by other States Parties of their rights or the performance of their obligations under this Convention.
>
> 3. Two or more States Parties may conclude agreements modifying or suspending the operation of provisions of this Convention, applicable solely to the relations between them, provided that such agreements do not relate to a provision derogation from which is incompatible with the effective execution of the object and purpose of this Convention and provided further that such agreements shall not affect the application of the basic principles embodied herein, and that the provisions of such agreements do not affect the enjoyment by other States Parties of their rights or the performance of their obligations under this Convention.
> ...
> 5. This article does not affect international agreements expressly permitted or preserved by other articles of this Convention.

The provision that other treaties may be preserved (Article 311(2)) or concluded (Article 311(3)) as long as they do not affect the rights or duties of UNCLOS parties assumes the priority of UNCLOS.[133] The derogation in Article 311(5), however, applies to a wide range of agreements.[134] An example is the *Convention for the Conservation of Southern Bluefin Tuna* (1993), a regional agreement between states fishing for highly migratory species as encouraged by UNCLOS Article 64. As discussed above, this agreement was relevant to an unsuccessful claim before an arbitral tribunal by Australia and New Zealand over Japan's

---

[133] Art. 311(3) draws on VCLT Art. 41, which allows for *inter-se* modification that does not, *inter alia*, 'affect the enjoyment by the other parties of their rights under the treaty or the performance of their obligations'. On the application of UNCLOS Art. 311 to trade-related marine conservation disputes brought under the WTO, see Richard McLaughlin, 'Settling Trade-Related Disputes Over the Protection of Marine Living Resources: UNCLOS or the WTO?' (1997) 10 *Georgetown International Environmental Law Review* 29, 58.

[134] Franckx notes that not less than one-sixth of the total number of articles contained in UNCLOS provide for this kind of derogation: Erik Franckx, 'The Protection of Biodiversity and Fisheries Management: Issues Raised by the Relationship Between CITES and LOSC' in David Freestone, Richard Barnes and David Ong (eds.) *The Law of the Sea: Progress and Prospects* (2006) 210, 220.

fishing of southern bluefin tuna.[135] The tribunal considered that the non-binding dispute settlement provisions in the 1993 agreement were intended to require the consent of parties before a tribunal had jurisdiction to resolve relevant disputes. As a result, the compulsory dispute settlement mechanisms of UNCLOS Part XV did not apply.

The potential for regional and other agreements to 'trump' the compulsory dispute settlement mechanisms within UNCLOS has led commentators to observe the potential for fragmentation within the law of the sea itself.[136] The different choices of dispute settlement forums and the potential for different tribunals to develop legal principles in different ways may result in substantive and procedural differences. This phenomenon was perhaps predicted by UNCLOS negotiators, who referred to the 'interrelated' problems of ocean space and the need for them to be 'considered as a whole', yet also affirmed that matters not regulated by UNCLOS 'continue to be governed by the rules and principles of general international law'.[137] This underlying paradox within the law of the sea reflects the overall situation of fragmentation in international law described in Chapter 1, and demonstrates again why a consideration of regime interaction for fisheries issues is so important.

The Fish Stocks Agreement also affirms that matters outside of its scope 'continue to be governed by the rules and principles of general international law'.[138] As described above, the Agreement seeks to implement UNCLOS, yet does not have an identical membership of states.[139] It encourages states to develop interdependencies between the law of the sea and other areas of international law by consultations with each other. For example, the Agreement refers to the need for consultation by states through competent RFMOs and IGOs before states take action within IGOs with competence with respect to living resources.[140]

The new Port State Measures Agreement similarly provides that it 'shall be interpreted and applied in conformity with international law

---

[135] See above n. 84 and surrounding text.
[136] Rosemary Rayfuse, 'The Future of Compulsory Dispute Settlement Under The Law of the Sea Convention' (2005) 36 *Victoria University of Wellington Law Review* 683.
[137] UNCLOS, Preamble.   [138] Fish Stocks Agreement, Preamble.
[139] See above n. 68 and accompanying text.
[140] Fish Stocks Agreement, Art. 8(6). See further on marine living resources, CITES and the Convention on Biological Diversity, considered below pp. 56ff.

taking into account applicable international rules and standards'.[141] Such standards are to include those established through the International Maritime Organization (IMO), as well as other international instruments.[142]

The FAO Code of Conduct provides that it is to be interpreted and applied 'in accordance with other applicable rules of international law' and 'in the light of' relevant international environmental laws.[143] In addition, it impliedly refers to CITES when it recognises that states should 'cooperate in complying with relevant international agreements regulating trade in endangered species'.[144]

With respect to international trade law, the FAO Code of Conduct appears to transplant the policy of the WTO covered agreements. Under the heading of 'responsible international trade',[145] the Code provides that its provisions 'should be interpreted and applied in accordance with the principles, rights and obligations established in the WTO Agreement'.[146] It endorses the idea that states should ensure that measures that affect international trade in fish should be transparent and, when applicable, based on scientific evidence.[147]

The Code deviates from the core principles of liberalised trade, however, by recognising that states and relevant IGOs should ensure that the promotion of international fish trade does not 'adversely impact the nutritional rights and needs of people for whom fish is critical to their health and well being and for whom other comparable sources of food are not readily available or affordable'.[148] This introduces a concept of 'food security' that appears in a number of FAO documents. Such a right is largely absent from the WTO covered agreements (apart from a preambular reference in the Agreement on Agriculture, which does not apply to fisheries) and, indeed, appears to conflict with the idea of comparative advantage that underpins efforts to enhance the aggregate welfare of states through trade liberalisation. In addition, the Code recommends that in developing and implementing trade laws, states 'facilitate appropriate consultation with and participation of industry

---

[141] Port State Measures Agreement, Art. 4.   [142] Ibid.
[143] FAO Code of Conduct, Art. 3.2(c). (The Code is to be interpreted and applied 'in the light of the 1992 Declaration of Cancun, the 1992 Rio Declaration on Environment and Development, and Agenda 21 adopted by the United Nations Conference on Environment and Development (UNCED), in particular Chapter 17 of Agenda 21, and other relevant declarations and international instruments'.).
[144] Ibid., Art. 11.2.9.   [145] Ibid., Art. 11.2.   [146] Ibid., Art. 11.2.1.
[147] Ibid., Art. 11.2.3; see also Art. 11.3.1.   [148] Ibid., Art. 11.2.15.

as well as environmental and consumer groups'.[149] Such approaches bring additional and possibly conflicting elements to the interdependencies between trade law and the law of the sea.

5 Further links: the UN Consultative Process

In addition to the recognition of interdependencies in various instruments of the law of the sea, there are various examples of institutional links. In particular, the UN has recognised the necessity for collaboration between relevant agencies. A coordinating role was temporarily undertaken by the Subcommittee on Oceans and Coastal Areas (SOCA) of the UN's Administrative Committee on Coordination. SOCA operated from 1993 to 2002.

In 1999, the UN General Assembly resolved to establish a consultative process to facilitate the General Assembly's review of developments in ocean affairs and the law of the sea.[150] The UN Consultative Process is aimed at identifying areas where coordination and cooperation at the intergovernmental and inter-agency levels should be enhanced. At its first meeting in 2000, delegates identified international coordination as one of their primary areas of concern. IGOs with mandates to take measures relating to these areas 'expressed their willingness to cooperate with international organizations and States, in particular sharing their experiences and information'.[151]

The UN Consultative Process has had a mixed reception. Delegates have considered that one of its main contributions has been in assisting the General Assembly negotiations on resolutions on 'sustainable fisheries',[152] although the social and 'development' aspects of fisheries have allegedly been neglected.[153] At times, delegates to the Consultative Process have been unable to agree on outcomes.[154] There is continuing disagreement between delegates as to whether it is a consultative body

---

[149] Ibid., Art. 11.3.2.
[150] See UN Resolution 54/33 (1999). Delegates could not agree on the group's title, which varied between the United Nations Informal Process on Oceans and the United Nations Informal Consultative Process on Oceans and the Law of the Sea (UNICPOLOS) (see summary in UN Doc A/64/131 (13 July 2009), para. 22); it is given the label here of UN Consultative Process.
[151] UN Doc A/55/274 (31 July 2000), para. 45.
[152] See e.g. seventh meeting (see UN Doc A/61/156 (17 July 2006), para. 17); eighth meeting (see UN Doc A/62/169 (30 July 2007) para. 12). These resolutions are referred to above at n. 41.
[153] See UN Doc A/64/131 (13 July 2009).
[154] See, e.g., inability to finalise outcomes at the sixth and eighth meetings, noted in UN Doc A/64/131 (13 July 2009), para. 10.

or a negotiating body, and indeed whether these functions are mutually exclusive.[155] By contrast, an International Union for Conservation of Nature (IUCN) workshop on high seas governance has recently called for the expansion of the Consultative Process into an intergovernmental steering body with special responsibility for marine areas beyond national jurisdiction.[156]

One outcome of the UN Consultative Process was advocacy for the establishment of a new coordinating mechanism for ocean issues after the dissolution of SOCA.[157] This led, in 2003, to the decision by the United Nations System Chief Executives Board for Coordination to establish a replacement inter-agency coordination mechanism. Called the Oceans and Coastal Areas Network (UN-Oceans),[158] it aims to strengthen the 'coordination and cooperation of United Nations activities related to oceans and coastal areas'.[159]

UN-Oceans is headed by the Executive Secretary of the Intergovernmental Oceanographic Commission of UNESCO (IOC) and since 2005 has provided reports to the UN Consultative Process. Its activities are mainly addressed at specific time-bound task forces. It has been involved in post-tsunami reconstruction, and the process for global reporting and assessment of the state of the marine environment. More recently, it has pursued a task force on marine biodiversity beyond areas of national jurisdiction, which includes DOALOS and the Convention on Biological Diversity (CBD) Secretariat as lead agencies,[160] and a task force on the protection of the marine environment from land-based activities, which involves the World Summit on Sustainable Development, the CBD, the IOC, the FAO and UNEP.[161]

---

[155] Ibid., paras. 44–72.
[156] IUCN, Workshop in High Seas Governance for the 21st Century, Co-Chairs' Summary Report, December 2007, 6, 17.
[157] Report of the work of the United Nations Open-ended Informal Consultative Process on Oceans and the Law of the Sea, UN Doc A/58/95 (2003) para. 24.
[158] For recognition of the need for such a mechanism, see World Summit Plan of Implementation (2002), para. 30(c). The UN Consultative Process welcomed the establishment of this group: see UN Doc A/59/122 (2004), para. 3(a).
[159] www.oceansatlas.org/www.un-oceans.org/Index.htm.
[160] See www.oceansatlas.org/www.un-oceans.org/TaskForces.htm; see further Warner, above n. 43, 211–14.
[161] DOALOS, UNDP, IMO, the World Bank and the International Seabed Authority also expressed interest in participating in this task force: see UN Doc A/62/169 (30 July 2007) para. 110. See further www.gpa.unep.org/.

Although UN-Oceans is recognised as potentially very useful in promoting coordination and cooperation within the UN system, some delegates to the UN Consultative Process have raised questions about UN-Oceans's internal procedures.[162] Information-sharing, reporting and transparency is one area of concern. For example, the posting of online material has been subject to delays, perhaps due to a failure to get consensus on content.[163] There has also been criticism about accessibility and participation within UN-Oceans. The decision by members of UN-Oceans to restrict task-force participation to UN actors has been questioned.[164] Alongside the need for UN-Oceans to carry out its work in an 'efficient, transparent, accountable and responsive manner', there have been suggestions that international organisations themselves should enhance cooperation and coordination though memorandums of understanding.[165]

Indeed, there is growing recognition of the need for IGOs to enhance cooperation and coordination within the organisations themselves. In the FAO, there is an encouragement of 'concerted action' with respect to the conservation of natural resources.[166] The mission of the Fisheries and Aquaculture Department is to work with its members and, within its mandate, to forge 'closer and more effective partnerships with national and international institutions, academia, the private sector and civil society to achieve long-term sustainable results in the fisheries sector'.[167] The example of a Memorandum of Understanding between the FAO and the CITES Secretariat is considered in detail in the next chapter. The remainder of this chapter considers the framework for rules affecting fisheries sustainability within international environmental law and trade law.

---

[162] See e.g. UN Doc A/62/169 (30 July 2007) paras. 112–13.
[163] A long-standing disclaimer on the website of UN-Oceans stated: 'The information presented on this UN-OCEANS web-site is under development. None of the information presented here has been approved by the Partners of UN-OCEANS. March 9, 2005.' See www.oceansatlas.org/www.un-oceans.org/Index.htm. UN-Oceans continually acknowledges the need to revitalise its website: see UN Doc A/62/169 (2007) para. 113; UN Doc A/63/174 (2008), para. 131; UN Doc A/64/131 (2009) para. 43.
[164] UN Doc A/62/169 (2007) para. 113.
[165] UN Doc A/63/174 (2008) para. 133. For further discussion, see the Memorandum of Understanding between CITES and the FAO considered in Chapter 4.
[166] FAO Constitution, Art. I.
[167] www.fao.org/fishery/about. See also COFI, above n. 106 and accompanying text.

B International environmental law

The protection of the marine environment has been an original theme of international environmental law,[168] as it developed through intergovernmental meetings, state practice and specific case law. Now, the body of laws that is conceived as constituting international environmental law includes a large number of Multilateral Environmental Agreements (MEAs). Some treaties address the over-exploitation of marine resources by addressing particular species such as whales.[169] Others address particular activities such as the use of drift-nets, or address species protection within particular regions.[170] Paramount among the instruments that seek to control the trade in endangered or threatened species is CITES, which is examined in the following section. Next, the Convention on Biological Diversity (CBD), which confirms norms relating to the need for precautionary approaches in the utilisation of marine resources, is discussed. The interdependencies between these two Conventions and the law of the sea are examined in the final section.[171]

1 Convention on the International Trade in Endangered Species

CITES is one of the oldest international Conventions in environmental law. It was developed in the early 1970s to combat trade in wildlife and plants, which was threatening certain species with over-exploitation. It

---

[168] See 1972 Stockholm Conference on the Human Environment, Report of the UN Conference on the Human Environment, UN Doc A/Conf.48/14 (1972), 11 ILM 1416 (Stockholm Report).

[169] See e.g. the International Whaling Convention (1946). The overlapping regimes for the protection of whales, which are not considered in the present study, are examined in Alexander Gillespie, *Whaling Diplomacy* (2005) and his 'Forum Shopping in International Environmental Law: the IWC, CITES, and the Management of Cetaceans' (2002) 33 *Ocean Development and International Law* 17; see also Peter Sand, 'Japan's "Research Whaling" in the Antarctic Southern Ocean and the North Pacific Ocean in the Face of the Endangered Species Convention (CITES)' (2008) 17(1) *RECIEL* 56.

[170] See the Wellington Driftnet Convention (1990). For an example of a regional approach, see the Inter-American Convention for the Protection and Conservation of Sea Turtles (which featured in the *US – Shrimp* dispute discussed in Chapter 5).

[171] For a discussion of the early influence of MEAs and soft-law environmental instruments on regulating the marine environment, see Warner, above n. 43, 68-96) (citing declarations of IUCN, UNEP and WWF (the World Conservation Strategy of 1980) and the World Commission on Environment and Development (Brundtland Report of 1987).

grew from a resolution of members of IUCN.[172] As such, it is cited in support of the argument that IGOs and NGOs, rather than states, can be the driving forces of regime formation.[173]

CITES deals with prominent species such as elephants and tigers, but also applies to marine species. It seeks to regulate trade in all species agreed to be listed by the parties.[174] Trade extends to 'export, re-export, import and introduction from the sea', the latter defined as 'transportation into a State of specimens of any species which were taken in the marine environment not under the jurisdiction of any State'.[175] As such, CITES presumably applies to the international trade in all marine species (whether taken from within EEZs or high seas) and domestic fishing from high seas.[176] CITES has 175 parties,[177] however the trade-sanctioning nature of its regulatory mechanism means that non-party states that have not assented to the protection of species are brought within its ambit.[178]

Like many environmental treaties, CITES utilises an institutional model whereby functions are carried out by a Secretariat and decisions are made by the states parties who meet every two or three years at a Conference of the Parties (COP).[179] The CITES Secretariat is provided by the Executive Director of the United Nations Environment Programme (UNEP).[180] The function of the Secretariat encompasses all manner of administrative arrangements under the Convention, including the provision of reports and listing details. The Secretariat may also make recommendations for the implementation of the aims and provisions

---

[172] The IUCN has state and non-state members; it has 80 state members out of a total of more than 1000 members: see www.iucn.org/about/union/members/. Its functions include the compilation of a 'Red List of Threatened Species', on which an increasing number of commercial fish species such as the Golden Line Fish are beginning to appear (see further www.iucnredlist.org/).

[173] Oran Young, 'The Politics of International Regime Formation: Managing Natural Resources and the Environment' (1989) 43 *International Organization* 349, 353-4.

[174] CITES, Art. I.   [175] Ibid.

[176] The interpretation of 'introduction from the sea' is the subject of some controversy: see further Chapter 4.

[177] www.cites.org/eng/disc/parties/chronolo.shtml.

[178] But note that even though trade with non-parties is subject to conditions, such trade can significantly undermine CITES objectives: Paul Matthews, 'Problems Related to the Convention on the International Trade in Endangered Species' (1996) 45 *ICLQ* 421, 428.

[179] CITES was in fact the first MEA to use the term 'Conference of the Parties': Robin Churchill and Geir Ulfstein, 'Autonomous Institutional Arrangements in Multilateral Environmental Agreements: A Little-Noticed Phenomenon in International Law' (2000) 94 *AJIL* 623, 630.

[180] CITES, Art. XII.

of the Convention, including the exchange of information of a scientific or technical nature.[181]

The regulatory mechanism of CITES operates through a licensing system that authorises the trade (import, export, re-export and introduction from the sea)[182] of selected species. Three appendices divide selected species into levels of protection. Appendix I lists species that are threatened with extinction. Trade in specimens of these species is only permitted in exceptional circumstances and under set conditions.[183] Fewer than ten species of fish are currently included in Appendix I.[184] Appendix II lists species that are not necessarily threatened with extinction, but may become so unless trade is subject to strict regulation 'in order to avoid utilization incompatible with their survival'.[185] Export permits are granted subject to similar conditions as required for Appendix I species: exporting countries need to show that the commercial trade is not detrimental to the wild population, and that the particular specimen was legally obtained.[186] Approximately ten fish species are currently listed in Appendix II, although there are far more marine species.[187] Appendix III contains species that are protected by national laws, where the protecting country has asked other CITES parties for assistance in controlling the trade.[188] The CITES licensing system is administered by national management authorities, which are required to be designated by each party. These management authorities are advised by designated scientific authorities.

Proposals for the listing of species or their transfer between or removal from appendices, which are known as 'amendments' to the appendices, can only be made by parties.[189] These proposals are facilitated by the Secretariat, which communicates the proposal to the parties

---

[181] Ibid., Art. XII (2)(h).
[182] Ibid., Art. I, see above n. 175 and accompanying text.   [183] Ibid., Art. III.
[184] www.cites.org/eng/app/appendices.shtml. These figures are for the fish (class *actinopterygii* and *sarcopterygii*) and include species of sturgeon, totoaba, catfish and coelacanth. Other marine species listed in Appendix I include cetaceans, such as several species of dolphins and porpoises, and marine turtles.
[185] CITES, Art. II(2).   [186] Ibid., Art. IV.
[187] www.cites.org/eng/app/appendices.shtml (including seahorses, the european eel and the humphead wrasse). A large number of other marine species listed in Appendix II include sharks, coral species and the remainder of the dolphin species not listed in Appendix I.
[188] There are few marine species listed in Appendix III. An example is a species of sea cucumbers in Ecuador: see www.cites.org/eng/app/appendices.shtml.
[189] CITES, Art. XV.

and makes recommendations. The Secretariat has special obligations when a proposal relates to a marine species, in which case it shall:

> consult inter-governmental bodies having a function in relation to those species especially with a view to obtaining scientific data these bodies may be able to provide and to ensuring co-ordination with any conservation measures enforced by such bodies.[190]

The scientific and other views of IGOs are therefore useful for, but not determinative of, the Secretariat's own recommendations regarding listing.

The adoption of a proposal to amend the appendices occurs with a two-thirds majority vote at the COP.[191] In voting on the proposal, parties consider certain criteria according to which they have resolved *inter alia* to adopt a precautionary approach 'so that scientific uncertainty should not be used as a reason for failing to act in the best interest of the conservation of the species'.[192] The listing criteria include the information that parties are required to assemble in their listing proposals, including data on utilisation and levels of trade, conservation and management.[193] Reservations may be entered by parties that do not agree with the listing of the species. They will be treated as a non-party with respect to that species, until the reservation is withdrawn.

In addition to delegates of the parties, the COP is also open to representatives from IGOs and NGOs that are 'technically qualified in protection, conservation or management of wild fauna and flora'.[194] The bodies must be organisations with legal personality and an 'international character, remit and programme of activities'.[195] These bodies may attend as observers (without the right to vote) unless one-third of the parties present object.[196]

In parallel to the listing process, various committees within CITES monitor implementation and effectiveness. The permanent Standing Committee reports to the COP,[197] and coexists as the senior committee to the Animals Committee, Plants Committee and Nomenclature Committee. Its chair is elected from among the regional members.[198]

---

[190] Ibid., Art. XV(2)(b).
[191] Ibid., Art. XV. With respect to Appendix III species, however, decisions to add or remove such species may be made at any time and by any party.
[192] CITES Resolution Conf. 9.24 (Rev. CoP14). ('CITES listing criteria (2007)'); see also Annex 4(A).
[193] Ibid., Annex 6, paras. 6, 8.   [194] CITES, Art. XI(7).
[195] CITES Resolution Conf. 13.8.   [196] CITES, Art. XI(7).
[197] CITES Resolution Conf. 11.1 (Rev. CoP14).   [198] Ibid.

The Animals and Plants Committees have a particular role in conducting reviews of significant trade in Appendix II species. The reviews were instituted due to concern about a lack of implementation of CITES management measures such as population assessments.[199] The relevant committees work in consultation with exporting states to support national measures to prevent detriment to wild populations by controlled trade. They examine biological, trade and other relevant information for Appendix-II species, assisted, where required, by the Secretariat. Such a review has occurred for at least one commercially exploited marine species listed in Appendix II. A queen conch species became the subject of review after concerns were expressed about inadequate management and control of the species. In undertaking the review, the CITES Animals Committee agreed on a series of recommendations for harvesting and exporting countries, which included encouragement to seek assistance from the FAO.[200]

Enforcement of CITES depends on the parties adopting appropriate legislation, designating scientific and management authorities and imposing penalties for violations by importers and exporters. There is, however, a lack of implementing legislation within the domestic legal systems of approximately half of CITES parties.[201] In theory, compliance with these provisions can be encouraged through litigation, although unlike UNCLOS there are no binding dispute settlement measures. Instead, disputes arising over the interpretation or application of CITES provisions are subject to negotiation between the parties,[202] and may be submitted to binding arbitration, such as at the Permanent Court of Arbitration (PCA), only with prior consent.[203] Although the PCA has established a specialised panel for environmental matters,[204] no CITES dispute has ever been submitted to it or to another arbitral tribunal.

Instead, the reviews of significant trade of Appendix II species, such as the queen conch example given above, help in monitoring compliance

---

[199] CITES Resolution Conf. 12.8 (Rev. CoP13).
[200] CITES Doc. AC19 WG3 Doc. 1 (2003) 35; see FAO response that it was 'preparing to provide assistance as required and to the extent allowed by the limited resources of the Organization': COFI:FT/IX/2004/3 at 3. See further Chapter 4.
[201] See CITES CoP15 Doc. 20, 5 Nov 2009, referring to Resolution Conf. 8.4 (Rev. CoP14), 'National Laws for Implementation of the Convention' (revised 3–15 June 2007).
[202] CITES, Art. XVIII(1).   [203] Ibid., Art. XVIII(2).
[204] See Permanent Court of Arbitration, 'Optional Rules for Arbitration of Disputes Relating to Natural Resources and/or the Environment' (16 April 2002) (www.pca-cpa.org/upload/files/ENV%20CONC.pdf).

and identifying problems. The CITES Secretariat or Committees may recommend specific measures, including the suspension of all trade in a relevant species by the non-complying party.[205] In addition, the CITES Secretariat provides advice and assistance to parties and identifies countries where there is an urgent need for legislative reform.[206]

2 Convention on Migratory Species

Like CITES, the Convention on Migratory Species (CMS) recognises that states need to cooperate to conserve and effectively manage migratory species. Also known as the Convention on the Conservation of Migratory Species of Wild Animals, or the Bonn Convention, the CMS entered into force in 1979 and has 113 parties.[207] It contains a list of endangered species, which parties agree upon at regular meetings. Unlike CITES, however, it does not seek to restrain or regulate international trade. Instead, the CMS concentrates on the taking of migratory species by the range states that feature in relevant migration routes, and imposes obligations on these states to protect or restore habitat for certain species.

There are two appendices to the CMS. Appendix I covers endangered migratory species, of which any taking by range states is prohibited.[208] There are certain limited exceptions for scientific purposes or for the accommodation of the needs of traditional subsistence users of the relevant species. Appendix II covers migratory species that have an unfavourable conservation status, or that have a conservation status 'which would significantly benefit from the international cooperation that could be achieved by an international agreement'.[209] Relevant range states endeavour to enter into agreements on the conservation and management of such species.[210] They are also encouraged to take action with a view to concluding agreements for any population or geographically separate part of the population of any species,[211] which they often do by developing less formal Memoranda of Understanding between states.

---

[205] See CITES CoP12 Doc. 26, 'Compliance with the Convention' (3–15 November 2002) (adopted at Santiago, Chile, 3–15 November 2002).
[206] National Legislation Project: see CITES Resolution Conf 8.4 (Rev. CoP14) noted above n. 201.
[207] *Convention on the Conservation of Migratory Species of Wild Animals*, 23 June 1979, 19 ILM 15 (1980). For parties, see www.cms.int/about/Partylist_eng.pdf.
[208] CMS, Art. III.   [209] Ibid., Art. IV.   [210] Ibid., Art. IV(3).   [211] Ibid., Art. IV(4).

By encouraging range states to enter into agreements, the CMS acts as a framework Convention. States are encouraged to nominate national authorities to implement the measures, and to provide for ongoing research, education and review of the agreement.[212] The CMS Secretariat is provided by the Executive Director of UNEP.[213] The Secretariat arranges meetings of the Conference of the Parties and a Scientific Council, which provides expert advice.[214] The CMS Secretariat also undertakes extensive reporting on the status of migratory species.[215]

Marine species are eligible for inclusion in the CMS appendices. The Convention implicitly extends to terrestrial, marine and avian migratory species.[216] The inclusion of threatened marine species in the CMS appendices can extend the conservation efforts afforded by the coverage of such species by CITES, because the CMS addresses the taking of species for both domestic consumption *and* international trade. Moreover, the CMS can cover the health and habitat of endangered marine species that are not traded *per se*, but that are threatened by the fishing and other activities of states. Marine turtles are an obvious example of species that suffer from the harvesting practices of states,[217] and there are many other species that are vulnerable as by-catch. On the other hand, the enforcement of CMS measures does not benefit from the possibilities of trade restraints, such as can be initiated by CITES. Moreover, disputes arising from the interpretation or application of the CMS are not subject to compulsory settlement. Like CITES, consent of the parties is required before a dispute may be submitted to binding arbitration.[218]

There have been a number of efforts to protect marine species under the CMS. Whales and dolphins feature in a number of agreements between parties.[219] The conservation and management of marine

---

[212] Ibid., Art. V.   [213] Ibid., Art. IX.
[214] Ibid., Art. VIII (Scientific Council); Art. IX (The Secretariat).
[215] See e.g. CMS report on the decline of whales, dolphins and porpoises from by-catch in fishing nets, 4 February 2010 (www.cms.int/reports/small_cetaceans/contents.htm).
[216] There are no restrictions as to the type of species covered by the CMS: see further www.cms.int/about/intro.htm.
[217] See further Chapter 5.   [218] CMS, Art. XIII.
[219] See e.g. Agreement on the Conservation of Cetaceans of the Black Sea, Mediterranean Sea and Contiguous Atlantic Area (ACCOBAMS), Agreement on the Conservation of Small Cetaceans of the Baltic, North East Atlantic, Irish and North Seas (ASCOBANS), Memorandum of Understanding for the Conservation of Cetaceans and their Habitats in the Pacific Islands Region and the Memorandum of Understanding concerning the Conservation of the Manatee and Small Cetaceans of Western Africa and Macaronesia.

turtles has also received widespread support by states.[220] After delay in agreeing on measures to implement protection for the Appendix II-listed shark species,[221] a Memorandum of Understanding on migratory sharks is likely to be finalised.[222]

3 Convention on Biological Diversity

In 1992, the same year in which states convened to discuss measures to enhance responsible fishing,[223] the United Nations Conference on Environment and Development (UNCED – or the Earth Summit) met in Rio de Janeiro. Two measures important to fisheries governance were agreed upon. First, Agenda 21, which calls for action for the twenty-first century to address the human impact on the environment, was adopted by the 178 countries present. Secondly, the Convention on Biological Diversity (CBD) was adopted; it now attracts a membership of 193 states.[224]

Agenda 21 was prepared with coordination by the UNCED Secretariat and the FAO. Chapter 17 of Agenda 21 proclaims that the marine environment is an integrated whole.[225] It sets out policies on a range of marine resource issues, including the sustainable use and conservation of marine living resources of the high seas and under national jurisdiction and the strengthening of international, including regional, cooperation and coordination. It recognises the essential nature of information exchange and calls on the development and sharing of analytical and predictive tools, such as stock assessment and bioeconomic models, and the establishment of appropriate monitoring and assessment programmes.[226] The use of precautionary approaches is

---

[220] See e.g. Memorandum of Understanding on the Conservation and Management of Marine Turtles and their Habitats of the Indian Ocean and South-East Asia, 1 September 2001.

[221] See Holly Edwards, 'When Predators Become Prey: The Need for International Shark Conservation' (2007) 12 *Ocean and Coastal Law Journal* 305, 337–40.

[222] See Third Meeting on International Cooperation on Migratory Sharks (SHARKS III) Under the Convention on Migratory Species (CMS), 8–12 February 2010, Manila, Philippines (Signature by The Philippines, Senegal, Togo, USA, Republic of Congo, Costa Rica, Ghana, Guinea, Liberia and Palau): www.cms.int/bodies/meetings/regional/sharks/sharks_meetings.htm.

[223] The International Conference on Responsible Fishing in Cancun, Mexico, 1992 led to the 'Declaration of Cancun', a precursor to the FAO Code of Conduct. The UN DOALOS and the FAO were both involved in the administration of the conference.

[224] See www.cbd.int/convention/parties/list/.   [225] Agenda 21, chapter 17, para. 17.1.

[226] Ibid., para. 17.86; see also para. 17.46. See also chapter 40 of Agenda 21, which requires states to report on sustainability indicators.

endorsed in particular areas,[227] following the approach of the Rio Declaration, which applies to threats of serious or irreversible damage, and requires that 'lack of full scientific certainty shall not be used as a reason to postpone cost-effective measures to prevent environmental degradation'.[228] Chapter 17 of Agenda 21 was reaffirmed by the World Summit in 2002.[229]

The CBD is aimed at 'the conservation of biological diversity [including from all marine sources and aquatic ecosystems], the sustainable use of its components and the fair and equitable sharing of the benefits arising out of the utilization of genetic resources'.[230] States parties must promote the protection of ecosystems and natural habitats both *in situ* and *ex situ*.[231] This extends to an obligation to regulate and manage biological diversity within areas of national jurisdiction, including marine environments. However, the obligations of parties to minimise threats to biodiversity *beyond* their national jurisdiction (ie on the high seas) are less concrete.[232]

Parties to the CBD regularly meet to review the implementation of the Convention at conferences of the parties (COPs). These conferences are attended by a range of governmental and non-governmental bodies. Such bodies must first meet three requirements: they must be 'qualified in fields relating to conservation and sustainable use of biological diversity', have informed the Secretariat of their wish to attend, and their attendance must not be objected to by more than one-third of parties present.[233] In addition, the states parties agree to facilitate the exchange of all publically available information 'relevant to the conservation and sustainable use of biological diversity'.[234]

In 1995, parties to the CBD created a work programme for the management of marine living resources, which emphasised *inter alia* the precautionary approach.[235] The programme has been adopted in later

---

[227] See para. 17.5(d), 17.21, 17.22 (re. coastal marine areas), para. 17.97 (re. climate change uncertainties and the marine environment) and para. 17.128 (re. small islands).
[228] Rio Declaration, Principle 15.
[229] World Summit Plan of Implementation (2002), para 30(b).
[230] CBD, Art. 1. The incorporation of marine sources and aquatic ecosystems into the definition of biological diversity is found in Art. 2.
[231] See CITES, Arts. 8, 9.
[232] See Christopher C Joyner, 'Biodiversity in the Marine Environment: Resource Implications for the Law of the Sea' (1995) 28 *Vanderbilt Journal of Transnational Law* 635, 650.
[233] CITES, Art. 23(5).   [234] Ibid., Art. 17.
[235] Jakarta Mandate on Marine and Coastal Biodiversity (1995).

conferences.[236] CBD parties have also endorsed an 'ecosystem approach' to the promotion of the conservation and sustainable use of biological diversity, which recognises as a 'functional unit' the interaction of 'a dynamic complex of plant, animal and micro-organism communities and their non-living environment'.[237] Such an attitude accords with a growing view within fisheries scientists and conservation biologists that prefers large-scale protections of marine ecosystems rather than traditional exploitation rates that achieve maximum sustainable yields.[238]

4 Interdependencies with other regimes

As well as recognising their own synergies by entering into a memorandum of understanding,[239] CITES and the CBD have considered their interdependence with other international Conventions, particularly with respect to the law of the sea.

The CBD recognises its co-existence with the law of the sea in Article 22, as follows:[240]

1. The provisions of this Convention shall not affect the rights and obligations of any Contracting Party deriving from any existing international agreement, except where the exercise of those rights and obligations would cause a serious damage or threat to biological diversity.
2. Contracting Parties shall implement this Convention with respect to the marine environment consistently with the rights and obligations of States under the law of the sea.

Some commentators consider this provision to be of profound importance for the management of marine biodiversity resources.[241] It apparently accommodates the qualified freedom of the high seas concept, which, as described above, allows for fishing rights but also requires states to cooperate in the conservation and management of living

---

[236] CBD Decision on Marine and Coastal Ecosystems (1998); World Summit Plan of Implementation (2002), para. 32(b).
[237] CBD, Art. 2; CBD, CoP5 Decision, 'Ecosystem Approach', Nairobi, 15–26 May 2000 (Decision V/6); CBD, CoP7 Decision, 'Ecosystem Approach', Kuala Lumpur, 9–20 February 2004 (Decision VII/11).
[238] Boris Worm et al., 'Rebuilding Global Fisheries' (2009) 325 *Science* (31 July 2009).
[239] See Memorandum of Cooperation between CITES and CBD Secretariats referred to in CITES Resolution Conf. 10.4(Rev.14); see further Rosie Cooney, 'CITES and the CBD: Tensions and Synergies' (2001) 10 *RECIEL* 259.
[240] CBD, Art. 22.   [241] Joyner, above n. 232, 651.

resources of the high seas.[242] The provision also incorporates general UNCLOS obligations to protect and preserve the marine environment.[243] Yet despite these apparent synergies, the CBD's aspirations for integrated environmental protection and the implementation of biodiversity protection measures in areas beyond national jurisdiction continue to be constrained by a lack of coordinating mechanisms and institutional capacity.[244]

CITES Article XIV provides:

> 4. A State party to the present Convention, which is also a party to any other treaty, convention or international agreement which is in force at the time of the coming into force of the present Convention and under the provisions of which protection is afforded to marine species included in Appendix II, shall be relieved of the obligations imposed on it under the provisions of the present Convention with respect to trade in specimens of species included in Appendix II that are taken by ships registered in that State and in accordance with the provisions of such other treaty, convention or international agreement.[245]

According to this provision, if an Appendix II-listed marine species is afforded 'protection' by another treaty or international agreement that had entered into force before CITES, any party to that agreement is relieved of its CITES obligations with respect to those species. 'Protection' is not defined. This provision is said to be related to the International Whaling Convention (1946) and activities of the International Whaling Commission.[246] Were it to apply to RFMOs, it would only apply to pre-1975 bodies and only if the constitutive agreements of the RFMOs contained provisions affording protection to the relevant marine species.

Expert consultations at the FAO on this issue have pointed to the Inter-American Tropical Tuna Commission (IATTC) and the International Commission for the Conservation of Atlantic Tunas (ICCAT) as examples of fisheries agreements that were earlier in time than CITES.[247] An unresolved issue is whether these fisheries

---

[242] See UNCLOS Arts. 116–120 regarding the conservation and management of the living resources of the high seas, discussed above n. 43 and surrounding text.
[243] See UNCLOS, Art. 192 referred to above n. 5 and surrounding text; see also Warner, above n. 43, 94.
[244] Warner, above n. 43, 95, 97.   [245] CITES, Art. XIV.
[246] Gillespie, above n. 169, 31; Sand, above n. 169, 61.
[247] FAO, *Fisheries Report No. 741* Report of the Expert Consultation on Implementation Issues Associated with Listing Commercially-Exploited Aquatic Species on CITES Appendices (25–28 May, 2004), para. 31.

agreements can be said to accord 'protection' to relevant species, and how such protection ought to be assessed.

By contrast, CMS refers to the rights and obligations of parties deriving from existing treaties, but provides that the CMS 'shall in no way affect' such rights or obligations.[248] There is no criterion of 'protection' to be satisfied in the event of overlapping or conflicting obligations between the CMS and another treaty.

CITES and CMS are both expressly without prejudice to the codification and development of the law of the sea by UNCLOS, which at the time was finalising negotiations on the EEZ, as described above. The identical provision in both CITES (Article XIV(6)) and CMS (Article XII) is:

> Nothing in the present Convention shall prejudice the codification and development of the law of the sea by the United Nations Conference on the Law of the Sea convened pursuant to Resolution 2750 C (XXV) of the General Assembly of the United Nations nor the present or future claims and legal views of any State concerning the law of the sea and the nature and extent of coastal and flag State jurisdiction.

Apart from these conflict clauses, CITES parties recognise the need for collaboration with fisheries organisations before marine species are listed. As described above, if proposals to amend the appendices are made that relate to marine species, the CITES Secretariat communicates the proposals to the other parties and consults IGOs 'having a function in relation to those species'.[249] The CITES Secretariat's consultation is aimed at obtaining scientific data and 'ensuring coordination with any conservation measures enforced by such bodies'.[250] The CMS Secretariat similarly assists a standing scientific council and liaises with parties and other international organisations 'concerned with migratory species'.[251]

5 Further links: UNEP

Provisions for the interdependencies between the law of the sea and international environmental law suggest that the legal regimes aimed at fisheries conservation and management were never intended to be autonomous. The negotiators of the instruments described above were aware of the interrelationships between regimes as they expanded and diversified the content of international law. Apart from the legal

---

[248] CMS, Art. XII(2). [249] CITES, Art. XV(2)(b).
[250] Ibid. See further Chapter 4. [251] CMS, Art. IX (4).

provisions, the negotiators also recognised the need to institutionalise collaboration between regimes.

Chief among the institutional innovations within international environmental law is the role of the United Nations Environment Programme (UNEP). One of the functions of UNEP is to promote coherence in environmental policy-making. It overseas a number of MEA secretariats, including CITES, the CMS and the CBD. UNEP also seeks to integrate environmental considerations in economic, trade and financial policies and practices.[252] For example, in the area of fisheries, it aims to enhance awareness of the impacts of fisheries subsidies, provide a forum for discussions and promote sustainable fishery practices.[253]

UNEP is mandated by the General Assembly to 'promote international cooperation in the field of the environment and to recommend, as appropriate, policies to this end'.[254] Decisions are made by a Governing Council, which meets every two years and reports to the General Assembly through the Economic and Social Council.[255] The Governing Council envisages participation of certain UN bodies and other IGOs. This extends to deliberation, but without the right to vote.[256] Accredited NGOs may attend the Governing Council as observers and submit statements under certain conditions.[257] The availability and degree of participation of external bodies may bring important perspectives to the interdependencies of these regimes, as will be considered in the following chapters.

C International trade law

International trade law includes the multilateral framework agreed under the auspices of the GATT and WTO and the bilateral or regional framework existing in specific trade agreements. This book focuses on the multilateral framework, although I acknowledge that bilateral and

---

[252] See World Summit Plan of Implementation (2002), paras. 97(c), 154; see also the work of the Economics and Trade Branch (UNEP-ETB), part of the UNEP Division of Technology, Industry and Economics, at www.unep.ch/etb/about/index.php.
[253] See further Chapter 3.
[254] UNGA Resolution 2997 (XXVII) of 15 December 1972, Art. I (2)(a); See also Agenda 21, paras. 38.21–38.23.
[255] The Governing Council is made up of 58 representatives elected by the General Assembly, for four-year terms, taking into account the principle of equitable regional representation.
[256] UNEP Rules of Procedure (1988), Rule 68.   [257] Ibid., Rule 69.

regional trade agreements are increasingly preferred by countries.[258] Before moving to an overview of relevant WTO provisions, it is therefore important to note the influence of regional trade agreements on the regulation of the global fish trade.

Regional trade agreements have a significant impact on the global trade of fish and fish products. For example, negotiations between the African, Caribbean and Pacific Group of States (ACP) group of countries and the EU to conclude economic partnership agreements are intended to allocate fishing access rights to the EU and accord preferential trade arrangements to the recipient states.[259] The growing concentration of the fishery trade within regions is relevant to this issue; FAO statistics from 2004 and 2005 demonstrate, for example, that more than 84 per cent of EU fishery exports went to other EU countries, and about 50 per cent of EU imports came from other EU countries.[260]

The relevance of both multilateral and regional trade agreements to trade-related aspects of fisheries can give rise to real complexity when disputes arise between states. For example, in a recent development in a long-running dispute over US measures on the importation, marketing and sale of tuna and tuna products, Mexico has alleged that the United States has acted inconsistently with its WTO obligations. The US, in response, has criticised the establishment of a WTO panel, arguing that the case should be adjudicated in the dispute settlement system set up under the North American Free Trade Agreement (NAFTA).[261] Clearly, the fragmentation of trade provisions (or to use a different metaphor, the 'spaghetti bowl' of regional trade agreements)[262] gives rise to similar challenges for regime interaction as the fragmentation of international law.

Even as such arrangements may reduce the scope of WTO rules on non-discrimination,[263] multilateral rules continue to have relevance. Preferential trade arrangements must satisfy WTO provisions that allow for the establishment of customs unions and free trade areas, grant

---

[258] See WTO, Consultative Board, *The Future of the WTO: Addressing Institutional Challenges in the New Millennium* (2004) 19–27.

[259] On EU fisheries access policy, see further above n. 34 and surrounding text.

[260] SOFIA (2006) 45.

[261] *United States – Measures Concerning the Importation, Marketing and Sale of Tuna and Tuna Products* (DS 381) panel composed 14 December 2009).

[262] A metaphor devised and popularised by Jagdish Bhagwati; see his *Termites in the Trading System: How Preferential Agreements Undermine Free Trade* (2008) 63.

[263] Ibid.

exceptions for special and differential treatment or allow for special waivers.[264] The EU's exemption from customs duties for canned tuna from ACP countries is an example.[265] WTO commitments such as obligations relating to import restrictions, technical trade barriers, subsidies and anti-dumping provisions, have major significance to fisheries and are the focus of the remainder of this chapter.

1 The WTO Agreements

The WTO is mandated to provide an institutional framework for the 'conduct of trade relations' among its 153 members.[266] WTO members agree to facilitate the production and trade in goods and services through a range of commitments. These obligations were updated during the Uruguay Round, which concluded in 1993, and are now set out in a number of agreements known as the 'WTO covered agreements'.

The WTO covered agreements are headed by the Marrakesh Agreement, which formally establishes the WTO and its Secretariat. According to the Marrakesh Agreement, the WTO provides a forum for trade negotiations, monitors national trade policies and handles trade disputes.[267] The dispute settlement mechanism of the WTO is significant in international law in that it has compulsory jurisdiction: WTO members have given prior consent to the establishment of panels in trade disputes. The only other standing international tribunal with mandatory dispute settlement is ITLOS, although as I described above its jurisdiction is ameliorated by many factors.[268] The WTO has separate legal personality and is accorded 'by each of its Members such legal capacity as may be necessary for the exercise of its functions'.[269]

The core instrument is the General Agreement on Tariffs and Trade 1947 (GATT).[270] WTO members agree to accord mutual non-discriminatory

---

[264] See GATT 1994, Art XXIV (customs unions and free trade areas); Enabling Clause (1979); Marrakesh Agreement, Art. IX.3 and 4 (waiver).
[265] See further Margaret A Young, 'WTO Undercurrents at the Court of Justice' (2005) 30 *European Law Review* 211.
[266] Marrakesh Agreement, Art. II. For membership, see www.wto.org (as at 23 July 2008).
[267] Ibid., Art. III.   [268] See above nn. 37 and 87 and surrounding text.
[269] Marrakesh Agreement, Art. VIII.1.
[270] The other two main agreements are the General Agreement on Trade in Services (GATS) and the Agreement on Trade-Related Aspects of Intellectual Property Rights (TRIPS). These are not reviewed here because they have minimal impact on the fisheries sector. The TRIPS Agreement might be expected to have a greater impact on the trade in fishery products in the future, however, due to the growth of aquaculture and potential uses of biotechnologies of fishery production: see further www.fao.org/fishery/topic/13275.

preferences to like products of other members[271] and that internal taxation and regulation should not afford protection to domestic production.[272] In addition, members agree to the general elimination of restrictions on the importation or exportation of any product.[273]

There have been a number of examples where GATT obligations have been violated by national fisheries policies. For example, a Canadian rule that required all salmon and herring caught in Canadian waters to be processed in Canada before export was found to be a quantitative restriction.[274] A US measure that banned tuna that had been caught in ways harmful to dolphin was found to be both a quantitative restriction and inconsistent with the national treatment requirement.[275] Perhaps the most famous example is the US measure that banned the import of shrimp from countries that allegedly failed to minimise the death of turtles in the shrimp harvesting process.[276] The *US – Shrimp* case prompted the WTO panel and Appellate Body to consider the relevance of a range of other international laws and is the focus of my analysis of the adjudication of trade disputes in Chapter 5.

Two further agreements relating to goods aim to eliminate protectionism. First, the Agreement on Technical Barriers to Trade (TBT Agreement) seeks to ensure that technical regulations and standards do not create unnecessary obstacles to international trade. These technical regulations include packaging and labelling requirements and the TBT Agreement expressly encourages the harmonisation of these requirements through the development and use of international standards. This agreement has been at issue in one WTO dispute settlement proceeding involving fish products. In *EC – Sardines*, an EC measure relating to the labelling of sardines products was found to violate the TBT Agreement because it had failed to adopt a relevant international standard that would have fulfilled its requirements.[277]

---

[271] This obligation, known as most-favoured-nation treatment or MFN, is found in GATT, Art. I.
[272] Ibid., Art. III.   [273] Ibid., Art. XI.
[274] GATT Panel Report, *Canada – Herring and Salmon*.
[275] *US –Tuna I*. See Ted McDorman, 'Fisheries Conservation and Management and International Trade Law' in Ellen Hey (ed.) *Developments in International Fisheries Law* (1999) 501. There is substantial overlap between Art. III and Art. XI: see e.g. Panel Report, *India – Automotive*, paras. 7.217–7.224.
[276] AB Report, *US – Shrimp*.   [277] AB Report, *EC – Sardines*.

Secondly, the Agreement on the Application of Sanitary and Phytosanitary Measures (SPS Agreement) aims to minimise the negative effects on trade of certain specific technical regulations and standards known broadly as 'SPS measures'.[278] Like the TBT Agreement, international standards are expressly recognised as useful in minimising the negative effects of SPS measures on international trade. Moreover, members are obliged to ensure that their SPS measures are scientifically justified, most commonly through conducting risk assessments.[279] This requirement was the subject of dispute in the *Australia – Salmon* case, which found that an Australian import prohibition on certain fresh, chilled and frozen ocean-caught salmon from Canada violated the SPS Agreement because it was not based on a risk assessment and it resulted in 'discrimination or a disguised restriction' on imports of salmon compared to imports of other fish and fish products such as herring and finfish.[280]

Additional agreements that impact upon national fisheries policies include the Agreement on Rules of Origin, which seeks to ensure that 'rules of origin are prepared and applied in an impartial, transparent, predictable, consistent and neutral manner'.[281] This Agreement governs the criteria used to define where a product was made. The origin of a product is important for trade policies that seek to discriminate against countries, which, as described above are permitted in the WTO in certain circumstances.[282] Yet defining the origin of fisheries products is extremely complex. If a can of tuna, for example, mixes tuna from catches from a range of places, or includes tuna that has been harvested in one country but canned in another, it is difficult to determine its country of origin for trade purposes. There have been no disputes on this point to date, but the intersection of trade and fisheries in rules of origin is receiving increasing attention.[283] Also addressing non-tariff barriers of potential relevance to the fisheries sector is the

---

[278] SPS measures are defined in the SPS Agreement, Annex A, and relate broadly to measures necessary to protect human, animal or plant life or health. See further Joanne Scott, *The WTO Agreement on Sanitary and Phytosanitary Measures: A Commentary* (2007).

[279] SPS Agreement, Art. 5.

[280] AB Report, *Australia - Salmon*, see esp. paras. 136, 158, 178.

[281] Rules of Origin Agreement, Preamble.

[282] See reference to preferential trade arrangements, exceptions for special and differential treatment and special waivers above n. 264 and surrounding text.

[283] Roman Grynberg and Natallie Rochester, 'Expert Opinion: Fixing Cotonou's Rules of Origin Regime' in Adil Najam, Mark Halle and Ricardo Meléndez-Ortiz (eds.) *Trade and Environment: A Resource Book* (2007) 107.

Agreement on Import Licensing Procedures, which recognises that the 'flow of international trade could be impeded by the inappropriate use of import licensing procedures'.[284]

Of further significance for the fisheries sector are provisions that regulate national action against 'dumping' practices.[285] Dumping occurs in certain situations when a company exports a product at a price lower than the normal domestic price, and the GATT condemns it if it causes, or threatens to cause, injury to domestic industries. Relevant provisions regulate how governments can retaliate through the levying of countervailing duties. These provisions are among the most heavily litigated at the WTO and often involve fisheries products. For example, a long conflict over the value of Norwegian farmed salmon culminated in a successful claim by Norway against the EC's anti-dumping measures.[286] Anti-dumping cases on shrimp have involved many WTO members from Europa, Asia and the Americas.[287]

The decision by WTO members to constrain their use of subsidies is another area of trade law that affects fisheries. Subsidies are often trade-distorting, and where they lead to perverse incentives to exploit finite resources, can be ecologically harmful as well. At other times subsidies can be helpful for fishing sustainability – such as where states provide financial aid for marine research programmes. The current framework to assess and 'discipline' domestic subsidies is contained in the Agreement on Subsidies and Countervailing Measures (SCM Agreement). A separate framework exists for agricultural products in the Agreement on Agriculture, which expressly excludes fisheries products.[288]

The SCM Agreement disciplines the use by the governments of WTO members of certain subsidies and regulates the action that WTO members may take to counter the negative effects of subsidies. Members can

---

[284] Import Licensing Procedures Agreement, Preamble.
[285] GATT, Art. VI; Anti-Dumping Agreement.   [286] Panel Report, *EC – Salmon*.
[287] See, e.g., dispute brought by Thailand against US anti-dumping measures on imports of frozen warmwater shrimp: AB Report, *US – Shrimp (Thailand)*. Brazil, Chile, China, the EC, India, Japan, Korea, Mexico and Vietnam were third parties. See also dispute brought by Ecuador in *US – Shrimp (Ecuador)*, which, in addition to Thailand, involved the same third parties except for Vietnam.
[288] Agreement on Agriculture, Art. 2, Annex 1. Stone discusses proposals to bring fish into the agriculture framework during the Uruguay Round, which failed due to resistance of some major fishing nations: Christopher Stone, 'Too Many Fishing Boats, Too Few Fish: Can Trade Laws Trim Subsidies and Restore the Balance in Global Fisheries?' in Kevin Gallagher and Jacob Werksman (eds.) *International Trade and Sustainable Development* (2002) 286, 295.

challenge offending subsidy programmes within the WTO dispute settlement system or levy countervailing duties after domestic investigations.[289] The SCM Agreement also provides for the notification of particular subsidies programmes to the WTO.[290] The SCM Committee is charged with examining these notifications every three years.[291] It may seek advice from a 'Permanent Group of Experts', which consists of five independent persons highly qualified in the fields of 'subsidies and trade relations', although this has never occurred.[292]

Countries currently provide subsidies, in very large amounts, to the fishing sector. The effects of many of these subsidies include the distortion of the global trade in fish and the over-exploitation of fish stocks. Yet the SCM Agreement currently disciplines very few of these subsidies. To remedy this, WTO members have agreed to 'clarify and improve WTO disciplines on fisheries subsidies'.[293] In doing so, WTO members and the WTO Secretariat have consulted with a range of participants from the FAO fisheries management regime and other regimes, as described in Chapter 3.

The negotiations to amend the SCM Agreement are part of a wider set of negotiations on changes to the WTO covered agreements. These negotiations are conducted in a framework called a 'single undertaking', which means that a range of subjects are separately negotiated but must be signed by the WTO membership as a whole. A Trade Negotiations Committee ('TNC') oversees several disparate negotiating groups which periodically report in plenary.[294] The single-undertaking character of the WTO negotiations means that, in effect, WTO members cannot enter reservations to any of the WTO covered agreements. A different situation exists for WTO agreements that are updated through express amendment approved by two-thirds of the members.[295]

The current round of negotiations under the single undertaking was launched in Doha in 2001.[296] The 'Doha Round', as it has become

---

[289] SCM Agreement, Part V.   [290] Ibid., Part VII.
[291] Ibid., Art. 26.1. The SCM Committee is established by Art. 24.   [292] Ibid., Art. 24.3.
[293] WTO Ministerial Declaration adopted on 14 November 2001 (WT/MIN(01)/DEC/1, 20 November 2001) (2002) ILM 746 ('WTO Doha Declaration (2001)'), para. 28.
[294] The TNC operates under the authority of the WTO General Council (discussed below n. 321 and surrounding text): see WTO Doha Declaration (2001), para. 46.
[295] Marrakesh Agreement, Art. X. In certain circumstances, such amendment will only apply to the WTO members that have agreed to it: see further Art. X.3. Note also that there are a number of 'plurilateral' trade agreements, such as the Agreement on Government Procurement, which are not part of the WTO covered agreements.
[296] WTO Doha Declaration (2001).

known, impacts upon several areas of fisheries policy besides subsidies. In one set of negotiations, members are updating their agreements on market access.[297] These negotiations have the potential to reduce tariffs for the fisheries sector. As such, they also threaten to reduce the comparative advantage of those WTO members who have secured preferential treatment for their fish products in the past.[298] Another negotiating issue under the Doha Round relates to environmental labelling.[299] These discussions have significant relevance for the fisheries products displaying labels that inform consumers about environmental impacts, such as many brands of canned tuna. These and other issues have potential impact on the trade in fish and fish products, and WTO members have expressed real commitment to addressing them. Yet WTO members have failed to agree on other key issues in the Doha Round, including, most famously, agricultural subsidies, and the negotiations have overrun their deadline by several years. Given the 'single undertaking' nature of the Doha Round, the likelihood of change is currently slight.

2 Interdependencies with other regimes

The WTO Agreements expressly recognise their interdependencies with other regimes, both normatively and institutionally. For example, trade relations between WTO members are to be conducted in the context of 'allowing for the optimal use of the world's resources in accordance with the objective of sustainable development',[300] which resonates with the 'optimum utilisation' ethos of UNCLOS and the 'sustainable utilisation' ethos of CITES.

In the GATT, exceptions to trade obligations are set out in Article XX. These exceptions apply *inter alia* to measures relating to the conservation of exhaustible natural resources (which includes marine living resources) and measures necessary to protect human, animal or plant

---

[297] Ibid., para. 16. Because fish and fish products are classified at the WTO as industrial goods, these negotiations fall under the non-agricultural market access negotiations (known colloquially as 'NAMA') rather than negotiations to update the Agriculture Agreement. Proposals to eliminate tariffs for select sectors of export interest to developing countries and LDCs have included fish and fish products: see WTO Doc. TN/MA/W/35.

[298] See e.g. under the EU's access agreements noted above n. 34 and surrounding text; for a further discussion of the political economy of such trade preferences, see Young, above n. 265.

[299] WTO Doha Declaration (2001), para. 32(iii).   [300] Marrakesh Agreement, Preamble.

life or health[301]. Measures, for example the prohibition on the import of shrimp harvested using devices known to harm turtles, must also conform to the *chapeau* of GATT Article XX, including the requirement that WTO members first attempt to find multilateral, non-trade based solutions for their environmental objectives.[302]

Equivalent provisions exist in other WTO covered agreements. For example, the TBT Agreement recognises members' rights to adopt technical regulations they consider necessary to protect 'human health or safety, animal plant life or health, or the environment',[303] as long as such regulations do not create unnecessary obstacles to trade. The SPS Agreement also reaffirms the right of members to take SPS measures necessary 'to protect human, animal or plant life or health' as long as the measures are not applied in a discriminatory manner or do not constitute a disguised restriction on international trade.[304] Moreover, in cases of insufficient scientific information, members may provisionally adopt SPS measures in certain circumstances.[305]

WTO members often seek to prove that their trade measures meet these exceptions by reference to other regimes. For example, the United States referred to a number of international environmental Conventions when it sought to demonstrate that its measures to protect sea turtles were measures related to the conservation of exhaustible natural resources and were thus in conformity with GATT Article XX.[306] The EC claimed that its measures on sardines were based on international standards produced by the Codex Alimentarius Commission

---

[301] Art. XX (b) and (g). On the conclusion of marine species as exhaustible natural resources, see GATT Panel Report, *US - Tuna (Canada)*, para. 4.9; GATT Panel Report, *Canada - Herring*, para. 4.4; AB Report, *US - Shrimp*, para. 131. Besides these provisions, Art. XX (a) on public morals may have relevance, particularly where there are cultural or indigenous aspects to fisheries and an expanded notion of security may also call into question Art. XXI. I am grateful to Christopher Tran for this point.

[302] AB Report, *US - Shrimp*, para. 166. Earlier incarnations of GATT Art. XX demonstrate the historic regard to the interplay between trade and environmental protection: see Steve Charnovitz, 'The WTO's Environment Progress' (2007) 10 *JIEL* 685 (noting in particular that the draft Charter of the International Trade Organization (Havana Charter), which was abandoned in favour of the GATT, contained parallel exceptions to Art. XX as well as a general exception for measures 'taken in pursuance of any inter-governmental agreement which relates solely to the conservation of fisheries resources, migratory boards or wild animals').

[303] TBT Agreement, Art. 2.2.   [304] SPS Agreement, Preamble; see also Art. 2.

[305] SPS Agreement, Art. 5.7. On the relationship between Art. 5.7 and the precautionary principle, see Scott, above n. 278, 126-8.

[306] AB Report, *US - Shrimp*; see further Chapter 5.

(Codex) in order to satisfy provisions of the TBT Agreement.[307] There has not yet been a case in which a WTO member has sought to defend its trade policies by explicitly invoking a specific trade obligation contained in an MEA such as CITES. A common view is that MEAs will likely prevail over WTO rules.[308] Interaction between regimes in WTO dispute settlement is considered in more depth in Chapter 5.

Further scope for the interaction between the WTO regime and other regimes arises through the possible use of non-violation complaints. Non-violation complaints are available if WTO members consider that benefits accruing to them directly or indirectly under the GATT are being nullified or impaired by the conduct of another WTO member.[309] Arguably such conduct could relate to the ratification by a WTO member of norms within an exogenous regime. For example, membership by a state of UNCLOS or the Fish Stocks Agreement could create reasonable expectations for its trading partners that the law of the sea will be complied with. If the state breaches its obligations under UNCLOS or the Fish Stocks Agreement, and this breach has market access implications (such as the reduction of access to that state's market from what would have been reasonably anticipated on the basis of its bound tariffs and UNCLOS or Fish Stocks Agreement-conforming fishing practices) then its trading partners could possibly bring a claim based on the nullification and impairment of their rights under the GATT.[310] Such a claim would require a WTO panel to consider laws from other regimes.

Other sources of WTO law are relevant to the interaction between the WTO and other regimes. For example, a GATT provision requires WTO members to accept certain determinations of exchange arrangements by the International Monetary Fund (IMF).[311] There is an SCM Agreement exemption for certain export subsidies that are in conformity with an

---

[307] AB Report, *EC - Sardines*. Note also AB Report, *Brazil - Retreaded Tyres*, where Brazil pointed to domestic and international environmental standards in support of the claim that its trade restriction on retreaded tyres from the EC was necessary (see esp. paras. 203-204).

[308] Pauwelyn, above n. 78, 351; Joost Pauwelyn, 'Recent Books on Trade and Environment: GATT Phantoms Still Haunt the WTO' (2004) 15 *EJIL* 575; see further Chris Wold, 'Multilateral Environment Agreements and the GATT: Conflict and Resolution?' (1996) 26 *Environmental Law* 841; Marceil Yeater and Juan Vasquez, 'Demystifing the Relationship Between CITES and the WTO' (2001) 10 *RECIEL* 271.

[309] GATT, Art. XXIII; DSU Art. 26; see Kyle Bagwell, Petros Mavroidis and Robert Staiger, 'It's a Question of Market Access' (2002) 96 *AJIL* 56.

[310] This hypothetical follows closely an example of an ILO Convention given by Bagwell, Mavroidis and Staiger, ibid., 73.

[311] GATT, Art. XV.2.

existing undertaking between members of the Organisation for Economic Co-operation and Development (OECD).[312] Under the declaration on 'Coherence in Global Economic Policy Making', members agree that the WTO should develop cooperation with IGOs responsible for monetary and financial matters.[313]

The interdependencies between the WTO and other regimes are subject to reform as part of the Doha Round. In particular, members have agreed to negotiations 'with a view to enhancing the mutual supportiveness of trade and environment'.[314]

One issue that has been selected for negotiation is the normative relationship between WTO rules and MEAs. Under the Doha Agenda, WTO members have agreed to negotiate on 'the relationship between existing WTO rules and specific trade obligations set out in multilateral environmental agreements (MEAs)'.[315] Out of the six MEAs that have been targeted as relevant to these discussions, CITES is one of the most prominent.[316]

Members have adopted different approaches to this agenda item. Some, like the EU and Switzerland, have called for discussions about the substantive relationship of 'mutual supportiveness' between MEAs and the WTO, and are in favour of setting down primacy for MEA obligations. Others, such as Australia and the United States, have instead focused on domestic procedures that enhance coherence between trade and environment policy. These countries have encouraged other WTO members to submit papers outlining their experiences in negotiating and implementing specific trade obligations in MEAs.

Importantly, the WTO/MEA negotiations do not affect WTO members that are not parties to the relevant MEAs. Paragraph 31(i) contains the following exclusion clause:

The negotiations shall be limited in scope to the applicability of such existing WTO rules as among parties to the MEA in question. The negotiations shall

---

[312] SCM Agreement, Art. 3.1(a) footnote 5; Annex I (k).
[313] WTO Declaration on Coherence in Global Economic Policy Making (1993).
[314] WTO Doha Declaration (2001) para. 31. [315] Ibid., para. 31(i).
[316] These six MEAs are: CITES, Montreal Protocol on Substances that Deplete the Ozone Layer (Montreal Protocol), Basel Convention on the Control of Transboundary Movements of Hazardous Wastes and their Disposal (Basel Convention), Cartagena Protocol on Biosafety to the Convention on Biological Diversity (Biosafety Protocol), Rotterdam Convention on the Prior Informed Consent Procedure for Certain Hazardous Chemicals and Pesticides in International Trade (PIC), Stockholm Convention on Persistent Organic Pollutants (POPs).

not prejudice the WTO rights of any Member that is not a party to the MEA in question.³¹⁷

This exclusion reflects the idea that the obligations of MEAs should not apply to WTO members that have not consented to them.³¹⁸ WTO members that are not parties to the relevant MEAs have been careful to ensure that such negotiations do not give rise to obligations from other regimes 'through the back door'. This concern that regime interaction should only occur when states are members of the relevant regimes (the 'consent to regime interaction' issue flagged in Chapter 1) recurs in several areas of fragmented fisheries governance, as I demonstrate in following chapters.

The exclusion for non-MEA parties from the WTO/MEA negotiations has led critics to doubt the utility of negotiating the issue. If the general view is that MEAs are likely to prevail over WTO rules, as described above,³¹⁹ then there is no need to negotiate; indeed, contrary to the intention of the countries that sought to include the issue in the Doha Round, the primacy of MEAs is undermined by the negotiations.³²⁰ Although WTO members appear divided on the issue of the relationship between specific trade obligations within the WTO and MEA regimes, there are institutional links that suggest a far more united view on the benefits of regime interaction, as described below.

3 Further links: the General Council and the CTE

Apart from legal provisions entrenching regime interaction, there are a number of institutional arrangements that seek to enhance interdependencies between the WTO and other regimes. WTO members, meeting through a body called the General Council,³²¹ are responsible for making 'appropriate arrangements' for effective cooperation and consultation with IGOs and NGOs.³²² This has led to a number of cooperative

---

[317] WTO Doha Declaration (2001) para. 31(i).
[318] The *pacta tertiis* principle is noted in Chapter 1 n. 98.
[319] See above n. 308 and accompanying text.
[320] Pauwelyn, 'Recent Books on Trade and Environment' above n. 308, 590.
[321] Marakkesh Agreement, Art. IV.2.
[322] Marrakesh Agreement, Art. V. Article V has been considered as evidence that the WTO 'has treaty-making power and that in principle the General Council exercises it': Pieter Jan Kuijper, 'Some Institutional Issues Presently Before the WTO' in Daniel Kennedy and James Southwick (eds.) *The Political Economy of International Trade Law: Essays in Honour of Robert E Hudec* (2002) 81, 108.

agreements with IGOs, including with the UN.[323] In addition, the General Council has laid down rules of procedure requiring observership for IGOs to be considered by each WTO committee on a case-by-case basis, taking account of factors such as the nature of the work of the IGO, its membership (including the number of WTO members who are members of the other IGO) and reciprocity of access.[324] The General Council has also established guidelines for cooperation with NGOs.[325]

The WTO Secretariat itself has concluded cooperative agreements with other international Secretariats such as an agreement with the UNEP Secretariat.[326] Some consider that these agreements are not intended as binding commitments,[327] a view that may impact upon their ability to improve regime interaction.

Another important institutional arrangement to enhance regime interaction is the Committee on Trade and Environment (CTE), which was established in the Uruguay Round.[328] Ministers noted the Rio Declaration, Agenda 21 and other developments in international law and desired to 'coordinate the policies in the field of trade and environment'. The CTE's mandate is *inter alia* to 'identify the relationship between trade measures and environmental measures, in order to promote sustainable development', and 'make appropriate recommendations' on the need for rules.[329]

The CTE is open to all WTO members. As part of its functions, it has marshalled a large amount of information about the operation of other regimes. For example, the Secretariat introduced the FAO Code of Conduct to the CTE in 1995.[330] The CTE received a communication from DOALOS about the Fish Stocks Agreement in 1997, was notified of the FAO IPOA-Capacity in 1999 and hosted a presentation by DOALOS in 2002.[331] Multiple background documents on the operation of MEAs have also been circulated, such as updates on the CBD[332] and

---

[323] In addition to the UN, the General Council has entered into agreements with the World International Property Organization (WIPO), the World Bank, the IMF, the World Organisation for Animal Health (OIE) and the International Telecommunications Union. Specialised councils within the WTO, namely the WTO's Services Council and the TRIPS Council, had important roles in discussing and adopting the texts of these agreements in the services and intellectual property sector: see Kuijper, above n. 322, 108.
[324] General Council Rules of Procedure (1996), Rule 11 and Annex 3.
[325] WTO Guidelines on NGOs (1996). [326] WTO–UNEP Cooperation Agreement (1999).
[327] Kuijper, above n. 322, 108. [328] WTO Decision on Trade and Environment (1994).
[329] Ibid. [330] WTO Doc. WT/CTE/W/15.
[331] WTO Docs. WT/CTE/W/62, WT/CTE/M/30, WT/CTE/W/126, WT/CTE/M/135.
[332] See e.g. WTO Docs. TN/TE/INF/9/Rev.1, WT/CTE/W/235.

CITES.³³³ In consultation with MEA Secretariats, the WTO Secretariat has prepared a significant study on the operation of fourteen MEAs, which includes a 'matrix' of details on the number of parties to the MEA, the number of WTO members party to the MEA, the openness of membership, the decision-making bodies, and provisions relating to amendments and protocols.³³⁴ Moreover, the CTE has invited a number of MEA Secretariats to participate in ad hoc information sessions and has conducted regional information sessions as part of the WTO's technical cooperation activities.³³⁵ The CTE grants observer status to a number of IGOs, including UNEP, the CBD Secretariat and the CITES Secretariat.³³⁶

The provision for institutional interactions between the WTO and other regimes has been targeted for reform during the Doha Round. In addition to negotiations on the WTO/MEA relationships discussed above,³³⁷ WTO members have agreed to negotiate on procedures for information exchange and the criteria for the granting of observer status.³³⁸ Members have also instructed the CTE on its work programme,³³⁹ restated the value of cooperation between the WTO and UNEP and other IGOs³⁴⁰ and agreed to negotiate on tariff and non-tariff treatment of environmental goods and services.³⁴¹

These aspects of the Doha negotiations are significant for the interaction between the WTO and other regimes in fisheries governance. As well as impacting upon the ongoing work of the WTO committees, the increased awareness of trade and environmental synergies influences how negotiations within the WTO take account of other regimes. This is the topic of the following chapter, which analyses the development of rules on fisheries subsidies at the WTO.

D Conclusions

This chapter provided a detailed overview of the laws and institutions relevant to the global governance of fisheries. It provided an analytical

---

³³³ See e.g. WTO Docs WT/CTE/GEN/7, WT/CTE/GEN/6, WT/CTE/GEN/5.
³³⁴ WTO Doc, WT/CTE/W/160/Rev.4.   ³³⁵ WTO Doc TN/TE/S/2/Rev.2 (paras. 5-14).
³³⁶ Ibid. (paras. 18-21).   ³³⁷ See above n. 314 and surrounding text.
³³⁸ WTO Doha Declaration (2001), para. 31(ii). WTO members have been criticised for failing to simply agree on information exchange and observership criteria at the Doha Ministerial, instead of embarking on further negotiation on the issue: see Steve Charnovitz, 'WTO Cosmopolitics' (2002) 34 *NYU Journal of International Law and Politics* 299, 334-5.
³³⁹ WTO Doha Declaration (2001), para. 32.   ³⁴⁰ Ibid., para. 6.   ³⁴¹ Ibid., para. 31(iii).

framework that divided analysis between (i) the law of the sea; (ii) international environmental law; and (iii) international trade law.

Within the law of the sea, the 'enclosure of the oceans' into exclusive economic zones was described. The resulting EEZ regime impacts on relations between coastal states and other states, including those with large numbers of vessels that fish in distant water, and generates a range of other laws, including access agreements that allow states to fish in other states' EEZs in return for trade preferences and other privileges. Areas outside national jurisdiction are the 'high seas', where the concept of freedom of fishing once reigned, but where states increasingly recognise their obligations as well as their rights. The chapter detailed high seas, fish stocks and regional regimes, including the Fish Stocks Agreement that implements provisions for straddling and highly migratory stocks. Complementary to these conservation and management ideals are norms developed within the FAO fisheries management regimes, including the FAO Code of Conduct for Responsible Fisheries and the FAO Compliance Agreement. These regimes have interdependencies with each other and with other regimes in international law, and further links are intended to be developed through an institutional initiative in the form of the UN Consultative Process.

In describing relevant international environmental laws, the chapter focused on the Convention on the International Trade in Endangered Species and the Convention on Biological Diversity. The chapter pointed to the institutional responsibility of UNEP in ensuring the synergies between environmental regimes and with other regimes in international law. The increasing relevance of the CITES regime for marine species has generated uncertainty about relationships with long-standing management efforts, particularly within the FAO, as will be described in Chapter 4.

The chapter then provided an overview of international trade law's relevance to fisheries, summarising WTO covered agreements, relevant normative overlaps and the role of bodies such as the General Council and the Committee on Trade and Environment. The Doha negotiations to clarify and improve disciplines on fisheries subsidies offer crucial prospects for reform of a global fishing industry that is currently structured around perverse and ecologically harmful incentives. How, in these reform efforts, the WTO takes account of other regimes of international law is the subject of the first chapter of Part II.

PART II
Selected Case Studies

3 The negotiation of WTO rules on fisheries subsidies

> *Recognizing* that their relations in the field of trade and economic endeavour should be conducted ... while allowing for the optimal use of the world's resources in accordance with the objective of sustainable development, seeking both to protect and preserve the environment...[1]

In Part II, I provide three detailed case studies of the interaction between regimes in global fisheries governance. In the current chapter, I analyse regime interaction during the *making* of new laws to promote fisheries sustainability. The second chapter of Part II concentrates on the *implementation* of existing laws within the CITES and FAO fisheries management regimes. The third chapter assesses the *adjudication* of disputes that arise in the WTO over import measures on fish products. Each of these case studies highlights different stages of international law to draw out relevant aspects of regime interaction. As mentioned in Chapter 1, the current literature on fragmentation of international law tends to focus only on the dispute settlement stage, when an international tribunal is called upon to adjudicate conflicting norms. The methodology of this book and the research presented in this Part seeks to expand upon that approach.

Although each case study is chosen to emphasise discrete points in the process of international law, my selected issue areas are obviously not so bounded. For example, I use the issue of fisheries subsidies in this chapter to demonstrate how regimes are taken into account when international law is made, but there may be a time when a dispute over fisheries subsidies is filed at the WTO. Similarly, although

---

[1] Marrakesh Agreement Establishing the World Trade Organization 1994, Preamble.

Chapter 4 assesses the implementation of existing law from the CITES and FAO fisheries management regimes, the constant updating by the parties of new CITES species could arguably be conceived as the making of new law. The dispute over import restrictions of fish products analysed in Chapter 5 has emerged after the WTO rules (and exceptions to them) were made and implemented. To ignore the interdependencies between the *stages* of international law could misrepresent the case studies and even skew the findings of the book.

To alleviate this problem, I have structured each chapter in Part II in a fairly uniform way, albeit with different levels of detail in corresponding sections. Each chapter of Part II starts by asking 'why', and gives a brief overview of the particular problem, either in restraining subsidies that lead to overfishing (Chapter 3), restricting trade in endangered marine species (Chapter 4), or banning imports that threaten exhaustible natural resources (Chapter 5). The chapters then ask 'how', and explore the differing regimes of relevance, be they the WTO, UNCLOS, FAO, CITES or other regimes. Next, the chapters question how the regimes 'know about' one another, and explore capacities for inter-regime learning. Chapter 3 on the fisheries subsidies negotiations is the most detailed on this point. Next, the chapters ask how regime interaction is entrenched. Chapter 4 on the FAO/CITES Memorandum of Understanding explores this in the most depth. Finally, the chapters query how disputes will be resolved, with the longest analysis appearing in Chapter 5. The implications of the answers to these questions are drawn together in Part III of the book, which considers the challenges and opportunities for regime interaction in international law.

To turn, then, to the issue of fisheries subsidies, which are currently subject to ongoing negotiations in the WTO's Doha Round. This chapter begins with an exploration of the connection between fisheries subsidies and overfishing. It then shows how the choice of forum to deal with fisheries subsidies has been heavily contested, with much opposition to any role for the WTO. The politics behind such forum struggles is important for any legal framework that attempts to facilitate regime interaction. Next, the chapter considers modes of inter-regime learning in the Doha negotiations. There have been problems for WTO negotiators to classify and discipline ecologically harmful subsidies when they have scant institutional or epistemic capacity in this regard. The chapter then uses the draft text produced by the Chair of the Rules Group to consider how regime interaction might be entrenched in the resulting rules in the face of opposition to any deference to principles of

THE NEGOTIATION OF WTO RULES ON FISHERIES SUBSIDIES   87

international fisheries management contained in existing regimes. Finally, the chapter investigates how disputes relating to the WTO subsidy rules and other regimes might be resolved. The future adjudication of subsidies rules faces significant challenges, and there is real uncertainty about whether organisations such as the FAO should have a role in reviewing the fishing practices of subsidising WTO members. The conclusions of this chapter anticipate the wider theoretical discussion of regime interaction in the book's final Part.

A Subsidies and overfishing

The reasons for the emerging possibilities for regime interaction described in this chapter are to be found in the link between subsidies, overfishing and the unsustainability of global fishing practices. This section commences by providing some background to this link in scientific and economic terms.

Scientists agree that one of the main factors that threaten the sustainability of global fisheries stocks is overfishing.[2] Overfishing occurs when there are excessive levels of harvesting capacity, particularly in areas without well-defined property or user rights such as the high seas.[3] 'Capacity' is a contested term,[4] but 'fishing capacity' has been defined as 'the amount of fish (or fishing effort) that can be produced over a period of time (such as a year or a fishing season) by a vessel or a fleet if fully utilised and for a given resource condition'.[5] Capacity grows not merely because of the number of vessels, but also because of their increasing size and sophisticated technological capabilities.

Amidst growing awareness that there is too much 'fishing capacity' in the oceans, there is increasing attention to the economics of the fishing industry. There is increasing recognition that marine fisheries operate

---

[2] See Chih-hao Hsieh et al., 'Fishing Elevates Variability in the Abundance of Exploited Species', (2006) 443 Nature 859-62 (responding to alleged uncertainty about whether the decline in fisheries is due to fishing, to environmental change, or to some combination of these effects).
[3] FAO Fisheries Technical Paper No. 433/1, J M Ward, J E Kirkley, R Metzner and S Pascoe, 'Measuring and Assessing Capacity in Fisheries: 1. Basic Concepts and Management Options' (2004).
[4] 'Capacity' has different meanings for fisheries technologists, fisheries scientists and fisheries managers: ibid.
[5] FAO Fisheries Report No. 613, Report of the Technical Consultation on the Measurement of Fishing Capacity, (FAO, 2000). Ward et al. note that this definition involves a combination of inputs (effort, boat numbers, etc) and outputs (catch): above n. 3.

globally at a net economic loss.[6] Economists have shown that government practices contribute to perverse incentives for vessel production. The fishing sector is heavily supported by states. Although the level of support is difficult to access and quantify, a range of studies indicate that subsidies could amount to a quarter of the value of revenues in the fisheries sector or even higher.[7] Based on a collection of notifications and data supplied by the countries to organisations such as the World Bank, Asia-Pacific Economic Cooperation (APEC), the OECD and the WTO, Japan is the leading subsidising state, providing between US$2 and $3 billion annually to its fishing industry.[8] Significant amounts are also paid out by the European Union.[9] The United States, Canada, Korea and China are other major subsidising countries.[10]

While subsidies can promote useful management and conservation techniques, many subsidies are directed at enhancing the capacity of fishing fleets and therefore risk overfishing. They might be paid because commercial fishing interests have secured special preferences from

---

[6] World Bank and the FAO, *The Sunken Billions: The Economic Justifications for Fisheries Reform* (2008) 1.

[7] For example, OECD countries are estimated to pay US$6 billion a year to support their fisheries sectors: see OECD, *Subsidies: a Way Towards Sustainable Fisheries?* Policy Brief (December 2005); developing countries also provided significant support to their fisheries sector: APEC, *Study into the Nature and Extent of Subsidies in the Fisheries Sector of APEC Member Economies* (2000). The methodology of these papers involved querying governments on their levels of financial support. Stone comments on a study undertaken for the FAO that adopted a different methodology by inferring the level of government support through an industry profile that subtracted gross costs from gross revenues on a global basis, and reported an apparent $54 billion deficit: see Christopher Stone, 'Too Many Fishing Boats, Too Few Fish: Can Trade Laws Trim Subsidies and Restore the Balance in Global Fisheries?' in Kevin Gallagher and Jacob Werksman (eds.) *International Trade and Sustainable Development* (2002) 286, 293–4.

[8] WWF, *Hard Facts, Hidden Problems, A Review of Current Data on Fishing Subsidies* (2001) 18 (incorporating notifications to the OECD and the World Bank, and including capital and infrastructure investments, insurance, foreign access payments and measures improving harvesting techniques).

[9] The EU has allocated EUR 3.8 billion in subsidies to its fisheries sector over the next seven years: 'Friends of Fish Denounce EU Aid Package', *Financial Times* (17 June 2006). For data on EU subsidies, which includes detailed information on port and vessel recipients, see www.fishsubsidy.org.

[10] WWF estimates the US to pay out $1 billion annually, and Canada to pay out over $800 million annually, based on OECD and APEC data. WWF refers to 'official reports of annual fishing subsidies in Korea' to estimate expenditure there of over $300 million, but notes discrepancies with WTO and OECD figures. China's APEC notifications amount to $50 million but WWF notes that a number of known subsidy programmes are not given monetary values: WWF, above n. 8, 18–19.

governments through lobbying. According to public choice theory, such lobbyists represent a 'concentrated minority' with more motive to act than the 'diffuse majority' representing conservation interests.[11] Governments may also decide to provide subsidies to the fishing sector to increase their capacity to compete against other states, especially in accessing fish from the high seas.

In response to the growing evidence of the link between subsidies, overfishing, and the unsustainability of global fishing practices, states have begun to develop strategies to reduce excess global fishing capacity and limit their subsidies payments. These strategies have been developed within different regimes. States first turned their efforts to developing voluntary policies within the FAO. As the following section explores, these efforts have met with mixed success, leading to reform efforts within a regime that can offer binding disciplines: the WTO.

B The forum shop: regimes of relevance

There are a number of regimes of relevance to efforts to restrain fisheries subsidies.[12] The dominant one of the 1990s was the FAO fisheries management regime. The 1995 FAO Code of Conduct for Responsible Fisheries[13] recognised that too many boats were fishing for too few fish. It expresses the general principle that states 'should prevent overfishing and excess fishing capacity', and that states should take measures 'to ensure that fishing effort is commensurate with the productive capacity of the fishery resources and their sustainable utilization'.[14] This is not just preventative: the Code recognises that excess fishing capacity needs to be 'eliminated'.[15] States' management measures are to provide that

---

[11] See further Christopher Carr and Harry Scheiber, 'Dealing with a Resource Crisis: Regulatory Regimes for Managing the World's Marine Fisheries' (2002) 21 *Stanford Environmental Law Journal* 45, 54 (and sources cited therein). See also Ostrom and others, Chapter 2, n. 24.

[12] Efforts have also been made in regional regimes, such as RFMOs. See e.g. Harry N Scheiber, Kathryn J Mengerink and Yann-huei Song, 'Ocean Tuna Fisheries, East Asian Rivalries, and International Regulation: Japanese Policies and the Overcapacity/IUU Fishing Conundrum' (2007) 30 *University of Hawai'i Law Review* 97, 119–36 (reviewing ICCAT and the Western and Central Pacific Fisheries Commission in particular). RFMO policies on overcapacity are not considered in this chapter.

[13] For background on the FAO Code of Conduct, see Chapter 2, n. 114 and surrounding text.

[14] FAO Code of Conduct, Art. 6.3; see also Art. 7.4.3.   [15] Ibid., Art. 7.1.8; see also Art. 7.6.3.

'the economic conditions under which fishing industries operate promote responsible fisheries'.[16]

Although the Code of Conduct does not mention subsidies expressly, two voluntary international plans of action call on states to reduce subsidies which contribute to the build-up of excessive fishing capacity: The IPOA for the Management of Fishing Capacity (IPOA-Capacity)[17] and the IPOA to eliminate illegal, unregulated and unreported fishing (IPOA-IUU Fishing).[18]

The FAO's IPOA-Capacity calls upon states to 'reduce and progressively eliminate all factors, including subsidies and economic incentives... which contribute, directly or indirectly, to the build-up of excessive fishing capacity'.[19] Supported by information on subsidies to be gathered by the FAO,[20] states are to achieve capacity reduction in four ways: (i) by conducting national, regional and global assessments of capacity; (ii) by preparing and implementing national plans; (iii) by strengthening RFMOs; and (iv) by undertaking immediate action for major straddling, highly migratory and high seas fisheries.[21] To date, only three states have submitted 'national plans of action' to the FAO: the United States, Indonesia and Namibia.[22]

A similar approach is reflected in the FAO's IPOA-IUU Fishing, which provides that the FAO will have a role in collecting relevant information and analysis on subsidies practices.[23] It calls on states to 'avoid conferring economic support, including subsidies, to companies, vessels or persons that are involved in IUU fishing', but only 'to the extent possible in their national law'.[24]

The voluntary approach to subsidies reduction has largely failed to change national policies and has failed to alleviate the overfishing and economic drain in the fishing sector. At the World Summit on Sustainable Development in 2002, states were reminded that to achieve

---

[16] Ibid., Art. 7.2.2. [17] See Chapter 2, n. 127 and surrounding text.
[18] See Chapter 2, n. 125 and surrounding text. [19] IPOA-Capacity, para. 26.
[20] Ibid., para. 45. [21] Ibid., para. 8.
[22] The US submitted its plan of action in August 2004. This plan focuses on the management of capacity in federally managed fisheries and not in fisheries under the states' jurisdiction or in purely international fisheries. The latter is considered by the United States to be addressed in the FAO IPOA-IUU Fishing. Namibia and Indonesia submitted their plans of action in 2007 and 2008 respectively. There are also two regional plans of action: from the Lake Victoria Fisheries Organization (LVFO) and the Inter-American Tropical Tuna Commission (IATTC). See further www.fao.org/fishery/ipoa-capacity/npoa/en.
[23] IPOA-IUU Fishing, para. 88. [24] Ibid., para. 23.

sustainable fisheries they must ratify or accede to UNCLOS and the Fish Stocks Agreement, effectively implement the FAO Code of Conduct, and urgently develop and implement national and regional plans of action.[25] For the elimination of subsidies, however, attention had moved from the law of the sea and FAO regimes. Instead, states were called upon to improve relevant parts of a legal regime that directly confronts the national policies of subsidising states: the World Trade Organization.[26]

1 The WTO's SCM Agreement

The WTO's SCM Agreement disciplines the use by WTO members of certain subsidies that are specific to enterprises or industries and regulates the actions other WTO members can take to counter the effects of such subsidies.[27] It categorises subsidies as prohibited, actionable, or non-actionable. These categories are often described by mixing a traffic light metaphor: the prohibited subsidies are 'red-box', actionable subsidies are 'amber-box' and non-actionable subsidies are 'green-box'. The green box, which includes subsidies for the adaptation of existing facilities to environmental sustainability, has lapsed under the existing SCM framework.[28]

WTO members have so far been unwilling to challenge the fisheries subsidies of another WTO member under the SCM Agreement. On the one hand, it is fairly certain that most government support to the fishing sector will be 'deemed' to be a subsidy.[29] Grants or low-rate loans for the construction or repair of fishing vessels are obvious examples. Even the rights obtained by WTO members to allow their distant-water fleets access to fish within a foreign country's EEZ seem to be covered.[30] On

---

[25] World Summit Plan of Implementation (2002), para. 31.   [26] Ibid.
[27] For introductory remarks on the SCM Agreement, see Chapter 2, pp. 73–75.
[28] SCM Agreement, Arts. 8, 9. The green box was adopted on a provisional basis for a period of five years: SCM Agreement, Art. 31. When it expired on 31 December 1999, a lack of consensus among WTO members meant that it was not renewed.
[29] A subsidy is 'deemed' to exist under the SCM Agreement if there is a 'financial contribution' by a government or public body that confers a benefit on any domestic industry: see SCM Agreement, Art. 1.1(1). The only industry exempt from the ambit of the SCM Agreement is the agricultural industry, which is separately covered in the Agreement on Agriculture (which expressly excludes fisheries): see further Chapter 2, n. 288.
[30] See description of access agreements in Chapter 2, n. 34 and accompanying text. The issue is more complex for access rights obtained through non-financial arrangements such as the provision of preferential trade incentives. For the view that these subsidies do fall within the current definition of SCM Art. 1, see David Schorr, 'Towards Rational Disciplines on Subsidies to the Fishery Sector: A Call for New International Rules and Mechanisms' in WWF, *The Footprint of Distant Water Fleets on World Fisheries* (1998).

the other hand, the breadth of the practice in granting fisheries subsidies may mean that WTO members are reluctant to expose themselves to scrutiny when pursuing claims.[31] Indeed, even if WTO members do not provide subsidies themselves, they may indirectly benefit from subsidies of other members, which are used to procure access rights for fleets to fish in distant water EEZs, and are unlikely to challenge such arrangements.

The absence of any WTO litigation on fisheries subsidies may, instead, be attributed to a lack of certainty about the applicability of the red-box and amber-box disciplines of the SCM Agreement, as currently framed. Current red-box subsidies are those subsidies that are contingent on export performance or the use of local content.[32] Because the majority of fisheries subsidies are granted by countries, such as Japan, which are net importers of fish, and are usually designed to encourage domestic fish supply rather than exports, the red-box test is said to be difficult to meet in many cases. Also ill suited is the red box's emphasis on *ex-ante* legal design rather than effects.[33]

The actionable subsidies of the amber box, on the other hand, are assessed according to their adverse effects on the trade interests of other WTO members. These subsidies, which must also be specific to a particular enterprise or industry,[34] are assessed according to injury to the domestic industry of another member, the nullification or impairment of benefits of another member or serious prejudice to the interests of another member.[35] The establishment of these adverse effects is difficult in the fisheries context, however, given that the likely distortions caused by fisheries subsidies will be in resource availability, rather than price for exporters, and that such effects are spread over heterogeneous fisheries products.[36]

A further deficiency for the disciplining of fisheries subsidies under the current framework relates to inadequate compliance with the SCM Agreement's notification and surveillance regime. WTO members are

---

[31] See Marc Benitah, 'Ongoing WTO Negotiations on Fisheries Subsidies' *ASIL Insight* (June 2004).
[32] SCM Agreement, Art. 3; see also Art. 2 and Art. 1.2.
[33] But note that the SCM Agreement prohibits subsidies that are contingent on export or domestic supply 'in law or *in fact*': ibid Art. 3.1 (emphasis added). As such, effects may become relevant; see further David Schorr, *Healthy Fisheries, Sustainable Trade: Crafting New Rules on Fishing Subsidies in the World Trade Organization* (WWF Position Paper and Technical Resource, 2004) 39.
[34] SCM Agreement, Arts. 1.2, 2.   [35] SCM Agreement, Art. 5.
[36] See generally WTO Docs. TN/RL/W/3 and TN/RL/W12.

obliged to notify the WTO annually of any subsidy granted or maintained,[37] yet fisheries programmes are apparently under-reported and the WTO lacks meaningful data for the fisheries sector.[38]

The lapsed green-box category of non-actionable subsidies is also inapplicable to fisheries. Even if it were to be renewed, it would probably remain inapplicable because of its focus on 'industrial' pollution problems rather than resource use.[39] This is a shortcoming because some fisheries subsidies have positive effects on environmental sustainability. For example, by contrast to the subsidies that lead to vessel overcapacity described above, some subsidies are directed at reducing capacity in a targeted fishery through the decommissioning of vessels (known as 'buybacks') or retirement of fishing licenses. These buyback programmes can lead to an overall reduction in capacity, although such programmes must be carefully administered to ensure that the retired vessels are not used to enhance capacity in other fisheries.[40] Notwithstanding that other regimes have encouraged states to provide resources to promote good management of their fishing sectors (such as the FAO Code of Conduct's aspiration for states to support fisheries research[41]), there is no provision in the SCM Agreement to exempt these programmes.

The apparent inapplicability of the SCM Agreement for disciplining trade-distorting and ecologically harmful fisheries subsidies, and for exempting those subsidies that are positive for sustainability, has led to proposals for reform. The following section details the Doha negotiations on this issue.

---

[37] SCM Agreement, Art. 25.
[38] See e.g. WTO Docs. TN/RL/W/3, para 8; WTO Doc. TN/RL/M/7, para 23; WTO Doc. TN/RL/M8.
[39] See Schorr, above n. 33, 45.
[40] In some conditions, buyback programmes can be futile or even counterproductive. Problems include (i) the 'leakage' of the vessels that are subject to the buyback to other fisheries; (ii) the existence of a 'latent capacity' of inactive licences or unused vessels which may become active after the buyback; (iii) the 'capital stuffing', or the use of profits from a buyback programme to enhance the power or capacity of remaining vessels; and (iv) the 'perverse incentives' where there is an increased fishing effort in anticipation of the compensation of a buyback programme: these are discussed by reference to domestic experience in a communication from the United States: WTO Doc. TN/RL/GEN/41. In the United States context, federal buyback programmes have required that bought-out vessels be scrapped or permanently withdrawn from domestic and foreign fishing: see US national plan of action for the management of fishing capacity submitted to the FAO in 2004, noted above n. 22, 21.
[41] See e.g. FAO Code of Conduct Art. 12.5.

2 The Doha negotiations

WTO members have recognised the inadequacy of the SCM Agreement in restraining fisheries subsidies. Discussions between members first took place in the Committee on Trade and Environment (CTE). The CTE was formed to 'identify the relationship between trade measures and environmental measures in order to promote sustainable development'[42] but has no rule-making function. At the launch of the Doha Round of negotiations, WTO members agreed to assign fisheries subsidies reform to the Negotiation Group on Rules ('Rules Group'),[43] which also deals with general subsidies reform, anti-dumping and regional trade agreements. WTO members stated:

> In the light of experience and of the increasing application of these instruments by members, we agree to negotiations aimed at clarifying and improving disciplines under the Agreements on Implementation of Article VI of the GATT 1994 and on Subsidies and Countervailing Measures, while preserving the basic concepts, principles and effectiveness of these Agreements and their instruments and objectives, and taking into account the needs of developing and least-developed participants ... In the context of these negotiations, participants shall also aim to clarify and improve WTO disciplines on fisheries subsidies, taking into account the importance of this sector to developing countries. We note that fisheries subsidies are also referred to in paragraph 31.[44]

This proposal directly acknowledges the environmental dimensions to the fisheries subsidies negotiations by referring to the reform agenda for the mutual supportiveness of trade and environment (contained in paragraph 31).[45] The link between the trade-distorting and environmentally harmful aspects of fisheries subsidies has been restated in the course of the negotiations. The Ministerial Declaration at the Hong Kong meeting of trade ministers in 2005 states that Ministers:

---

[42] See further Chapter 2, pp. 80–1.
[43] The Rules Group reports to the Trade Negotiations Committee. On the TNC, see Chapter 2, n. 294 and surrounding text.
[44] WTO Doha Declaration (2001), para. 28.
[45] Ibid., para. 31. ('With a view to enhancing the mutual supportiveness of trade and environment, we agree to negotiations, without prejudging their outcome, on: (i) the relationship between existing WTO rules and specific trade obligations set out in multilateral environmental agreements (MEAs) ... (ii) procedures for regular information exchange between MEA Secretariats and the relevant WTO committees, and the criteria for the granting of observer status; (iii) the reduction or, as appropriate, elimination of tariff and non-tariff barriers to environmental goods and services. We note that fisheries subsidies form part of the negotiations provided for in paragraph 28.')

recall [their] commitment at Doha to enhancing the mutual supportiveness of trade and environment, *note* that there is broad agreement that the Group should strengthen disciplines on subsidies in the fisheries sector, including through the prohibition of certain forms of fisheries subsidies that contribute to overcapacity and over-fishing, and *call on* Participants promptly to undertake further detailed work to, *inter alia*, establish the nature and extent of those disciplines, including transparency and enforceability. Appropriate and effective special and differential treatment for developing and least-developed Members should be an integral part of the fisheries subsidies negotiations, taking into account the importance of this sector to development priorities, poverty reduction, and livelihood and food security concerns.[46]

The Doha negotiations were due for completion by December 2006 but have overrun. The fisheries subsidies negotiations have proceeded with a comparatively greater degree of consensus than other negotiations such as those relating to agricultural subsidies and tariffs. Given the single-undertaking character of the negotiations, an agreement on fisheries subsidies issues will probably not be resolved until the conclusion of the entire round.

In November 2007, the Chair of the Rules Group, Ambassador Guillermo Valles Games, circulated a draft consolidated text of the proposed disciplines ('Chair's text').[47] The Chair's text is in the form of an annex ('Annex VIII') of the SCM Agreement and is reproduced in the Appendices.[48] The Chair's text has not been agreed by members, and differences remain over the need for effective rules and the need to address development priorities, poverty reduction and food security concerns.[49] The Chair has subsequently tabled a 'roadmap' of key issues rather than a new draft text, on which negotiations are now proceeding.[50]

Two main positions have been taken by the Rules Group participants. The first position, associated with an informal grouping of WTO members self-named 'Friends of Fish',[51] is based on a conviction of the link between enhanced subsidy disciplines and trade, environmental and

---

[46] WT/MIN(05)/DEC (18 December 2005), Annex D, para. 9; see also Doc. TN/RL/W/195.
[47] WTO Doc. TN/RL/W/213 'Draft Consolidated Chair Texts of the AD and SCM Agreements' (30 November 2007) 87–93.
[48] See Appendix A below.
[49] WTO Doc. TN/RL/W/232 (Annex C – Fisheries Subsidies) (28 May 2008).
[50] WTO Doc. TN/RL/W/236, 85–94 (19 December 2008).
[51] Membership varies according to time and the content of submissions. Members have included Australia, Chile, Ecuador, Iceland, New Zealand, Peru, Philippines and the United States.

development needs.[52] These countries have proposed a general prohibition of all fishery subsidies that support fishing enterprises with limited exceptions. Exceptions to this prohibition, to be identified and defined during the negotiations, would relate to subsidies that are expressly concerned not to encourage overfishing, such as subsidies to decommission vessels and support social programmes to retrain fish workers for other industries.

The second position, which has been taken by Japan, Korea, Taiwan and others, contests the link between subsidies and environmental damage, and instead asserts that inadequate fisheries management is the main cause of unsustainable fishing.[53] These countries contend that subsidies are not dangerous to fishery resources if the fishery is properly managed, and that reliance on the work of fisheries management regimes is sufficient. Explicit in this approach is the deference to other regimes in international law. In the early stage of the negotiations, some WTO members even directly opposed new WTO disciplines on the basis of competing mandates from UNCLOS, FAO and other regimes.

3 Opposition to the role of the WTO

Opposition to the role of the WTO in restraining harmful fisheries subsidies has been directed to the WTO's mandate, expertise, alleged 'structural bias' and a perceived need for coherence in trade policy.

(a) Mandate

The 2001 Doha negotiations coincided with a decision of the FAO Committee on Fisheries that the FAO should take the lead role in the coordination of work on fisheries subsidies and the relationship with responsible fisheries.[54] Given that membership between the 153-strong WTO and 137-strong FAO Committee on Fisheries substantially overlaps,[55] this decision was presumably agreed by many states that are also WTO members. Within the WTO itself, countries such as Japan, Korea, Taiwan and a group of developing-country island states originally lobbied against the development of fisheries subsidies

---

[52] This perspective has been elaborated in a number of papers to the CTE and later the Rules Group. See e.g. WTO Docs. TN/RL/W/3, TN/RL/W/12, TN/RL/W/21, TN/RL/W/58, TN/RL/W/77, TN/RL/W/154, TN/RL/W/166, TN/RL/W/169, TN/RL/W/196 and TN/RL/GEN/145.
[53] See e.g., WTO. Docs TN/RL/W/11, TN/RL/W/17, TN/RL/W/52, TN/RL/W/69 and TN/RL/W/97.
[54] FAO, COFI, Report of the 24th Session of the Committee on Fisheries (2001) Item 8.
[55] See further Chapter 2, pp. 47, 70.

disciplines and claimed that deference should be given to existing regimes.[56]

In support of their claims that the WTO was inappropriately mandated to deal with fisheries, these members pointed to the capacities of other international forums.[57] For example, they claimed that the FAO, RFMOs and MEAs offer a set of mechanisms more responsive to the problem of threatened fish species.[58] UNCLOS was said to safeguard the rights of certain small coastal states to grant subsidies.[59] One WTO member suggested that the World Summit was a more appropriate forum for subsidies leading to overcapacity and IUU fishing and that the Rules Group should limit its attention to trade-distorting subsidies.[60] This questioning of the WTO's mandate has extended to doubts about the role of the Rules Group vis-à-vis other WTO forums. Some WTO members have asserted that the Rules Group should leave the fishery subsidies discussions to the CTE[61] and even the Committee on Trade and Development.[62]

(b) Expertise and judicial competence

A related argument that the WTO is not the right forum to discipline fisheries subsidies is that such rules are outside the WTO's existing institutional expertise and the competence of its judicial branch. Concerns about the WTO's expertise have been expressed by several states and also by commentators who claim to be 'friends of the WTO'.[63]

One view is that subsidy rules aimed at the fisheries sector would lead to the fragmentation of the subsidies regime and the entire WTO

---

[56] See e.g. positions of Antigua and Barbuda, Barbados, Belize, Dominican Republic, Fiji, Grenada, Guyana, Jamaica, the Maldives, Papua New Guinea, St Kitts and Nevis, St Lucia, Solomon Islands, and Trinidad and Tobago, set out in WTO Docs. TN/RL/W/136, TN/RL/GEN/57/Rev.2, esp. para. 12. See also debates on the need to 'preserve the basic concepts and principles' of the SCM Agreement at the Hong Kong Ministerial of 2005: see (2005) 9:40 *Bridges Digest* (23 November 2005).

[57] WTO. Doc. TN/RL/M/8.

[58] WTO Doc. TN/RL/GEN/57/Rev.2, para. 8; see also TN/RL/W/136.

[59] WTO Doc. TN/RL/W/136, p. 2. This claim is reproduced in Roman Grynberg and Natallie Rochester, 'The Emerging Architecture of a World Trade Organization Fisheries Subsidies Agreement and the Interests of Developing Coastal States' (2005) 39 *JWT* 503, 522.

[60] See WTO. Doc. TN/RL/M/17.

[61] WTO Docs. TN/RL/W/52 (Japan), TN/RL/W/97 (Korea), TN/RL/GEN/57/Rev.2 (Antigua and Barbuda *et al.*).

[62] WTO Doc. TN/RL/M/10.

[63] Seung Wha Chang, 'WTO Disciplines on Fisheries Subsidies: A Historic Step Towards Sustainability?' (2003) 6 *JIEL* 879.

system.[64] Japan and Korea have claimed that the Rules Group should consider reforms to subsidies in the light of the existing SCM Agreement and on a cross-sector basis.[65] In addition, some countries have voiced concern about the technical and administrative competence of the WTO in dealing with potential disputes. The concern is that a WTO panel called upon to resolve a subsidies claim could be adjudicating upon a fisheries management issue with no perceivable trade effects.[66] These delegates have insisted that the FAO and appropriately mandated RFMOs are more competent to deal with these issues.

(c) Bias

Related to the argument that the WTO is not the right forum to address fisheries subsidies because it does not have adequate expertise is the concern that the WTO is inappropriately biased to deal with ecological problems of overfishing. NGOs that campaign about the ecological risks of overfishing and extinction have been wary of the danger that the subsidy disciplines will lead the WTO to 'cross the thin green line' into environmental issues.[67] Greenpeace, for example, has called for the Convention on Biological Diversity instead of the WTO to act on subsidies.[68] This fear about the WTO adjudicating on environmental issues has arisen in other contexts. In response to the aspects of the Doha negotiations involving MEAs and the WTO, some commentators argue that the ICJ and, improbably, the ILC (which has no adjudicatory competence) should adjudicate on trade matters.[69] These positions are based on a belief that the trade regime is biased against environmental issues.

(d) Coherence and effectiveness of trade policy

Apart from complaints about mandate, expertise and bias, different views exist about the 'appropriateness' of the WTO to rule on environmental issues such as ecologically harmful fisheries subsidies. These

---

[64] See WTO Doc. WT/CTE/GEN/10 para. 7; see also Chang, ibid., 918.
[65] See, e.g., WTO Docs. TN/RL/W/11 (Japan), TN/RL/W/17 (Korea), TN/RL/W/52 (Japan), TN/RL/W/69 (Korea) and TN/RL/W/97 (Korea).
[66] See, e.g., WTO Doc. TN/RL/GEN/57/Rev.2, para. 12. [67] Schorr, above n. 33, 27.
[68] Juergen Knirsch et al., *Deadly Subsidies: How Government Funds are Killing Oceans and Forests and Why the CBD Rather Than the WTO Should Stop this Perverse Use of Public Money* (Greenpeace International (undated, circa 2006)).
[69] Stefanie Pfahl, 'Is the WTO the Only Way?' Briefing Paper for Greenpeace International and Friends of the Earth, (undated, circa 2005).

views centre on problems with 'coherence' and 'effectiveness'. For example, Bagwell, Mavroidis and Staiger consider that WTO rules should only relate to market access policies, which they see as policies 'reflecting the competitive relationship between imported and domestic products'.[70] Tariff measures, which affect producers and consumers, are the main example of market access policies. This theory leads to the interpretation that the WTO's role in 'expan[ding] production of trade in goods and services' requires a laissez-faire attitude to subsidies. Indeed, economists have considered that disciplining subsidies may be counterproductive to a market access framework because they reduce the incentive for WTO members to make tariff commitments.[71]

Other economists offer general prescriptions about 'non-trade' policy areas such as issues relating to the environment or investment. Lloyd, for example, considers that rules relating to the environment should be kept outside the WTO because such rules will not be effective at correcting inter-country externalities.[72] Although he concedes certain exceptions to this general conclusion,[73] Lloyd suggests that attention be given to the 'non-trade' regimes to improve their capacity to respond to relevant policy issues.[74]

4 Responses

There are a number of arguments in support of the WTO's role in restraining harmful fisheries subsidies, which are based on economic and political science literature as well as the favoured tool for most international lawyers: the text of the legal agreements themselves.

(a) Policy fit

One response to forum queries is to establish and follow a model to determine whether a policy area is right for the forum. For example,

---

[70] Kyle Bagwell, Petros Mavroidis and Robert Staiger, 'It's a Question of Market Access' (2002) 96 *AJIL* 56.
[71] See further Mitsuo Matsushita, Thomas Schoenbaum and Petros Mavroidis, *The World Trade Organization: Law, Practice, and Policy* (2006) 333.
[72] Peter Lloyd, 'When Should New Areas of Rules be Added to the WTO?' (2005) 4 *World Trade Review* 275.
[73] For example, Lloyd suggests that CITES and the trade in hazardous waste both respond to market failures that result from the international trade in goods, and therefore meet the WTO's objectives: ibid., 286.
[74] This mirrors suggestions e.g. by Jagdish Bhagwati, 'Afterword: The Question of Linkage' (2002) 96 *AJIL* 126. See also suggestions to build up a World Environment Court noted in Chapter 1, n. 64.

Charnovitz offers a complex model of three categories with eight separate 'frames' through which forum questions can be asked.[75] He claims that a policy area is appropriate for inclusion in the WTO if it can be justified by frames in all three categories; hence his idea of 'triangulating' the WTO. Charnovitz first categorises the WTO as a forum to further inter-state relations. As such, its agenda should promote (i) cooperative openness; (ii) harmonisation; (iii) fairness; or (iv) risk reduction. Charnovitz secondly considers the WTO's role in addressing domestic politics. According to this category, its agenda should promote (v) self-restraint; or (vi) coalition-building. Under the frame of self-restraint, governments enter the WTO to pre-empt the adoption of trade-distorting policies encouraged by special interest groups.[76] Charnovitz describes a third category that focuses on the WTO's position vis-à-vis other international organisations. Here, the appropriateness of the WTO including particular issues in its agenda will depend on an assessment of (vii) trade functionalism; or (viii) comparative institutionalism. Calling for the WTO to be restricted to trade functions is hard to reconcile even with current subsidies rules and other WTO rules such as the TRIPS Agreement.[77] The frame of comparative institutionalism, on the other hand, leads to different conclusions because it is less concerned with functionality and more concerned with how an organisation can adapt to attaining particular goals.[78] This frame emphasises the effectiveness of the WTO's dispute resolution system vis-à-vis other regimes.

The disciplining of fisheries subsidies meets all three of Charnovitz's categories. In the category relating to inter-state relations, it falls squarely within the fairness frame, because it involves action against a government programme that has adverse effects in foreign countries. In terms of Charnovitz's domestic politics category, the disciplining of fisheries subsidies falls within the frame of self-restraint because it enables WTO members to counter the strong lobbying influence of domestic fishing interests.[79] Charnovitz's third category of the WTO's position vis-à-vis international organisations is met because the WTO's

---

[75] Steve Charnovitz, 'Triangulating the World Trade Organization' (2002) 96 *AJIL* 28.
[76] Ibid., 43–4.   [77] Ibid., 50.
[78] See also Frederick Abbott, 'Distributed Governance at the WTO-WIPO: An Evolving Model for Open-Architecture Integrated Governance' (2000) 3 *JIEL* 63.
[79] For some indication of the influence of Spanish fishing interests on EU policy-making, see Margaret A Young, 'WTO undercurrents at the Court of Justice' (2005) 30 *European Law Review* 211.

subsidies disciplines compare well with non-binding or poorly enforced instruments from other regimes such as the FAO's IPOA-Capacity.

(b) Textual basis

A common lawyerly response to questions of forum is to consult the constitutive texts of the relevant IGO. The WTO has a wide textual mandate, according to which it acts as a forum for the 'conduct of trade relations' among members.[80] The expansive term 'trade relations' is undefined. Yet WTO members recognise in the Preamble to the Marrakesh Agreement that

> their relations in the field of trade and economic endeavour should be conducted with a view to ... expanding the production of and trade in goods and services, while allowing for the optimal use of the world's resources in accordance with the objective of sustainable development, seeking both to protect and preserve the environment ...

This textual mandate appears to warrant the WTO's role in disciplining fisheries subsidies. It allows the WTO to revise its rules according to an evolving understanding of ecological issues. In the context of fisheries, it also mandates the WTO to discipline subsidies that contribute to overcapacity and overfishing. Indeed, in response to the questioning of the WTO's mandate to deal with fishery subsidies, other Rules Group delegates have repeated their commitment to target a sector with major environmental and economic problems,[81] and have pointed to the special case posed by the fisheries sector, in which subsidies affect the access to the resource as well as markets.[82] Countries that originally disputed the WTO's mandate, such as Japan, Korea and Taiwan, have subsequently accepted its power to deal with the issue.[83]

(c) Ongoing indeterminacy

A third response to forum queries puts into doubt the quest for coherence and effectiveness of trade policy. For example, even if the disciplining of subsidies is not part of traditional economic trade theory, Langille demonstrates the connection between tariff policies that protect domestic producers and subsidies to domestic producers that achieve the same result. He writes, '[o]nce the logic of the project of eliminating tariffs is accepted, the subsidies project cannot be

---

[80] Marrakesh Agreement, Art. II.
[81] See, e.g., (2005) 9:40 *Bridges Digest* (23 November 2005). [82] WTO Doc. TN/RL/M/2.
[83] See Rules Group meeting of 28 November 2004, summarised in WTO Doc. TN/RL/M/18.

avoided.'[84] According to Langille, this insight demonstrates the impossibility of divorcing economic principles from political issues, such that '[f]air trade is free trade's destiny.'[85] Indeed, such recognition is implicit in the GATT national treatment provision,[86] through which WTO members have agreed not to use internal measures so as to afford protection to domestic production.[87] The national treatment provision prevents WTO members from undermining or reducing the value of their tariff concessions by adopting other policies, such as subsidies, that have trade-restricting effects.

Apart from demonstrating the diffuse connections between economic and non-economic policies, the fisheries subsidies issue demonstrates the politicised nature of mandate queries. Some WTO members have perceived the positions of Japan and others to be forum-shifting strategies that attempt to move the debate away from rule-setting institutions. Indeed, several negotiating states considered this to be Japan's objective when it argued that the CTE, which is not mandated to negotiate new WTO rules, should resume control over fish subsidies discussions.[88] A similar strategy could be said to accompany some states' exclusive support for RFMOs and the FAO, which are more limited in their powers of implementation and enforcement than the WTO. Key to this perspective is the superior enforceability of the WTO disciplines vis-à-vis the often voluntary rules of other forums. This perceived effectiveness creates incentives for states wishing to avoid new rules to engage in mandate struggles.[89]

The effort by some countries to avoid being subject to WTO fisheries subsidies disciplines is not based on a rejection of fisheries governance. Unlike the Doha Round disagreement about entrenching relationships between WTO and MEAs,[90] which was manifest in states which were

---

[84] Brian Langille, 'General Reflections on the Relationship of Trade and Labor (Or: Fair Trade Is Free Trade's Destiny)' in Jagdish Bhagwati and Robert E Hudec, *Fair Trade and Harmonization: Prerequisites for Free Trade?* (1996) 231, 235.
[85] Ibid., 236.   [86] GATT, Art. III.
[87] See Henrik Horn and Petros Mavroidis, 'Still Hazy after All These Years: The Interpretation of National Treatment in the GATT/WTO Case Law on Tax Discrimination' (2004) 15 *EJIL* 39.
[88] See WTO Doc. TN/RL/M/3.
[89] Forum struggles have been documented in other regimes involving marine species: see further Chapter 2, n. 169. On the relevance of binding and non-binding norms, see Gregory Shaffer and Mark A Pollack, 'Hard vs Soft Law: Alternatives, Complements and Antagonists in International Governance' (2010) 94 *Minnesota Law Review* 706.
[90] WTO Doha Declaration (2001) para 31(i) (see further pp. 78–79).

THE NEGOTIATION OF WTO RULES ON FISHERIES SUBSIDIES   103

not parties to relevant MEAs, the forum controversies over fisheries subsidies have been prompted by WTO members that are already active within the law of the sea and FAO fisheries management regimes. The concern of these states does not depend on issues of uniform membership. Rather than WTO members that have *not* consented to international laws surrounding fisheries, it is the countries with vested interests in current management regimes under the law of the sea that have been the most vocal in contesting the WTO's mandate. This suggests that the motivation of states to ask, 'Should the WTO make new rules on fisheries subsidies?' is caused not by a lack of consent to fisheries governance but rather by a need for states to maintain control of existing regimes. In other words, states have already indicated their *consent* for multilateral action to curb overfishing; they are not concerned about obligations 'through the back door' but rather about controlling the form that such multilateral action takes. Japan and others have argued that the management of fisheries is the correct approach to achieving sustainability in the fisheries trade, rather than the disciplining of fisheries subsidies.

(d) From forum shopping to regime interaction

Concerns about whether the WTO is an appropriate forum to discipline fisheries subsidies have not been solely directed to its mandate. As described above, NGOs such as WWF have voiced concern about a possible institutional bias that the WTO might bring to the regulation of subsidies. This concern is supported to some degree by empirical research demonstrating certain 'ecological insensitivities' within the trade and commercial spheres.[91] For example, in evaluating particular ecological dilemmas, Perez has located specific 'environmental blindspots' which, though not able to be generalised, provide information about the dominance of certain perspectives within different institutions and discursive networks. Drawing on systems theory and using examples from international construction law, transnational environmental litigation, selected jurisprudence of the WTO and GATT, and the SPS and TBT Agreements, Perez argues that the trade/environment conflict engenders a diverse range of attitudes, ranging from, for example, an engineering ethos, a mercantilist bias, and a preoccupation with

---

[91] On this argument, see Oren Perez, *Ecological Sensitivity and Global Legal Pluralism: Rethinking the Trade and Environment Conflict* (2004) (identifying how structural attributes of certain systems affect their sensitivity to environmental issues).

scientific principles. Perez calls for a framework of study to deconstruct the resulting 'multiplicity of ecological insensitivities'.

Although Perez's institutional recommendations are directed to WTO adjudication rather than the negotiation of WTO rules,[92] his insight into the biases of different regimes supports an analysis of the WTO's incorporation of environmental perspectives in the framing of fishery subsidy rules. This shift from questions of forum to questions of institutional adaptability can also be observed in a number of country submissions. For example, the majority of submissions address the WTO's lack of competence and experience in the fishing sector by proposing institutional reform and dynamic interaction with other regimes. Japan and Korea have also abandoned their argument that subsidies disciplines should not be separated according to sectors.[93] A 'linkage model' that seeks merely to ascertain the current institutional expertise of the WTO fails to give due regard to the scope for flexibility and revisability of the WTO's competences in assessing whether fisheries subsidies ought to be disciplined under its auspices. In this regard, the question of the 'right forum' for fisheries subsidies is dangerously under-inclusive. A neglected part is the consideration of whether the WTO can adequately learn from other institutions and bodies to develop the requisite expertise in fisheries.

In conclusion, concerns about the WTO's expertise and possible bias, like concerns about its mandate, are not simply resolved by ascertaining its appropriateness as a forum. Instead, such questions give rise to a consideration of the interaction between the WTO and other regimes. In this way, forum struggles in the context of fragmentation are less concerned with achieving coherence and more concerned with preventing policy issues from being buried, or preventing institutions from being insensitive to unfamiliar issues. As such, questions about mandate must be supplemented by questions about regime interaction.[94] For example, is the WTO able to learn from and work with other international organisations in framing and implementing the rules? Who is able to participate in rule-making? The rest of this chapter examines the process of regime interaction in the framing and implementation of fishery subsidy disciplines.

---

[92] See Chapter 5, n. 244 and accompanying text.
[93] The decision not to pursue this argument perhaps reflects an acknowledgment by Japan and Korea that the subsidies regime of the WTO is already fragmented, with the notorious agricultural sector dealt with quite separately; cf Chang, above n. 63, 918.
[94] See further Chapter 6.

C Inter-regime learning

This section explores the modes of inter-regime learning during the WTO negotiations.[95] It starts by considering how the WTO members themselves act as conduits and share information and experience within the Rules Group, and queries whether this sole reliance on states is adequate in allowing the WTO to take account of existing fisheries interests and perspectives. The section then examines scope for participation and observership by external bodies in the Rules Group. Finally, deliberations within other forums such as non-hierarchical workshops hosted by UNEP and other bodies are investigated.

1 WTO members as conduits

WTO members have been conduits for information and learning from other regimes. Rules Group participants have shown initiative in directly reporting on their relevant experience in other IGO forums both in proposals and in their written and oral replies to submissions during the negotiations.

For example, the negotiating proposals have relied heavily on the work of IGOs to classify the types of fisheries subsidies paid by countries. In one of the first communications from the Friends of Fish, the group offered for discussion the classifications of fish subsidy programmes used by a range of international organisations such as APEC, the FAO and UNEP.[96] Later proposals drew on other IGOs in the formulation of express exceptions to the general prohibition on fisheries subsidies.[97] New Zealand referred to the work of the OECD, the FAO and UNEP in proposing that subsidies relating to 'fisheries management' be exempted from the general prohibition.[98] The OECD had divided these subsidies into sub-categories of research, creation, implementation and enforcement of fisheries management programmes, and these sub-categories were subsequently endorsed in publications of the FAO and UNEP. New Zealand proposed that this classification be directly reproduced in a list of permitted exemptions to prohibited subsidies, although reference to the IGOs themselves was not included in the proposed list.[99]

---

[95] See further the three modes of learning described in Margaret A. Young, 'Fragmentation or Interaction: The WTO, Fisheries Subsidies, and International Law' (2009) 8 *World Trade Review* 477.
[96] WTO Doc. TN/RL/W/58.   [97] WTO Doc. TN/RL/GEN/36.   [98] Ibid.   [99] Ibid.

The alternative negotiating positions have gone even further in relying on other IGOs. After conceding the WTO's mandate to constrain fisheries subsidies, Japan, Korea and Taiwan have maintained their emphasis on the non-WTO management of fisheries.[100] These countries have proposed a general permissibility for fisheries subsidies with certain limited exceptions.[101] In categorising the limited and exhaustive list of prohibited subsidies, Japan has drawn expressly on the work of IGOs.[102] According to Japan's proposal, for example, subsidies that relate to 'IUU fishing' would be expressly prohibited. The definition of IUU fishing was taken expressly from the FAO's International Plan of Action on IUU Fishing.[103] Japan also proposed that certain subsidies for vessel construction be allowed if the fisheries were properly managed under other regimes,[104] a position that attracted strong criticism from other WTO members.[105] In a further twist on the reliance by a WTO member on its experience in another regime, Korea reminded the Rules Group that the OECD negotiations to discipline steel subsidies had failed because OECD participants were not able to agree on exceptions.[106]

The attempts by WTO members to balance the needs of developing countries have also referred to other IGOs. Proposals to allow states to maintain certain subsidies for 'small-scale', 'artisanal' and 'subsistence' fishing have led to further questions of definition and classification, as well as controversies about whether exceptions for such subsidies should extend to developed as well as developing countries.[107] Rules Group participants have requested the WTO Secretariat to compare definitions from a range of IGO, NGO and national sources.[108] In providing this information, the WTO Secretariat was careful to include a disclaimer that its research on IGO definitions had no legal implications for the negotiations.

The learning and information-sharing described here is a form of regime interaction that relies solely on the initiative of the trade

---

[100] WTO Doc. TN/RL/W/164, para. 5–6. [101] See, e.g., WTO Doc. TN/RL/W/159. [102] Ibid.
[103] WTO Doc. TN/RL/GEN/114. The IPOA-IUU Fishing is discussed in Chapter 2 n. 125 and accompanying text.
[104] WTO Doc. TN/RL/W/164 (Japan).
[105] See WTO Doc. TN/RL/M/18, esp. para. 6 ('One Participant noted that the sponsor's definition of a properly managed fishery covered most, if not all, of the 75 per cent of fisheries that were currently overexploited').
[106] Reported at (2004) 8:38 *Bridges Digest* (2 November 2004).
[107] WTO Docs. TN/RL/W/172; TN/RL/GEN/92 (para. 4). [108] WTO Doc. TN/RL/W/197.

delegates of the WTO members. This fits with a traditional, state-centric model of international law. According to this model, trade delegations seek to ascertain and represent the consolidated views of their respective states in crafting fisheries subsidies disciplines. The model does not inquire into the capability of WTO members to ensure that there is adequate coordination of national policy in taking into account the experiences and preferences of their domestic stakeholders.

However, relying solely on WTO members may be inadequate to promote inter-regime learning. There may be major questions about how states devise and implement domestic participatory models to ensure stakeholders are included in trade policy-making.[109] This will be a matter for individual WTO members – and some will be more inclusive than others.[110] Outside the fisheries subsidies context, evidence suggests that there are ongoing problems in including the perspectives of domestic constituents in WTO negotiating positions.[111] The concept of 'national interest', at least as put forward in the fisheries subsidies negotiations, is an aggregate that does not sufficiently account for the positions of *all* parties affected by overfishing. Even beyond the trade context, major deficiencies in achieving domestic policy coordination in fisheries issues have been observed.[112] Perhaps because of the relative unimportance of fisheries issues on some national policy agendas, there are failures in coordinating national information and positions. The assumption that there is adequate stakeholder participation and policy coordination within the states and customs territories that make up the WTO appears to be misplaced.

In addition, there is a practical need in the fisheries context for information and expertise on global, as well as intra-state, behaviour. Given that failures in achieving fisheries sustainability often occur because of insufficient restraint by a multitude of states, an overarching perspective on fisheries is needed, rather than a reliance on individual accounts from WTO members themselves. As such, an investigation of

---

[109] Recall the stakeholders identified in the qualified summary of those affected by fisheries governance that I offered in Chapter 1: see p. 28.
[110] See e.g. Brian Hocking, 'Changing the Terms of Trade Policy Making: From the "Club" to the "Multistakeholder" Model' (2004) 3 *World Trade Review* 3 (considering domestic participatory models in the context of the EU and Canada).
[111] Valentin Zahrnt, 'Domestic Constituents and the Formulation of WTO Negotiating Positions: What the Delegates Say' (2008) 7 *World Trade Review* 393.
[112] Noted, e.g., by UN Consultative Process: see UN Doc A/55/274, Annex I, para. 15.

inter-regime learning needs to move beyond the members of the Rules Group itself. The next section examines the ability of non-WTO members to participate in the Rules Group.

2 Participation and observership by others

At the beginning of the Rules Group negotiations, the need to reach out to relevant IGOs and NGOs was identified. For example, the United States urged the Rules Group to learn from relevant IGOs and NGOs:

> In considering the possible structure of these improved disciplines, the Rules Group should explore ways to draw upon information about the state of fisheries stocks and similar expertise in other organizations, including development of relationships with the UN Food and Agriculture Organization and regional fisheries management organizations. The Group could also find ways to obtain the views of non-governmental groups and individuals with expertise, including the fisheries industry and environmental conservation groups.[113]

This proposal to 'enrich' the negotiations with perspectives from inter-governmental and non-governmental organisations was met with criticism by other Rules Group participants because of perceived effects on the WTO's character as a 'member-driven organisation'.[114]

According to the formal rules of the Rules Group, only WTO members may actively participate in the negotiations on fisheries subsidy rules. This applies both to the Rules Group and to the CTE, which first considered the issue of fishery subsidies. Although WTO members have voiced general concern about the limited accessibility and participation within Doha Round negotiating groups, they have focused reform efforts on the inadequate participation of WTO members, rather than of external bodies.[115] The exclusion of external participants includes both IGOs and NGOs. Specific guidelines addressed to NGOs in 1996 emphasised that NGOs could not be 'directly involved in the work of the WTO or its meetings'.[116]

---

[113] WTO Doc. TN/RL/W/77, para. 8.
[114] WTO Doc. TN/RL/M/7 ('Regarding expertise from other governmental or non-governmental organizations, the sponsor did not intend to detract from the intergovernmental nature of the WTO, but was of the view that such expertise could enrich the discussions').
[115] The issue of whether there is full participation between *all* WTO members themselves is an issue of wider controversy at the WTO, although it has not surfaced directly in the fisheries subsidies negotiations. This is usually referred to as a matter of 'internal transparency'; see further WTO Doha Declaration (2001) para. 49.
[116] WTO Guidelines on NGOs (1996); see also Chapter 2, n. 325 and accompanying text.

THE NEGOTIATION OF WTO RULES ON FISHERIES SUBSIDIES   109

While participation is not generally permitted, there is at least scope in some WTO committees for external bodies to observe proceedings. Regular WTO committees operating outside of the Doha negotiating framework can grant observer rights to an IGO after considering the nature of the IGO's work, its membership (including the number of WTO members who are members of it) and whether it extends reciprocal observership to the WTO.[117] On this basis, the CTE has granted observer status to a large number of IGOs with interests in fisheries, including UNEP.[118] Such groups have been able to observe the initial discussions of the fish subsidies issues. Indeed, commentators have considered that the attendance by FAO representatives at CTE meetings influenced states' commitment to fisheries subsidies reform.[119]

The grant of observer status has not extended, however, to the 'issue-identification' and 'text-based negotiations' phase of the development of fishery subsidy rules. The Rules Group is closed to observers. The decision by Rules Group members to deny access to observers was not inevitable: like the regular WTO committees, the Doha negotiating groups each have powers to adopt rules on observership.[120] The Doha negotiating groups on agriculture and services granted observer status to the IGOs that already participated in the work of the regular committees.[121]

Improving access to observers by revising the General Council's observership criteria is currently the subject of Doha negotiations.[122] These negotiations have been allocated to the specially constituted 'Special Session of the CTE', a separate group from the CTE. These discussions have so far made little progress. A common response of members is to refer to the fact that the granting of observer status is an issue of systemic importance across all WTO committees and councils, and to argue that it ought be resolved in the TNC and the General Council.[123] Other members

---

[117] General Council, 'Rules of Procedure for Meetings of the General Council', WT/L/161 (25 July 1996), Rule 11 and Annex 3; see also Chapter 2, n. 324.
[118] For a list of observers, see WTO Doc. WT/CTE/INF/6/Rev.4.
[119] Olav Schram Stokke and Clare Coffey, 'Institutional Interplay and Responsible Fisheries: Combating Subsidies, Developing Precaution' in Sebastian Oberthür and Thomas Gehring, *Institutional Interaction in Global Environmental Governance* (2006) 127, 136. The existence of the FAO IPOA-Capacity was communicated formally to the CTE in 1999: WT/CTE/W/126.
[120] WTO Doha Declaration (2001) paras. 45–52.
[121] This comparison was made by several members at the CTE Special Session: WTO Doc. TN/TE/R/4, see para. 115.
[122] WTO Doha Declaration (2001) para. 31(ii), referred to in Chapter 2, n. 338.
[123] See, e.g., WTO Doc. TN/TE/R5, paras. 103, 109.

disagree that developing criteria for observer status is directly linked to the more generic issues of observer status in international organisations considered in the General Council.[124]

The strategy of members opposed to granting observership has thus been to question the forum in which decisions on observership can be made. This has occurred not only in the CTE but in the Rules Group itself. At the beginning of the Doha negotiations, the Chair of the Rules Group reportedly engaged in informal consultations with participating WTO members about observership.[125] Some members stated that they did not wish to permit any observers when the issue remained unresolved in *other* WTO forums. As a consequence, the question of observers disappeared from the Rules Group agenda in early 2005, and the status quo of closed access remained.

The Rules Group's formal procedures for participation and observership can be contrasted to professed optimism within the WTO Secretariat about working with other intergovernmental organisations in the framing of the fisheries subsidies rules. WTO Director-General Pascal Lamy, for example, claims that cooperation between the WTO and other organisations 'depends very largely on goodwill and a common sense of problem solving rather than strict mandates'.[126] This optimistic and inclusive attitude, which has been manifest in other public speeches of the Director-General,[127] is a strong contrast to the exclusivity of the Rules Group described above.

Given the restrictions on formal interaction between the Rules Group, intergovernmental organisations and NGOs, it would seem that inter-regime learning must occur in alternative, informal settings. The next section considers alternative procedures where knowledge-production about subsidies and sustainability allows stakeholders, including WTO members, to deliberate upon fisheries subsidies issues outside of the institutional strictures of the Rules Group.

---

[124] Ibid., para. 106.   [125] WTO Doc. TN/RL/M/1; see also TN/RL/1.
[126] Consultation by author with Director-General Lamy during online chat discussion hosted by www.wto.org, 16 November 2007. ('The WTO today, and the GATT for fifty years before it, has always had a smaller membership than most other IGOs, but that has never stopped us from working closely with them.') The Director-General also referred to the WTO declaration on economic policy coherence, noted in Chapter 2, n. 313 and accompanying text.
[127] For other accounts of the Director-General emphasising such unity, see e.g. Pascal Lamy, 'The Place of the WTO and its Law in the International Legal Order' (2006) 17 *EJIL* 969.

3 Other forums including UNEP

Learning and information-sharing across regimes may occur outside of the formal structures established by WTO members, and instead through open deliberation by a range of actors who will be affected by the rules. These 'stakeholders' include environmental NGOs, fishing organisations and epistemic communities of experts, as well as states and intergovernmental organisations.[128]

In contrast to formal restrictions within the Rules Group, there are several examples in the fisheries subsidies negotiations of intergovernmental organisations and NGOs promoting deliberation through informal institutional linkages, reporting and information-sharing. The WTO Secretariat has led some of these initiatives. For example, the WTO Secretariat has organised special symposia as part of the quest for synergies between trade and environment,[129] and has conducted regional seminars for members with the participation of international organisations.[130] Officials from the WTO have attended FAO expert consultations and meetings.[131] The WTO Secretariat has issued reports on the activities of intergovernmental organisations and NGOs, mainly in servicing the CTE discussions,[132] but also for the Rules Group.[133] There is also coordination between the CTE and the Rules Group.[134] More recently, representatives from the FAO Secretariat have attended Rules Group negotiations in an informal capacity.[135]

The WTO Secretariat website facilitates this deliberation to some degree.[136] Most importantly, the website disseminates the results of the Rules Group negotiations. Thus, although the Rules Group is formally closed, there is scope for external participants to access its decision-making

---

[128] See Chapter 1, pp. 24–29.
[129] See, e.g., 'WTO Symposium on Trade and Sustainable Development within the Framework of Paragraph 51 of the Doha Ministerial Declaration' on 10–11 October 2005, which included panellists from the FAO, UNEP and a developing country NGO: see www.wto.org/english/tratop_e/envir_e/sym_oct05_e/sym_oct05_e.htm#part2.
[130] See e.g. WTO Doc. WT/CTE/W/216.
[131] WTO Doc. WT/CTE/W/189; See also the 'side-event' of the FAO Sub-Committee on Fish Trade, Bremen, 2–6 June 2008.
[132] See e.g. WTO Doc. WT/CTE/W/167/Add.1 (reviewing activities of APEC, FAO, OECD and UNEP as well as WWF and ICTSD).
[133] See, e.g., WTO Doc. TN/RL/W/197.
[134] See, e.g., attendance by Director of the Rules Division to CTE: WTO Doc. WT/CTE/GEN/10.
[135] Interview by author with staff member of the Food and Agriculture Organization (Rome, 2 July 2008); see also FAO, *Fisheries Report No. 830* (2007) para 18.
[136] The website is part of the WTO's outreach promoted by its guidelines on NGOs; see Chapter 2, n. 325; see also www.wto.org/english/forums_e/ngo_e/ngo_e.htm.

because country submissions are mostly posted online.[137] Secretariat summaries of Rules Group meetings are also made public, although they are usually limited in detail and there is sometimes a delay of several months between meeting and publication. Reporting by other organisations suggests, however, that the Secretariat responds to requests for general information and briefings.[138] In addition, the website includes a portal for non-state submissions which is accessible for members as well as the general public, although the papers are not systematically catalogued.[139]

Outside of the WTO, other opportunities for informal information-sharing and learning have emerged. UNEP, as part of its stated strategy to ensure national coordination between trade, environment and sectoral ministries,[140] has offered a series of workshops,[141] which it often co-convenes with NGOs including WWF and ICTSD. Participants have come from national governments (including officials from environment, trade, and fisheries agencies), intergovernmental organisations, RFMOs, NGOs and academic institutions. Such consultations are often timed to coincide with closed sessions of the Rules Group.[142]

Rather than striving for consensus, the workshops have encouraged open experimentation and learning. In addition to information about subsidy effects and fisheries issues, the workshops have provided information on the laws and standards that operate within other international regimes.[143] For example, participants have discussed the relative breadth of acceptance of particular norms, such as the near universal application of the FAO Code of Conduct as compared to the very limited enforcement of the FAO IPOA-Capacity. This has allowed WTO members to learn about and interrogate intergovernmental organisations and standards that may be relevant to the final disciplines. Written conclusions from the workshops, which tend to demonstrate plurality of perspectives rather than agreement, are distributed directly to Rules

---

[137] Cf Informal country ('job') submissions, which are internal and not available.
[138] This is in keeping with the General Council's commitment to enhancing contacts with NGOs and other organisations through Secretariat consultations: See its guidelines on NGOs, noted in Chapter 2, n. 325.
[139] The NGO portal includes papers on fisheries subsidies submitted to the WTO by a range of NGOs from a little-known Filipino development NGO to Greenpeace International.
[140] See further Chapter 2, n. 252 and surrounding text.
[141] See e.g. www.unep.ch/etb/events/2007fish_symposium.php.
[142] See e.g. reported at (2006) 10:10 *Bridges Weekly Trade News Digest* 5.
[143] See e.g. David Schorr and John Caddy, *Sustainability Criteria for Fisheries Subsidies: Options for the WTO and Beyond* (commissioned by UNEP and WWF) (2007).

Group participants by certain WTO members[144] and are publicly disseminated via the UNEP and WWF websites.

The influence of informal deliberation by stakeholders in the fisheries subsidies negotiations is difficult to prove empirically. According to my interviews with trade delegates, WTO Secretariat staff and FAO representatives, the informal consultations have been regarded as extremely useful opportunities for learning about fisheries subsidies issues. In addition, an analysis of the written Rules Group submissions reveals a strong influence. Some of the proposals have closely followed the positions of external stakeholders such as the WWF.[145] This is reflected in comments by the WTO Secretariat itself, which acknowledges that NGOs can influence the negotiating positions of Rules Group participants.[146] Indeed, the Director-General has attributed the inclusion of fisheries subsidies in the Doha agenda to the influence of civil society.[147]

The activities described above suggest that a community of stakeholders has addressed fisheries subsidies issues at the WTO notwithstanding that they have been denied formal rights to participate and observe negotiations in the Rules Group. Interdisciplinary workshops and informal links between Secretariats and other groups are spatially and temporally linked to the Rules Group negotiations and have had a material influence on proposals. The implication of these forums of learning for regime interaction in international law is examined in Part III of the book. The remainder of this chapter, however, examines the proposed text of the subsidy disciplines and speculates upon regime interaction in the ongoing implementation and adjudication of the fisheries subsidies rules.

D Entrenching interaction

The proposed WTO rules, as currently contained in the Chair's text,[148] incorporate norms from other regimes and entrench collaborative

---

[144] See e.g. formal submission by New Zealand in WTO Doc. TN/RL/W/207.
[145] See, e.g., Brazil's concept of 'patently at risk' fishery in WTO Doc. TN/RL/GEN/79/R3 as similar to the WWF Position Paper noted above n. 33, 130).
[146] See e.g. comment of Director of the Rules Division about the impact of environmental NGOs on the subsidies negotiations in WTO Doc. WT/CTE/GEN/10 ('Over time, these views from civil society may eventually find their way into the negotiating positions of certain participants').
[147] Pascal Lamy, 'Civil Society is Influencing the WTO Agenda', Keynote address to the WTO Public Forum on 4 October 2007: see www.wto.org/english/news_e/sppl_e/sppl73_e.htm.
[148] See above n. 47; The Chair's text is reproduced at Appendix A below.

procedures in three main and related areas: the notification of subsidies, their classification, and the use of benchmarks for fisheries management. This is not the first time that WTO members have crafted rules that require ongoing interaction or collaboration with non-WTO regimes, and this section examines the proposals for fisheries subsidy disciplines in the context of practice resulting from other parts of the SCM Agreement and the SBS and TBT Agreements.

Before turning to these areas of entrenched regime interaction, it is equally important to note the absence of potential interaction in some areas – where WTO members have refuted aspects of the law of the sea and FAO fisheries management regimes. The Chair's text does not endorse *all* related standards. For example, the Chair's text rejects the major jurisdictional concept of these regimes: by attributing subsidies to members regardless of vessel flags or the application of rules of origin to the fish involved,[149] the text purposively avoids enduring problems of 'flag' jurisdiction in the law of the sea, as well as difficulties in assessing the origin of fish products.[150]

As well as exposing areas of express differences between the regimes, this observation reinforces the influence of inter-regime learning on the subsidy rules. The decisions on what standards to include and exclude from the Chair's text, although difficult to prove, may well have been made after trade delegates learned from the experiences of fisheries delegates, international organisations and NGOs.

1 Notifications

The most basic area of fisheries subsidies reform that draws on links with IGOs is the proposed rules relating to notifications. As described above, WTO member compliance with current notification requirements for subsidy programmes is extremely low, and surveillance is inadequately enforced.[151] Inadequate notifications are particularly controversial in the area of agricultural subsidies.[152] The perceived lack of transparency in subsidy notifications has led to the establishment by an NGO of an online searchable database of WTO subsidy notifications.[153]

Rules Group participants have noted that the current implementation of fishery subsidy notifications can benefit from increased interaction with bodies that are maintaining existing databases. For example,

---

[149] Ibid., Art. IV.2. [150] On problems of rules of origin in a fisheries context, see p. 72.
[151] See above n. 38. [152] See World Trade Organization *World Trade Report* (2006).
[153] See Global Subsidies Initiative (GSI) Database at www.globalsubsidies.org.

Chile has suggested that notifications be complementary to subsidy notifications in other regimes, in particular the FAO.[154] Brazil has gone even further to propose that members notify the Committee about, *inter alia*, the conservation status of fisheries harvested by any subsidised fishing vessels 'according to the criteria established by the FAO'.[155]

The United States, on the other hand, has structured proposed notifications according to the relevant member's own domestic management system, 'including measures in place to address fishing capacity and fishing effort and the biological status of managed stocks', for which they would be obliged to maintain an 'enquiry point' in order to maintain transparency.[156] The United States has approved of this structure in the TBT context,[157] and has retreated from earlier suggestions about making notification requirements complementary with other IGOs.[158] The European Union has envisaged that members would notify their fishery-related subsidies to the Permanent Group of Experts (PGE) established under the existing SCM framework,[159] which would then be responsible for reporting on members' subsidisation practices.[160]

The Chair's text requires members to notify their subsidies to the SCM Committee,[161] perhaps due to current deficiencies in subsidy notifications to the FAO.[162] The SCM Committee is to review these notified subsides.[163] Regime interaction is relevant to this process, however. The FAO gathers information and conducts its own review of members' stock assessments and fisheries management systems,[164] in tandem with the work of the SCM Committee, as is discussed further below.

2 Classifying subsidies using fisheries standards

The Chair's text contains a list of negotiated prohibited subsidies ('red-box') as well as exempted subsidies ('green-box'). The provisions utilise existing fisheries data, norms and institutions in a variety of ways to classify fisheries subsidies.

One method is to expressly adopt an FAO standard. The proposed red box includes a prohibition on 'subsidies the benefits of which are

---

[154] WTO Doc. TN/RL/W/115.   [155] WTO Doc. TN/RL/GEN/79/R3 (proposed Art. 5.2).
[156] WTO Doc. TN/RL/GEN/145 (proposed Art. 7).   [157] See TBT Agreement, Art. 10.1.
[158] For early US suggestions, see WTO Doc. TN/RL/W/77, para. 7.
[159] The PGE is described in Chapter 2 n. 292 and accompanying text.
[160] WTO Doc. TN/RL/GEN/134 (proposed Art. 5).
[161] Chair's text, above n. 47, Art VI.1.
[162] On the role of the FAO in collecting relevant information and the inadequacy of state responses, see above nn. 20–24 and accompanying text.
[163] Chair's text, above n. 47, Art. VI.7.   [164] Ibid., Art. V.1, footnote 86.

conferred on any vessel engaged in illegal, unreported or unregulated fishing'.[165] The provision directly incorporates the definition of IUU fishing from the FAO's international plan of action.[166] According to this definition, illegal fishing includes fishing by vessels flying the flag of states that are parties to a relevant RFMO but that 'operate in contravention of the conservation and management measures adopted by [the RFMO] and by which the states are bound, or relevant provisions of the applicable international law'. As a consequence, the implementation of the WTO subsidy rules depends upon an assessment of WTO members' compliance with relevant rules of an RFMO or other applicable international law.

WTO lawyers will be familiar with this form of express reference to other regimes in the classification of WTO-consistent behaviour. For example, the SPS and TBT Agreement rely on international standard-setting bodies to determine whether WTO members can be exempt from disciplines because their trade measures are 'necessary' for a legitimate domestic objective. According to the SPS Agreement, WTO members' measures are presumed to be 'necessary' if they conform to international standards promulgated by listed groups such as Codex or are preceded by proven scientific risk assessments.[167] According to the TBT Agreement, members' measures are presumed to be exempt from disciplines if they are 'in accordance' with international standards set by an open list of international standard-setting bodies.[168] Such bodies are often private producer-led organisations whose aim is to lower trade barriers; as long as they are open to WTO members, they can qualify as standard-setting bodies under the TBT Agreement.

The reference to international standards in the TBT and SPS Agreements has led to a number of criticisms. The standards may have a determinative role in WTO disputes even if they were not intended to be binding in forums such as Codex.[169] This problem is exacerbated if a standard has been agreed without consensus between relevant

---

[165] Ibid., Art. I.1(h), footnote 81. ('The terms "illegal fishing", "unreported fishing" and "unregulated fishing" shall have the same meaning as in paragraph 3 of the International Plan of Action to Prevent, Deter and Eliminate Illegal Unreported and Unregulated Fishing of the United Nations Food and Agricultural [sic] Organization.')
[166] As introduced in Chapter 2, n. 125 and surrounding text.
[167] SPS Agreement, Art. 3.2.   [168] TBT Agreement, Art. 2.
[169] See esp. TBT Agreement, Art. 2.4; AB Report, *EC – Sardines* WT/DS231/AB/R (circulated 26 September 2002).

participants in the standard-setting process.[170] As such, a WTO member may be required to be in conformity with a particular standard even if it voted against the adoption of the standard in the relevant forum and even if the standard is not binding in that forum.[171]

Similar criticisms have arisen during the Rules Group negotiations. Members have expressed concern about the representativeness and effectiveness of the organisations within the law of the sea regimes.[172] There was a perception that WTO members viewed the FAO with more suspicion than standard-setting bodies under the TBT and SPS Agreements. Out of the relevant fisheries organisations, the FAO was generally considered by Rules Group members to be the most credible organisation to set external standards.[173] Yet members remained distrustful of FAO standards because they were not universally accepted and because some standards did not apply to all relevant species. On the other hand, a requirement that standards be adopted by consensus would risk a complete failure of the international community to ever agree on fisheries management and conservation standards.[174]

One trade delegate speculated that to incorporate FAO standards into the subsidy rules would necessitate the improvement of the development of the standards.[175] However, giving the FAO a mandate and the resources to redevelop certain standards was considered to risk exposing the WTO rules to that forum's different set of economic interests and political pressures.[176]

In line with these concerns, delegates have proposed an alternative method of subsidy classification that does not rely on other regimes. As a respondent in much of the relevant case law, the EU is perhaps particularly sensitive to the potential application of standards from other international regimes. Suspicious of referring expressly to the

---

[170] Codex, the OIE and the International Plant Protection Convention IPPC have historically adopted standards by consensus, with some notable and controversial exceptions: see further Doaa Abdel Motaal, 'The "Multilateral Scientific Consensus" and the World Trade Organization' (2004) 38 *JWT* 855.

[171] See AB Report, *EC – Sardines* (WT/DS231/AB/R) (circulated 26 September 2002) para. 225 (relating to a Codex standard for the description of 'sardines'); see also AB Report, *EC – Hormones* WT/DS26/AB/R, WT/DS48/AB/R (circulated 16 January 1998) (relating to a Codex hormone-treated beef standard, which had been adopted by secret vote given the lack of agreement between the United States and the European Union).

[172] Interview by author with staff member of the Rules Division, WTO Secretariat (Geneva, 24 August 2005).

[173] Ibid.   [174] Ibid.

[175] Interview by author with delegate of WTO member (Geneva, 18 August 2005).

[176] Ibid.

law of the sea and FAO regimes in the proposed subsidy disciplines, it has proposed a catch-all prohibition which would prohibit subsidies where the benefits 'are conferred on any fishing vessel or fishing activity affecting fish stocks that are in an *unequivocally overfished condition*'.[177] The italicised phrase does not refer to any other international regimes. Yet this form has been heavily criticised by other WTO members, who have expressed concern that the 'unequivocally overfished' condition is too vague and obscure, and leaves an adjudicating body with too little guidance and too much discretion in interpreting the term.[178]

A third method of standard-setting appears in the proposed 'green-box' exemptions, which refer to international fisheries management regimes, systems and programmes. The Chair's text exempts a range of subsidies, including those used for (i) compliance with 'fisheries management regimes aimed at sustainable use and conservation';[179] (ii) vessel decommissioning or capacity reduction programmes provided that fisheries management systems are in place;[180] and (iii) 'special and differential treatment' provisions for developing countries.[181]

The special and differential treatment builds on earlier proposals that sought to rely on non-WTO regimes to ensure that exempted subsidies maintained standards of sustainability. For example, Brazil initially proposed to allow developing countries that were part of an RFMO to grant capacity-enhancing subsidies, as long as the developing country's fishing capacity did not exceed the sustainable level of exploitation as defined by the particular RFMO.[182] Argentina initially proposed that developing countries wishing to qualify for special and differential treatment should implement national fisheries management systems 'in keeping' with the FAO Code of Conduct.[183] The EU, on the other hand, sought to contain the issue of sustainable special and differential treatment entirely within the WTO. It proposed that developing countries be exempted from subsidy disciplines as long as their fishing

---

[177] Chair's text, above n. 47, Art. I.2. (my emphasis).
[178] (2007) 11:44 *Bridges Weekly Trade News Digest*. Similarly, the Chair's text grants an exemption for otherwise prohibited subsidies that benefit vessel construction and operating costs 'in the exceptional case of natural disaster relief' (Chair's text, above n. 47, Art. I.1). This exemption is limited to a point in time at which fishing capacity is restored and 'sustainable'. This point in time is to be 'established through a science-based assessment of the post-disaster status of the fishery' (footnote 77 to Art. I.1).
[179] Chair's text, above n. 47, Art. II(b)(3).   [180] Ibid., Art. II(d)(4).
[181] Ibid., Art. III.   [182] WTO Doc. TN/RL/GEN/79 (see proposed Art. 6).
[183] WTO Doc. TN/RL/GEN/138 (see proposed Art. X.3)

capacity did not increase to an extent that it presented an 'impediment to the sustainable exploitation of fishery resources worldwide'.[184]

Instead, the Chair's text provides that developing countries may subsidise vessel construction but only if the vessels are used within the member's relevant EEZ, if the fish stocks within that EEZ have been scientifically assessed 'in accordance with relevant international standards', and if the stock assessment has been subject to peer review by the FAO.[185] In addition, international organisations such as the FAO are to provide technical assistance to developing countries.[186] The final overarching safeguard to ensure that special and differential treatment and other exempted subsidies do not lead to overfishing is contained in the condition that any member seeking an exemption 'shall operate a fisheries management system ... designed to prevent overfishing'.[187] The conditionality regarding fisheries management systems further entrenches ongoing regime interaction, as is described below.

3 Conditionality through benchmarking and peer review

To safeguard fisheries sustainability, the Chair's text proposes an overarching condition that all WTO members seeking exemptions for their subsidy practices operate a particular kind of fisheries management system.[188] These management systems are to be based on 'internationally-recognized best practices for fisheries management and conservation' and are to include regular science-based stock assessments and capacity and management measures. Such best practices are said to be reflected in a non-exhaustive list of five international instruments: (i) the Fish Stocks Agreement; (ii) the Code of Conduct; (iii) the Compliance Agreement; (iv) technical guidelines; and (v) plans of action.[189] Accordingly, instruments developed within the law of the sea and FAO regime will have a contingent role in the WTO subsidies regime.

This group of five fisheries instruments is relevant to other proposed exemptions and in the assessment of actionable subsidies. Subsidies relating to the procurement of access rights to foreign fisheries in developing-country WTO members are not prohibited provided that the terms of access include provisions designed to prevent overfishing 'based on internationally-recognized best practices ... as reflected in'

---

[184] WTO Doc. TN/RL/GEN/134 (see proposed Art. 6.1).
[185] Chair's text, above n. 47, Art. III.2(b)(3).
[186] Ibid., Art. III.4; see further clarification by Chair Valles Galmes noted above n. 50 and surrounding text.
[187] Chair's text, above n. 47, Art. V.1.   [188] Ibid., Art. V.1.   [189] Ibid., Art. V.

this group of five fisheries instruments.[190] The 'amber-box' test for actionable subsidies, which may be subject to member complaints based on their adverse effects,[191] also refers to the group of five fisheries instruments as relevant evidence. Information about the subsidising member's implementation of 'internationally-recognized best practices for fisheries management and conservation' will assist in establishing whether a member is causing, through the use of a subsidy, depletion of straddling or migratory fish stocks or stocks in which another WTO member has identifiable fishing interests.

One of the difficulties of using these five instruments as benchmarks is the potential for WTO members to adopt management systems *on paper* without adequately implementing or enforcing them. To address this, the Chair's text requires members to 'adopt and implement' domestic legislation and judicial enforcement mechanisms with full transparency. Information about these and other aspects of the fisheries management system is to be made publicly accessible and subject to peer review.

Review of WTO members' fisheries management systems takes place in three ways. First, members are to notify the existence and operation of their fisheries management systems, including the results of stock assessments, to the FAO.[192] The FAO will determine the specific information it requires and then subject the management system to peer review before the subsidy is granted. The Chair's text recognises that not all WTO members are members of the FAO. In that event, it provides that the relevant WTO member's fisheries management information shall be notified to 'another relevant international organisation'.[193] Secondly, information about the fisheries management system, including the outcome of the FAO peer review, is to be notified to the SCM Committee,[194] an internal committee made up of member representatives that currently reviews subsidy notifications.[195] Thirdly, WTO members are to maintain a domestic 'enquiry point', to which WTO members and other interested parties may direct their scientific or trade questions.[196]

---

[190] Ibid., Art. III.3.
[191] Ibid., Art. IV. See further Marc Benitah, 'Five Suggestions for Clarifying the Draft Text on Fisheries Subsidies (2008) 12:1 *Bridges Trade and Biological Resources News Digest* 21, 22.
[192] Chair's text, above n. 47, Art. V.1.   [193] Ibid., Art. V.1 (footnote 86).
[194] Ibid., Arts. V.1, VI.4.   [195] SCM Agreement Art. 24.
[196] Chair's text, above n. 47, Art. V.2. The enquiry point arrangement may have implications for the future allocation of the burden of proof for disputes: see

The role of the FAO as a peer reviewer of relevant aspects of WTO members' fisheries management systems appears to be a useful harnessing of the FAO's expertise, particularly on stock assessments. There is some evidence that the FAO was involved in the inclusion of this role in the Chair's text,[197] suggesting that the regime interaction in the framing of the rules had a substantive effect on the development of ongoing interaction in the rules' implementation. However, some WTO members have expressed suspicion about the proposed collaboration between the WTO and other regimes. Taiwan, which is not a member of the UN, apparently objects to the involvement of an international organisation to which not all WTO members belong.[198] Norway, supported by some developing-country members, would delete the peer review function and require simple notifications to the FAO or other relevant organisation.[199] Other WTO members have noted their 'systemic concerns' with another intergovernmental organisation 'passing judgment' on the adequacy of WTO members' management systems.[200]

In response to these concerns, the Chair of the Rules Group has emphasised the continuing role of the SCM Committee in maintaining a dialogue with the FAO, and proposed that the peer review process be considered as similar to the existing Trade Policy Review Mechanism (TPRM).[201] The Chair's emphasis on the continuing role of the SCM Committee in maintaining an inter-regime dialogue can be compared to a set of existing and potential arrangements with another rather controversial intergovernmental organisation: the OECD.

EC – *Sardines* and commentary in Robert Howse, 'The Sardines Panel and AB Rulings – Some Preliminary Reactions' (2002) 29(3) *Legal Issues of Economic Integration* 247, 254; see also Henrik Horn and Joseph Weiler, '*European Communities – Trade Description of Sardines*: Textualism and its Discontent' in Henrik Horn and Petros Mavroidis (eds.) *The WTO Case Law of 2002* (2003) 248, 275.

[197] (2008) 12:4 *Bridges Weekly Trade News Digest* 5.
[198] Aside from Taiwan and the other member customs territories, Brunei Darussalam and Singapore are the only WTO members that are not members of the FAO. There are several FAO members that are not members of the WTO.
[199] (2008) 12:11 *Bridges Weekly Trade News Digest* 6.
[200] WTO Doc. TN/RL/W/232 (Annex C – Fisheries Subsidies), C-60.
[201] (2008) 12:11 *Bridges Weekly Trade News Digest* 6. In the TPRM process, the WTO member under review produces a domestic trade policy report, which is then reviewed by the WTO Secretariat, which creates its own independent report. The General Council of the WTO (acting as a 'Trade Policy Review Body') then scrutinises both reports with a view to enabling 'regular collective appreciation and evaluation' of the member's domestic trade policies: see Annex 3 of the Marrakesh Agreement Establishing the World Trade Organization 1994, esp. para. C.

The OECD has a special role in the granting of existing exemptions for export subsidies that are otherwise prohibited. The SCM Agreement exempts export subsidies for certain export credit practices if they are in conformity with provisions of a certain undertaking or 'successor undertaking' on official export credits.[202] The infamous Annex I(k) provision refers to an arrangement concluded in the OECD in 1978 and revised most recently in 2010.[203] The arrangement is self-defined as a 'Gentlemen's Agreement among the Participants'.[204] Current participants are Australia, Canada, the EC, Japan, Korea, New Zealand, Norway, Switzerland and the United States. Other OECD members and non-members may be invited to become participants by the current participants.[205]

The term 'successor undertaking' allows the participants of the OECD Export Credit Arrangement to continually update the terms of the exemption as it applies to themselves and to the rest of the WTO membership. In the face of a challenge that it was unreasonable for a subgroup of WTO members to 'perpetually legislate on behalf of the overwhelming majority of the membership' in this way,[206] a WTO panel has considered that such a delegation is justifiable in order to share necessary expertise.[207] The Panel noted, however, that if the participants to the OECD Export Credit Arrangement 'were to abuse their power to modify the scope of the safe haven, the recourse of other WTO members would be to renegotiate' the relevant provision of the SCM Agreement.[208]

Indeed, the Rules Group has since examined the operation of the exemption for official export credit practices, as part of its mandate for general subsidies reform. The Chair's text contains a proposal that

---

[202] SCM Agreement, Art. 3.1(a), footnote 5; Annex I (k). For a detailed description of its operation, see Janet Koven Levit, 'A Bottom-Up Approach to International Lawmaking: The Tale of Three Trade Finance Instruments' (2005) 30 *Yale Journal of International Law* 125, 157–67.
[203] OECD, 'Arrangement on Officially Supported Export Credits' (2010 revision): See OECD Trade and Agriculture Directorate TAD/PG(2010)2 (28 January 2010), www.oecd.org/.
[204] Ibid., Art. 2. (Surprisingly, the revisions have not amended this rather outdated expression.)
[205] Ibid., Art. 10.
[206] Panel Report, *Brazil – Export Financing Programme for Aircraft* WT/DS46/R (circulated 14 April 1999), para. 5.84. Brazil had separately complained that developing countries 'are not members of the OECD. They have no voice in the OECD': see para. 4.98.
[207] Ibid., para. 5.88.   [208] Ibid., para. 5.89, footnote 86.

the undertaking be notified to the SCM Committee.[209] It further stipulates that the SCM Committee shall examine the notified undertaking on the request of a WTO member. This role will add to existing institutional arrangements, whereby the WTO Secretariat has observer status in the OECD export credit group, the OECD Secretariat has observer status within the SCM Committee, and panels may seek information during disputes.[210]

The existing and proposed collaborative practices of the WTO and OECD Secretariats change the position of the OECD Export Credit Arrangement as a benchmark of conduct in the SCM Agreement. These practices require a level of transparency and openness in the OECD procedures to frame and update the benchmark, as scrutinised by the SCM Committee.

A similar development has occurred in the context of the TBT and SPS Agreements, probably as a response to the perceived problems in the interpretation of standards by adjudicating bodies.[211] The TBT Committee, for example, has encouraged standard-setting bodies to operate with open, impartial and transparent procedures.[212] The SPS Committee has adopted a peer review role of member activities,[213] while the WTO Secretariat, seeking to address deficiencies in compliance with SPS notification obligations, has conducted workshops on transparency and created an online searchable database.[214]

Some commentators would take scrutiny of international standards even further. Scott compares the WTO adjudicating bodies' stance on

---

[209] Draft Consolidated Chair Texts of the AD and SCM Agreements, WTO Doc. TN/RL/W/213 (31 November 2007), 76 (amendment to Annex I (k)).

[210] I note, however, that the consultative power of the PGE or panel does not appear to have been utilised in the following disputes involving Annex I (k): Panel Report, *Brazil – Export Financing Programme for Aircraft* WT/DS46/R (circulated 14 April 1999); AB Report, *Canada – Measures Affecting the Export of Civilian Aircraft* WT/DS70/AB/R (circulated 2 August 1999); Panel Report (circulated 14 April 1999); Panel Report, *Canada – Export Credits and Loan Guarantees for Regional Aircraft* WT/DS222/R (circulated 28 January 2002).

[211] See above n. 171 and surrounding text.

[212] Decision of the Committee on Principles for the Development of International Standards, Guides and Recommendations with relation to Articles 2, 5 and Annex 3 of the Agreement (2002): WTO Doc. G/TBT/1/Rev.8. For critical assessment of the Decision, see Robert Howse, 'A New Device for Creating International Normativity: The WTO Technical Barriers to Trade Agreement and "International Standards"' in Christian Joerges and Ernst-Ulrich Petersmann (eds.) *Constitutionalism, Multilevel Trade Governance and Social Regulation* (2006) 383, 392–4.

[213] For an analysis of the SPS Committee acting as agent of peer review, see Joanne Scott, *The WTO Agreement on Sanitary and Phytosanitary Measures: A Commentary* (2007) 50ff.

[214] WTO Doc. G/SPS/R/47.

standards with that of the European Court of Justice (ECJ).[215] She claims that the Appellate Body could learn much from the EU, namely that 'the authority enjoyed by [standards] may be regarded as contingent, not absolute'.[216] She demonstrates the considerable scope of the ECJ to scrutinise relevant standard-setting procedures, particularly with regard to transparency and access to documents.[217] On this basis, she suggests that WTO adjudicating bodies could take on an oversight role for standards if those standards 'have exhausted their own internal reserves of authority' through adoption without consensus or some other procedural deficiency.[218] A far-reaching implication of Scott's work is that the question of whether a standard is 'inappropriate' or 'ineffective' for a particular domestic objective under the TBT Agreement[219] might be assessed according to the process by which the standard came into being.[220] A further implication is that disputing bodies ought to consult with standard-setting organisations during disputes about these procedures.[221]

The procedures discussed above emphasise transparency, accessibility and ongoing scrutiny of norms and benchmarks from international regimes, and are suggestive of emerging factors for WTO members to ensure appropriate regime interaction, a theme developed in Part III of the book. Such factors are important not only in the fishery subsidies negotiations and resulting disciplines but also for the proposed regime interaction in the settlement of subsidies disputes.

E Settling disputes

Proposals for the settlement of fisheries subsidies disputes involve ongoing links between the WTO and other IGOs. The Chair's text adopts two procedures for dispute settlement: one for disputes about prohibited subsidies, and the other for disputes arising under other provisions.[222]

---

[215] Joanne Scott, 'International Trade and Environmental Governance: Relating Rules (and Standards) in the EU and the WTO' (2004) 15 *EJIL* 307.
[216] Ibid., 311.   [217] Ibid., 317.   [218] Ibid., 354.
[219] TBT Agreement, Art. 2.4.   [220] Scott, above n. 215, 332.
[221] Relevant provisions for consultation by panels depart from the DSU: see SPS Agreement, Art. 11.2; TBT Agreement, Arts. 14.2–14.4; Annex 2. In the *EC – Sardines* case, the EC claimed on appeal that the Panel had erred in failing to consult Codex. This is assessed in Chapter 5, n. 202 and accompanying text.
[222] This section does not consider the settlement of disputes arising over fisheries subsidies that under certain conditions could be brought to a law of the sea tribunal according to UNCLOS, Part XV (see further Chapter 2).

For disputes about prohibited subsidies, panels are constituted according to the current procedure of the SCM Agreement. As such, the panel may request the assistance of the Permanent Group of Experts (PGE), whose conclusions shall be accepted without modification.[223] The PGE is made up of five experts 'highly qualified in the fields of subsidies and trade relations'[224] who are elected by the SCM Committee.[225] Perhaps because of the binding nature of its conclusions, the PGE has yet to be consulted in the history of the SCM Agreement. For this reason, and given the absence of any requirement of fisheries or environmental expertise, it may be unlikely that the PGE is consulted in fisheries subsidies disputes.

For disputes arising under other provisions in the fisheries subsidies disciplines, including under green-box provisions and special and differential treatment, the Chair's text provides that panels are to be constituted according to the DSU.[226]

In both cases, if the dispute raises 'scientific or technical questions related to fisheries', the panel must seek advice from 'fisheries experts'.[227] These experts are to be chosen by the panel in consultation with the disputing parties. The panel also has discretion to consult with international organisations.[228]

The provision for panel consultations with experts and international organisations is virtually identical to the dispute settlement provision

---

[223] Chair's text, above n. 47, Art. VIII.1, referring to SCM Agreement Art. 4.
[224] SCM Agreement, Art. 24.3; See further Chapter 2, n. 292.
[225] SCM Agreement, Art. 24.3.
[226] Chair's text, above n. 47, Art. VIII.1, referring to SCM Agreement, Art. 30.
[227] Chair's text, Art. VIII.4 ('Where a dispute arising under this Annex raises scientific or technical questions related to fisheries, the panel should seek advice from fisheries experts chosen by the panel in consultation with the parties. To this end, the panel may, when it deems it appropriate, establish an advisory technical fisheries expert group, or consult recognized and competent international organizations, at the request of either party to the dispute or on its own initiative.') Rather than using Brazil's suggested language that the panel 'may' seek advice from fisheries experts, the Chair's text reproduces the United States' language of 'should': see WTO Doc. TN/RL/GEN79R3 (proposed Art. 7.1) cf TN/RL/GEN/145 (proposed Art. 10).
[228] Ibid. On one reading, draft Art. VIII.4 gives the panel discretion to consult with an advisory fisheries expert group or relevant international organisation on scientific or technical matters. On another reading, the paragraph merely refers to the discretionary power of the panel to consult on the selection of fisheries experts. This is the interpretation preferred by the United States, which modified its initial proposal to clarify that FAO and other international organisations could be asked to provide assistance merely in identifying appropriate experts, and would not be called upon to provide expert opinions as organisations: WTO Doc. TN/RL/GEN/145 (annotation to proposed Art. 10).

adopted by the SPS Agreement,[229] which was drafted after substantial disagreement between negotiators. The Cairns group of agriculture-producing WTO members wished to ensure that any consultation was in a personal capacity, because it wanted to minimise the role of international standardising bodies in dispute settlement.[230] Subsequent application of the SPS Agreement reveals that there is some uncertainty about whether international organisations should be consulted only on the specific selection of experts[231] or on a wider set of issues.[232]

One of the ways that SPS negotiators sought to balance the role of international standards was to remind WTO members of their unimpaired rights to resort to the good offices or dispute settlement mechanisms of these international agreements.[233] Such a reminder is reproduced in the Chair's text,[234] a provision that sits rather uneasily with the SCM Agreement's stipulation that no other specific action can be taken against a subsidy of another member.[235]

The provision for experts can also be compared with a tribunal established under the law of the sea regime.[236] In disputes concerning the interpretation or application of UNCLOS, ITLOS (or another relevant tribunal) can, on the motion of the parties or on its own motion, choose no fewer than two experts who will sit but have no right to vote.[237] For fisheries disputes, these experts are to be chosen preferably from a list drawn up and maintained by the FAO.[238]

WTO members could have gone much further in designing inter-regime enforcement. An early proposal from WWF, for example, promoted much closer interaction between a WTO dispute settlement panel and international organisations.[239] WWF proposed that a permanent group, comprised of experts from the FAO, UNEP, RFMOs and NGOs, would sit at the WTO and would have the authority to make

---

[229] SPS Agreement, Art. 11. [230] See Motaal, above n. 170, 864 and citations therein.
[231] See AB Report, *Australia – Salmon*, WT/DS18/AB/R (circulated 20 October 1998); AB Report, *EC – Hormones* WT/DS26 and DS48/AB/R (circulated 16 January 1998); AB Report, *Japan – Agricultural Products II* WT/DS/76/AB/R (circulated 22 February 1999); Panel Report WT/DS76/R (circulated 27 October 1998).
[232] Panel Report, *EC – Biotech* WT/DS291/R, WT/DS292/R, WT/DS293/R, (circulated 29 September 2006), para 3.1. But note that this consultation was limited in other ways, as considered in Margaret A. Young, 'The WTO's Use of Relevant Rules of International Law: An Analysis of the Biotech Case' (2007) 56 *ICLQ* 907.
[233] SPS Agreement, Art. 11.3. [234] Chair's text, above n. 47, Art. VIII.5.
[235] See SCM Agreement, Art. 32.1. [236] See further Chapter 2, esp. pp. 37–38; 43–46.
[237] UNCLOS, Art. 289. [238] Ibid., Art. 289; Annex VII, Art. 2.
[239] Schorr, above n. 33, 105.

binding rulings on matters outside of the traditional competence of the WTO.[240] This proposal reflects other suggestions for shared adjudicative capacities made in the general trade and environmental context.[241]

The WWF suggestion was not accepted by the Rules Group, but even current arrangements could promote authority-sharing with other regimes to some extent. For example, if a panel adjudicating on an alleged prohibited subsidy is required to assess whether subsidies are benefiting fishing activity for stocks that are in an 'unequivocally overfished condition',[242] it could consult the PGE for a binding determination. The PGE has the discretion to draw on a wide range of sources for assistance;[243] depending on the PGE's initiative, this could include consultations between the PGE and international organisations, scientific experts and NGOs. In addition, there are a number of other methods by which a WTO adjudicating body can seek information from other regimes in resolving fisheries subsidies disputes. This includes information from *amicus* briefs and from the submissions of the disputing parties themselves.[244]

The institutional provision for panel consultation on fisheries issues is of course interrelated with the normative status of the fisheries regimes in the disciplines. For example, for a broad-based standard, such as the 'equivocally overfished' standard proposed by the EU, dispute settlement bodies retain full discretion to interact with other regimes. For standards that directly incorporate the norms of international organisations, such as the FAO international plan of action on IUU fishing, adjudication inevitably requires WTO adjudicating bodies to refer to those regimes.

Regime interaction in the adjudication of potential violations may thus depend on the character of the particular standard at issue. Paradoxically, for broad-based standards, the absence of express references to the norms or actors from other regimes may make regime interaction even more essential. Institutional provisions for consultation and fact-finding in dispute settlement may make up for a lack of normative support in allowing adjudicating bodies to take account of

---

[240] Ibid.
[241] See e.g. Perez, above n. 91, 98 and citations therein with respect to UNEP acting as a representative on WTO Panels.
[242] Chair's text, above n. 47, Art. I.2.
[243] The PGE may consult with and seek information from any source it deems appropriate: SCM Agreement Art. 24.5.
[244] These procedures are explored in Chapter 5.

the issues related to fisheries conservation and management that are currently outside the institutional competence of the WTO. Arguably, the less focused the standard, the more scope for regime interaction.

WTO jurisprudence suggests that dispute settlement bodies will draw upon other regimes to assess standards that are apparently objectively verifiable. This technique was implicitly endorsed by the Appellate Body in *EC – Tariff Preferences*. The AB assessed the EU's conformity with paragraph 3(c) of the Enabling Clause, which specifies that 'differential and more favourable treatment' provided under the Enabling Clause:

> ... shall in the case of such treatment accorded by developed contracting parties to developing countries be designed and, if necessary, modified, *to respond positively to the development, financial and trade needs of developing countries.* [Emphasis added][245]

The AB upheld India's complaint that the EU's GSP policy was not justified under the Enabling Clause because the policy discriminated arbitrarily between developing countries. It held that the Enabling Clause did allow different tariff preferences to be granted to different GSP beneficiaries but only if such preferences responded positively to the particular 'development, financial or trade need[s]' and were made available to 'all beneficiaries that share that need' on the basis of an objective standard. Of relevance to the present study is that the AB expressly recognised that such an objective standard could be ascertained by reference to multilateral instruments adopted by IGOs.[246]

If a WTO panel adjudicating upon a fisheries subsidies dispute adopted the approach of the AB in *EC – Tariff Preferences*, its assessment of whether subsidies should be prohibited because relevant stocks are in an 'unequivocally overfished condition' might therefore involve the norms of fisheries organisations that appear expressly in other parts of the Chair's text. Yet such regime interaction leads to further challenges. Consider, for example, a broad-based standard that does not expressly refer to the conservation and management obligations set out in the Fish Stocks Agreement. Currently, only 77 countries are parties to the Fish Stocks Agreement. If a dispute is brought against a

---

[245] Enabling Clause (1979), para. 3(c).
[246] AB Report, *EC – Tariff Preferences*, para. 163 ('Broad-based recognition of a particular need, set out in the *WTO Agreement* or in multilateral instruments adopted by international organizations, could serve as such a standard'); see also para. 163, footnote 335.

subsidising WTO member that is not a party to the Fish Stocks Agreement, that member will likely argue that a WTO adjudicating body should not draw on relevant FSA norms. Moreover, although the Compliance Agreement and the Code of Conduct are mutually compatible, there may be conflicting provisions between them and the Fish Stocks Agreement or norms from RFMOs. There may also be conflicts between evidence from NGOs such as WWF and fishing industry groups. There is a need for guidance to enable WTO adjudicating bodies to rank sources and assess the credibility of all of these stakeholders. This issue is discussed in greater depth in Chapter 5's examination of regime interaction in the context of WTO adjudication.

The paradox between broad-based and regime-dependent standards is apparent in the continuing shift by WTO members between including direct reference to fisheries management regimes and leaving standards open-ended. To leave standards open-ended risks giving the WTO dispute settlement bodies full discretion in incorporating norms from other regimes. Yet to incorporate standards from exogenous regimes in the disciplines themselves requires an agreement between WTO members about the status of the specific standard-setting bodies, which still suffer from problems of credibility. These concerns demonstrate that the controversies surrounding the WTO's new role in fisheries subsidies disciplines will remain long after the law-making stage. For fisheries subsidies disputes, the onus may well be on WTO dispute settlement bodies to assess WTO member conduct by interacting, on their own discretion, with other regimes. Part III of this book seeks to provide them with an appropriate framework to do so.

F Conclusions

State subsidies currently paid to the fishing sector lead to an oversupply of vessel capacity, and a resultant risk of overfishing, trade distortion and fishery collapse. This chapter has examined the making of international laws to restrain such subsidies. The choice of law has been made in the context of a fragmented international legal system, where a number of regimes relevant to fisheries already co-exist. The FAO fisheries management regime contains voluntary instruments aimed at reducing subsidies. Yet the FAO Code of Conduct and associated plans of action (IPOAs) have had little apparent effect in curbing subsidies. In contrast, the WTO's SCM Agreement provides for binding rules. Yet despite evidence of significant subsidising behaviour, a claim against fisheries subsidies has never been brought before a WTO panel, and there is a need to clarify and improve the SCM Agreement. To add to this complexity, there are some states that consider subsidies to be harmless as long as fisheries are properly managed. For these states, the Fish Stocks Agreement, UNCLOS, RFMOs and the FAO are most relevant.

Within this 'forum shop', there have been significant challenges to law-making. The chapter demonstrated that the choice of forum to deal with fisheries subsidies has been heavily contested, with much opposition to any new laws developed within the WTO. Four core concerns centred on whether the WTO was appropriately mandated, whether the WTO had requisite expertise and judicial competence in the fisheries sector, whether the WTO was biased against the ecological dimensions of the issue and whether the new disciplines would threaten coherence and effectiveness of trade policy.

Much of this opposition was grounded on concerns that regimes remain within their mandates and competences, which accords with traditional concerns about self-contained regimes. The chapter showed instead that the broad textual and functional mandates of the WTO comfortably include the disciplining of fisheries subsidies. More importantly, the chapter suggested the high political and economic stakes behind forum struggles. Rather than a technical issue of legal precision, the fragmented legal order reflects the conflicting interests in the utilisation and trade of fish and fish products. Moreover, appropriate levels of government support are one of the most contested questions of global capitalism. Some nations clearly wished to avoid the possibility of binding dispute settlement by the WTO, which, as I have discussed, has superior enforcement powers to other relevant regimes. These nations were not opposed to governance of fisheries *per se* – they were active participants of the law of the sea regimes – but they differed significantly in their forum preferences.

Instead of closing these debates by suggesting an a priori delimitation of competence, the chapter suggested a need for open and inclusive engagement on the issue. This finding contrasts with a narrow conception of the liberal trade project that some of the scholarly literature on 'linkage' tends to reinforce.[247] Within the context of the making of law,

---

[247] On the limitations of the 'linkage' literature, see e.g. Jeffrey Dunoff, 'Rethinking International Trade' (1998) 19 *University of Pennsylvania Journal of International Economic Law* 347, 384; see also Lang (cited in Chapter 1 n. 80), 538.

the finding prompts an investigation into the ability of the WTO to learn from and entrench interaction with other regimes. On a more theoretical level, the finding prompts an investigation into a framework for regime interaction in international law. This is the endeavour of Part III of the book.

Moving the focus, then, away from 'forum' questions and towards inter-regime learning, the chapter investigated the ability of the WTO to adapt and incorporate perspectives from exogenous regimes. The fisheries subsidies negotiations have relied heavily on WTO members to act as conduits of information. Yet while many trade delegates have taken pains to report on their experience in other forums, the assumption that they have all coordinated their national fishery policy and consulted with stakeholders is rather a bald one. In addition, trade delegates are ill equipped to report on global behaviour.

Inter-regime learning in the negotiations has been further hampered by the decision to insulate the Rules Group from external participants or observers, due to the oft-cited wish of WTO members to remain a 'member-driven organisation'. The breach has been filled, however, by activities of IGOs, including the WTO Secretariat itself, the FAO and UNEP, and NGOs. These alternative avenues for learning and collaboration have been apparently influential in negotiating proposals. This may suggest that 'public–civil society partnerships' influence the development of WTO law.[248] It also confirms UNEP's role in encouraging 'mutual supportiveness' of trade and environment policies.[249]

A conception of the deliberations conducted by UNEP, WWF and others is augmented by theories of governance and law. Part III of the book undertakes a theoretical investigation into models of governance that rely less on representative functions of states and IGOs and more on deliberative capacities of advocacy networks and stakeholder groups. Ideas of 'democratic experimentalism' and 'post-sovereign' governance may impact upon the framework of regime interaction in significant ways.[250]

The chapter continued the inquiry into regime interaction by investigating the way interaction would be entrenched in the disciplines,

---

[248] Cf Shaffer's work on public – private partnerships, which relates to corporate actors in the WTO: see Gregory Shaffer, *Defending Interests: Public – Private Partnerships in WTO Litigation* (2003).

[249] See Chapter 2, n. 252 and accompanying text.

[250] See e.g. Michael Dorf and Charles Sabel, 'A Constitution of Democratic Experimentalism' (1998) 98 *Columbia Law Review* 267; see further Part III.

using for analysis the draft text circulated by the Chair of the Rules Group. It described three areas for ongoing collaboration with other regimes in the implementation of the rules: in the notification of subsidies, their classification, and the use of benchmarks for fisheries management. The proposals to coordinate subsidy notifications reveal a mistrust of data from other IGOs. The proposals to classify 'good' and 'bad' subsidies by using standards from other regimes have also been deeply controversial. There are no universal definitions of which fisheries subsidies are harmful and which enhance sustainability, and proposals reflect underlying preferences for alternative forums. The use of fisheries standards brings a different rationale to the WTO subsidies disciplines, which have been critiqued as economically irrational and incoherent.[251] The proposed approach focuses disciplines on subsidies that are harmful to the achievement of agreed global goods. The rub, however, is how such agreement occurs.

The Chair's text adopts different approaches to entrenching regime interaction: either standards expressly incorporate norms from external regimes (such as the FAO IPOA-Capacity), or they are broad-based and objectively verifiable without reference to external regimes, or they refer generally to other regimes by suggesting benchmarks for good fisheries management systems. For the latter, WTO members must achieve international best practices in fisheries conservation and management in order that their subsidies qualify for green-box and special and differential treatment. A non-exhaustive list of five main instruments gives examples of these best practices: the Fish Stocks Agreement, the FAO Code of Conduct, the Compliance Agreement, technical guidelines and plans of actions.

The adjudication of fisheries subsidies disputes in the WTO will involve interaction with other regimes, and the chapter detailed the prospective role of the dormant Permanent Group of Experts and other innovative arrangements. Comparisons with the SCM Agreement's provision on export credit practices and the TBT and SPS Agreements' reliance on external standards demonstrated the substantial relationship between normative and institutional regime interaction in the settlement of disputes. For example, if FAO technical guidelines and plans of action, which are constantly developed and evolving, are normatively important, a WTO panel may require institutional assistance in understanding them by consulting IGOs. Moreover, the comparisons

---

[251] See e.g. Bagwell, Mavroidis and Staiger, above n. 70.

demonstrate challenges in according authority to exogenous standards, especially if such standards are not adopted by the same countries that make up the WTO. Such problems may be assuaged by ensuring that the relevant WTO bodies, including the SCM Committee and adjudicating bodies, have scope to contest the standards by inquiring into their internal procedures and accountability. Any attendant risk of managerialism, which is linked to the increased role and authority of technical experts and institutional bodies,[252] must be confronted. These ideas will be revisited as part of my inquiry into a legal framework for regime interaction.

In sum, this chapter has provided an examination of current efforts to make international laws constraining fisheries subsidies. It is the result of empirical engagement with trade delegates, Secretariat staff and NGOs and detailed analysis of country submissions and external reports, sometimes minutely described. The aim is not just to offer a narrative of forum shopping, inter-regime learning and entrenchment of future regime interaction, although the documentation will no doubt be useful for some. As discussed in Chapter 1, my aim is to theorise regime interaction in the face of conflicting ideas of international law: which include sovereignty, legitimacy and accountability. The technical details of the fisheries subsidies negotiations, and of the role and authority of relevant stakeholders and experts, help shape a meta-understanding of law-making in a fragmented international legal order. This is a theme that is developed in Part III of the book. Before moving to that Part, the next chapter offers a further case study of regime interaction, which arises from the restriction of trade in endangered marine species.

---

[252] Cf the definition of managerialism offered by Martti Koskenniemi, 'International Law: Constitutionalism, Managerialism and the Ethos of Legal Education' (2007) 1 *European Journal of Legal Studies* (online) ('A *managerial* approach is emerging that envisages law beyond the state as an instrument for particular values, interests, preferences ... law turns into rules of thumb or soft standards that refer to the best judgment of the experts in the [trade, or human rights, etc] box'); see also Martti Koskenniemi, 'The Fate of Public International Law: Between Technique and Politics' (2007) 70 *Modern Law Review* 1.

4 The restriction of trade in endangered marine species

> RECOGNIZING the primary role of sovereign States, FAO and regional fisheries management organizations in fisheries conservation and management...[1]

As commercially significant and heavily traded species such as basking sharks and bluefin tuna become threatened with depletion, questions arise about the role of existing regimes and novel interactions.

This chapter explores the efforts within international law to restrict the trade in endangered marine species. The increasing trade in fish and fish products has accompanied a drastic decline in numbers of certain marine species. Stalling and reversing the depletion of these species gives rise to interaction between two prominent regimes: the FAO fisheries management regime and the CITES regime that regulates the international trade in endangered species. The Secretariats of CITES and the FAO have worked closely to address the potential large-scale depletion of selected marine species, including some that are commercially valuable and heavily traded, such as sharks, queen conch and the bluefin tuna.

Rather than initiating new law, as was done in the regime interaction considered in the previous chapter, efforts to restrain international trade in marine species have involved the implementation of *existing* legal commitments. The implementation of these commitments requires collaboration and coordination between the FAO and CITES regimes. Countries have proposed to entrench this interaction through

---

[1] The controversial proposed introduction to the draft Memorandum of Understanding between the FAO and the CITES Secretariat (endorsed by the FAO Sub-Committee on Fish Trade, 2004) which was subsequently deleted from the final text.

a Memorandum of Understanding (MOU) between the FAO and CITES Secretariats.

On the one hand, the MOU promises to enhance coordination and cooperation between the two regimes. On the other, the MOU is an instrument with which states forum shop. This chapter investigates the drafting of the MOU, when opposition to CITES' role played out in attempts to give the FAO express primacy. Like the drafting of rules on fisheries subsidies considered in the previous chapter, issues of mandate, expertise, bias and coherence in fisheries management all featured in the lead-up to the MOU's joint signature by the FAO and CITES Secretariat.

After introducing the problem of endangered marine species, and the regimes that vie for relevance in efforts to combat the threats to such species, the chapter analyses the legal status and evolution of the MOU. It examines how interaction is entrenched in the MOU, in terms of both the written instrument and subsequent supporting practice. This analysis assists the broader discussion of regime interaction in Part III of the book.

A Endangered marine species

The emerging possibilities for regime interaction described in this chapter are due to the increasing ecological threat of depletion of certain marine species caused by international trade. Of many examples of marine species at risk, this section provides background scientific and economic data on sharks, seahorses and, perhaps the most prized of all fish species, the bluefin tuna.

Sharks, the largest fish in the sea and the apex predators in the marine food chain,[2] are caught and traded in increasing numbers for their meat and fins. Shark meat consumption has grown steadily, with consumers often unaware that they are buying and eating shark species due to varying market names.[3] Shark meat is now produced by a wide range of countries.[4] Exports were valued at US$131.5 million in 1997,

---

[2] The two largest fish are the whale shark and the basking shark respectively; both species are now recognised as under threat: see below n. 31 and surrounding text.
[3] For example, in Australia and the UK, picked dogfish and other species are marketed as 'flake': *FAO Fisheries Technical Paper No 389*, Stefania Vannuccini, 'Shark Utilization, Marketing and Trade', 1999, para. 6.1.1.
[4] FAO statistics indicate that production of fresh, frozen and cured chondrichthyan meat and fillets increased from nearly 18 000 tonnes in 1976 to 34 500 tonnes in 1986 and 69 300 tonnes in 1997: ibid., para. 6.1.3.

with major exporters including Spain, USA, Japan, UK, Canada, Taiwan, New Zealand and Indonesia.[5] In addition to shark meat, shark fins have become a highly traded and valued fish product.[6] The rise of average income levels in China has led to increased demand for the traditional delicacy of shark-fin soup. China, India and Indonesia are the largest producers of shark fins and 1997 exports were valued at US$90.4 million.[7]

Sharks have suffered sudden and drastic population decline. As well as the expanding catches of sharks due to the huge demand for shark meat and fins, sharks are often killed as by-catch in tuna and swordfish longline fisheries. They are also affected by marine pollution and by loss of their prey through other instances of overfishing. Population numbers are often unable to recover because sharks have slow growth and reproduction rates.[8] Over 135 shark species have been listed on the IUCN Red List, and over twenty are listed as critically endangered or endangered.[9]

Seahorses are another commercially valuable marine species that is threatened due to overfishing, by-catch vulnerability and habitat loss.[10] Seahorses are harvested and traded for use in traditional medicine and aquariums, and are caught as by-catch in shrimp trawlers. Seahorses have an unusual biology, in which offspring are dependent on the male of the species for a prolonged period, and their site and mate fidelity makes them particularly vulnerable to depletion.[11] There are at least two endangered or critically endangered species and several vulnerable species of seahorse on the IUCN Red List.[12]

---

[5] Ibid., para. 6.1.3 ('Exports of fresh, frozen and cured chondrichthyan meat and fillets have grown considerably from 17 600 tonnes, worth US$22.1 million in 1976 to approximately 58 600 tonnes, valued at US$131.5 million, in 1997').

[6] FAO statistics, which only partially record the relevant data due to incomplete reporting by countries, indicate that world production of shark fins has increased from 1 800 tonnes in 1976 to 6 030 tonnes in 1997: ibid., para. 6.2.8.

[7] Ibid.

[8] See Terence Walker, 'Can Shark Resources be Harvested Sustainably?' (1998) 49 *Marine & Freshwater Research* 553 (reviewing biological characteristics and noting different productivity rates of species).

[9] 'Shark Species Face Extinction Amid Overfishing and Appetite for Fins' *Guardian* (18 February 2008) 11. See further www.iucnredlist.org/.

[10] Pamela S Turner, 'Struggling to Save the Seahorse' *National Wildlife* (6 January 2005).

[11] N C Perante et al., 'Biology of a Seahorse Species, Hippocampus Comes in the Central Philippines' (2002) 60 *Journal of Fish Biology* 821.

[12] www.iucnredlist.org/.

Bluefin tuna are the 'ultimate political fish' for Japan and many North Atlantic fishing nations,[13] as well as southern nations such as New Zealand and Australia. There are two species, both highly migratory: the northern bluefin tuna (taken mainly in the Atlantic and also in the Pacific) and the southern bluefin tuna (from the southern Indian and Pacific Oceans). Both are highly valued and traded, especially for the Japanese sushi market, where one fish can command JPY16.3 million – about US$180,000.[14] Japan imports 90 per cent of bluefin tuna,[15] and is the world's largest consumer of the fish. Bluefin tuna are long-lived and slow to reproduce, and are generally considered to be severely depleted. The southern bluefin tuna is thought to be at risk of commercial extinction without significant conservation measures, and the northern bluefin tuna is thought to be depleted in the Atlantic and overfished in the North Pacific Ocean.[16]

This overview of the status of sharks, seahorses and bluefin tuna suggests a growing risk of depletion for many commercially fished and traded marine species. Depletion of such species has both environmental and economic consequences. Marine biodiversity and ecology will clearly be affected, particularly if important predator species such as sharks are lost. Economically, whole fishing industries could collapse. Ecotourism operations such as whale shark viewing, which have increasing economic significance for developing countries (and potentially little environmental impact), will also suffer.[17] The crisis has led to multiple suggestions about appropriate legal and regulatory responses at the international level, and about how environmental, trade and fisheries regimes can interact.

---

[13] Josué Martínez-Garmendia and James L Anderson, 'Conservation, Markets, and Fisheries Policy: The North Atlantic Bluefin Tuna and the Japanese Sashimi Market' (2005) 21 *Agribusiness* 17.
[14] Robin McKie, 'Push to Ban Trade in Endangered Bluefin Tuna' *The Observer* (14 February 2010).
[15] Harry N Scheiber, Kathryn J Mengerink and Yann-huei Song, 'Ocean Tuna Fisheries, East Asian Rivalries, and International Regulation: Japanese Policies and the Overcapacity/IUU Fishing Conundrum' (2007) 30 *University of Hawai'i Law Review* 97, 111 (citing the Japanese Ministry for Agriculture, Forest and Fishing).
[16] Carolyn Deere, *Net Gains: Linking Fisheries Management, International Trade and Sustainable Development* (2000) 1991, and sources cited therein.
[17] Note, for example, the proposed or existing ecotourism operations based on Whale Shark viewing in Western Australia (Ningaloo Reef), KwaZulu Natal (South Africa), Mozambique, Philippines, Seychelles, Maldives, parts of the Caribbean, and Gulf of California (Mexico): see CITES Prop. 12.35.

B The forum shop: regimes of relevance

There are a number of regimes of relevance to the problem of endangered marine species. Where such species swim through the high seas and straddle EEZs, UNCLOS and the Fish Stocks Agreement come into play, as do relevant RFMOs. The FAO fisheries management regime, the CMS, the CBD and CITES all contain measures that respond broadly to the view that the threat to marine species is a symptom of poor fisheries management.[18]

The Fish Stocks Agreement aims to enhance UNCLOS provisions relating to conservation and management of the high seas by obliging states to adopt necessary conservation and management measures,[19] yet it is hampered by weak implementation by states and RFMOs.[20] Problems with compliance and 'flag-hopping' are sought to be addressed by the FAO Compliance Agreement.[21]

The FAO fisheries management regime attempts to improve poor fisheries management more broadly by invoking voluntary instruments such as the FAO Code of Conduct and International Plans of Action (IPOA). The FAO IPOA-Sharks, for example, provides a framework for shark conservation and management.[22] Yet the IPOA does not provide sanctions or incentives for states to implement its recommended measures, and there has been a lack of progress in the development by states of national plans to implement the IPOA.[23]

The Convention on Migratory Species (CMS) provides for voluntary measures by 'range' states that are part of a route of listed migratory marine species. For example, there are a number of shark species listed in CMS Appendix II, and participating states are expected to implement a non-binding agreement to conserve and manage migratory sharks that swim through their waters.[24] Similarly, the CBD seeks to protect marine

---

[18] The new Port State Measures Agreement, noted esp. at p. 43, is expected to have a major role in addressing the depletion of endangered marine species, especially through IUU fishing. In addition, a number of RFMOs, including species-specific organisations such as the ICCAT, are increasingly called upon to address the issue. These are not considered in the current chapter.
[19] See pp. 41–3.   [20] See pp. 43–6.
[21] See pp. 47–8.   [22] The FAO IPOA-Sharks (1999) is noted in Chapter 2, n. 122.
[23] See further Holly Edwards, 'When Predators Become Prey: The Need for International Shark Conservation' (2007) 12 *Ocean and Coastal Law Journal* 305, 308; 323–4.
[24] As noted in Chapter 2, n. 222 and accompanying text.

biodiversity resources, which it does by obliging coastal states to protect ecosystems and natural habitats in areas of national jurisdiction.[25]

In response to concerns about a lack of compliance and enforceability of voluntary conservation and management measures, the use of trade measures offers a possible solution. States have turned to the trade mechanisms of CITES to protect certain species. The remainder of this section considers the increased role of CITES, amidst significant opposition, in the conservation and management of marine species.

1 Listing marine species in the CITES Appendices

As described in Chapter 2, the CITES regime regularises the trade in certain species through a licensing system that authorises their import, export, re-export and 'introduction from the sea'. Species threatened with extinction are listed in Appendix I and their trade is only permitted in exceptional circumstances. Appendix II contains species that may become threatened with extinction unless trade is subject to strict regulation. To export these species, exporting states must first obtain advice from their national scientific authorities that such export will not prove detrimental to the survival of the species. Exporting states' management authorities must be satisfied that the specimen was not obtained illegally, and that importing states have valid import permits.[26] Appendix III contains species that are subject to domestic protections because countries have sought assistance from other CITES parties to control external trade. With a coverage of 175 parties, the CITES membership is slightly smaller than the FAO's 191 members,[27] although it dwarfs any one regional fisheries management organisation.

CITES Appendix II, in particular, has potential to enhance fisheries sustainability for a wide range of species. Commercial trade in Appendix II species is only permitted if such trade is not detrimental to the wild population. The requirement that management authorities produce 'non-detriment' certificates in order to trade in listed species is potentially useful in combating IUU fishing.[28] The ongoing review of

---

[25] The areas outside national jurisdiction are less certain: see Christopher C Joyner, 'Biodiversity in the Marine Environment: Resource Implications for the Law of the Sea' (1995) 28 *Vanderbilt Journal of Transnational Law* 635.
[26] CITES, Art. IV; see further Chapter 2, n. 186 and accompanying text.
[27] Note that there are 135 members of the FAO Committee on Fisheries: see Chapter 2, n. 105.
[28] See, e.g., WWF, TRAFFIC, IUCN, 'Regulating International Trade in Commercial Marine Fisheries Products' (February 2002), 4.

significant trade of vulnerable Appendix II species by the CITES Animal Committee, as was done for the queen conch, is valuable for monitoring threats to marine species.[29]

The role of CITES in addressing problems of fish stock sustainability was foreshadowed by CITES parties in 1994 when the ninth COP recognised the conservation threat that international trade poses to sharks.[30] In the twelfth COP in 2002, parties adopted proposals to restrict trade in several commercially exploited marine species amidst vigorous opposition.[31] The Philippines' and India's proposal to list the whale shark and the United Kingdom's proposal to list the basking shark were adopted by the CITES parties; Indonesia, Iceland, Japan, Korea and Norway made subsequent reservations to these Appendix II listings. The whale shark listing was particularly significant because the two countries that proposed it – India and the Philippines – are, together with Taiwan, the significant sites for the catching, landing or trading of the species.[32] The countries may well have decided to seek CITES listing due to the significant ecotourism opportunities in their territory for whale shark viewing.[33] The meeting also provided that the CITES Animals Committee would have involvement in reviewing the FAO's IPOA on Sharks, and parties agreed to list all species of seahorses, which attracted similar levels of controversy.[34]

At the following CITES COP in 2004, the Irrawaddy dolphin was transferred from Appendix II to Appendix I[35] and the great white shark[36] was added to Appendix II. Amongst multiple proposals to expand CITES' coverage of marine species at the COP in 2007,[37] the European eel was

---

[29] See Chapter 2, n. 200 and accompanying text.   [30] CITES Resolution Conf. 9.17.
[31] See generally Chris Wold, 'Natural Resource Management and Conservation: Trade in Endangered Species' (2002) 13 *Yearbook of International Environmental Law* 389.
[32] Vannuccini, above n. 3, Appendix II: Commercially Important Shark Species by Country (by Sei Poh Chen).
[33] For comments on the Philippines' conflicting interests in whale shark meat and ecotourism, see Wold, above n. 31, 393.
[34] See generally Wold, above n. 31, 393-4. Subsequent reservations to the seahorses listings were entered into by Indonesia, Japan, Norway and South Korea.
[35] Japan, Norway, and Gabon opposed this proposal. Japan entered a reservation to the listing.
[36] The listing of the great white shark, which already appeared in Appendix III, was proposed by Australia and Madagascar. Palau entered a reservation in 2004 and Iceland, Japan and Norway entered a reservation in 2005.
[37] Proposals included extending CITES coverage to two other sharks, and certain species of coral, cardinelfish, sawfish and lobster: CITES Press Release, 'CITES Conference to Consider New Trade Rules for Marine, Timber and Other Wildlife Species' (May 2007).

added to Appendix II. The bluefin tuna was suggested for listing at the 2010 COP, which the European Commission, among others, supported,[38] but the proposal was rejected after strong opposition from Japan and other countries.[39]

2 Opposition to the role of CITES

Opposition to the role of the CITES in restricting the trade in endangered marine species has been directed to CITES' mandate, expertise, alleged 'structural bias' and the risk of conflicting rules and procedures in fisheries management.

(a) Mandate

States have expressed concern about the legal applicability of CITES with respect to marine species. Concern about CITES' mandate has focused on alleged uncertainties surrounding the express protection for marine species in the Convention text. The Convention allows parties to restrict the trade in any species provided that the listings are agreed by the requisite majority of parties voting at the COP.[40] 'Trade' is defined as the 'export, re-export, import and introduction from the sea'.[41] The phrase 'introduction from the sea' is further defined as 'transportation into a State of specimens of any species which were taken in the marine environment not under the jurisdiction of any State'.[42] CITES requires that states that introduce from the sea specimens of species listed in Appendices I and II must verify that the introduction will not be detrimental to the survival of the species involved.[43]

The inclusion of marine species in CITES has been the subject of considerable controversy dating from its original inception. During the original negotiations leading to CITES, Japan proposed the deletion of all references to 'introduction from the sea',[44] apparently due to concerns that whaling would become subject to CITES protections

---

[38] See (2009) 9:16 *Bridges Trade BioRes* (18 September 2009); (2010) 10:2 *Bridges Trade BioRes* (5 February 2010). Sweden and Kenya both sought unsuccessfully to list the Atlantic bluefin tuna on CITES in 1992 and 1994: See Erik Franckx, 'The Protection of Biodiversity and Fisheries Management: Issues Raised by the Relationship Between CITES and LOSC' in David Freestone, Richard Barnes and David Ong (eds.) *The Law of the Sea: Progress and Prospects* (2006) 210, 216.

[39] (2010) 10:5 *Bridges Trade BioRes* (19 March 2010).   [40] CITES, Art. XV.   [41] Ibid., Art. I(c).
[42] Ibid., Art. I(e).   [43] Ibid., Art. III(5) and Art. IV(6), (7).
[44] CITES, *travaux preparatoires*, Summary Record – Tenth Plenary Session, Tuesday, 20 February 1973 (Doc. SR/10 (Final) 5 March 1973) 3, cited in Franckx, above n. 38, 224–5.

rather than its preferred forum, the International Convention for the Regulation of Whaling.[45] Australia, by way of compromise, proposed the inclusion of a clause to give deference to existing agreements that afforded protection to marine species.[46]

Although the parties to CITES agreed that it should extend protection to threatened marine species, the parties were unsure about how to conceptualise the way listed marine species entered international trade given that negotiations about sovereign coastal boundaries were ongoing. As described in Chapter 2, an EEZ of 200 miles was not finalised until the third Conference on the Law of the Sea was completed in 1982,[47] and CITES expressly stated that it was without prejudice to these negotiations.[48] In 1979, the CITES COP adopted a resolution that stated that the jurisdiction with respect to marine resources in maritime areas adjacent to the coast of CITES parties 'is not uniform in extent, varies in nature and has not yet been agreed internationally'.[49] Participating states remained uncertain about whether 'introduction' occurred when a fishing vessel took a fish specimen on board (so that the flag state was the state of introduction) or whether introduction occurred when the fish was landed in a port. The practice of processing fish on board added to this complexity.[50] These concerns led to the 2002 recommendation of the Sub-Committee on Fish Trade that FAO members adopt a work plan to investigate possible problems.[51]

(b) Expertise

Some FAO members have expressed concerns that the CITES regime has insufficient expertise in fisheries management and that these and other factors will lead to increased costs of implementation of CITES' listings. The concerns have focused on alleged difficulties in listing marine species, as compared to land-based species. Information and understanding about the threat to marine species is important not only in decisions to add a particular marine species to one of the appendices, but in non-detriment findings and other assessments on viability.

---

[45] Statement by Delegate of Japan on 'Introduction from the Sea' (Doc. PR/11, 21 February 1973) 2–3, cited in Franckx, ibid., 224.
[46] CITES, Art. XIV(4), (5) (these provisions are noted in Chapter 2, n. 246 and accompanying text).
[47] See Chapter 2, p. 33.   [48] CITES, Art. XIV(6).
[49] CITES Resolution Conf. 2.8 (1979) (now included in Conf. 11.4 (Rev. CoP12).).
[50] *FAO Fisheries Report No. 746* (2004) Report of the Expert Consultation on Legal Issues Related to CITES and Commercially-Exploited Aquatic Species, para. 18.
[51] FAO, *Report No. 673* (2002), para. 22 and Appendix F. See also Franckx, above n. 38.

As described in Chapter 2, the CITES COP is aided in its work by a number of committees that provide scientific and technical expertise.[52] Although the Animals Committee has expertise in biology, its experience in marine matters has been minimal.[53] In addition, it is difficult for CITES parties and the committees to access information about marine species, especially for migratory species and species on the high seas, making decisions about proposals for listing and compliance measures very difficult.[54] This lack of information hampers the assessment of viability of trade for Appendix II species, which, as described above, is conditional on the production of a 'non-detriment' finding. CITES relies on states to provide non-detriment findings, yet the relevant national scientific authorities are often under-resourced or even non-existent.[55]

(c) Bias

Apart from concerns about mandate and expertise, states have raised concerns that CITES is institutionally biased against the sustainable utilisation of fisheries. These can be grouped broadly into two complaints with the CITES regime: first, some states make allegations about a structural bias against science; second, some states make claims about a structural bias against economic development. The latter group of complaints focus particularly on an alleged bias against 'utilisation' of marine species and consequent insensitivities towards the needs of developing countries.

The first set of concerns relate to CITES' use of science. Some states have argued within the FAO that the listing of species in the Appendices is not sufficiently rigorous. As described in Chapter 2, the listing of species in the Appendices is conducted according to particular listing criteria.[56] These criteria include scientific methods to assess population levels for allegedly threatened species. Within the FAO, states have expressed concern that these methods do not adequately account for the natural fluctuations of fish stocks or the requirement in UNCLOS

---

[52] See above Chapter 2, n. 197.
[53] Laura Little and Marcos A Orellana, 'Can CITES Play a Role in Solving the Problems of IUU Fishing?: The Trouble with Patagonian Toothfish' (2004) 16 *Colorado Journal of International Environmental Law and Policy* 21, 77.
[54] Ibid.
[55] James B Murphy, 'Alternative Approaches to the CITES 'Non-Detriment' Finding for Appendix II Species' (2006) 36 *Environmental Law* 531, 534.
[56] See CITES listing criteria (2007) in Chapter 2, n. 192.

for the use of the best scientific advice.[57] A related concern is that the process of transferring species within the Appendices is too onerous. CITES parties can vote to transfer species from Appendix I to Appendix II ('down-listing') and remove species from the Appendices ('deletion') in circumstances when species numbers have recovered, but must exercise precaution in cases of scientific uncertainty.[58] Within the FAO, some states have expressed concern that this approach will make it very difficult to down-list commercially exploited marine species.[59] They have asserted that proposals for down-listing are more onerous than the listing itself, and that this could lead to several commercially exploited marine species being wrongfully retained in the Appendices.

The second set of concerns that has been expressed within the FAO is that the CITES regime is institutionally biased against the concept of sustainable utilisation in favour of its primary objective of preventing a species from becoming endangered.[60] This implies that in listing a species in Appendix II, CITES' conservation objectives are insufficiently addressed to socio-economic factors. This concern has been framed, in particular, in the context of the employment, income and food security needs of developing countries, which might be disproportionately affected by a CITES listing.[61]

(d) Coherence in fisheries management

A dominant theme in the complaints against CITES' role is that there will be a lack of coherence in fisheries conservation and management, and the potential for conflicting rules and practices. For example, some FAO members that are coastal states have pointed to the possibility that their rights and obligations under UNCLOS will be different, and abrogated, by a CITES listing of a species.[62]

Some states have pointed to conflicts between a CITES listing of marine species and alleged UNCLOS concepts of freedom of fishing in

---

[57] See, e.g., *FAO Fisheries Report No. 667* (2001) Second Technical Consultation on the Suitability of the CITES Criteria for Listing Commercially-Exploited Aquatic Species, Namibia, 22–25 October 2001.
[58] CITES, Art. XV; Annex 4 of Resolution Conf. 9.24 (Rev CoP14).
[59] FAO Expert Consultation on Legal Issues (2001), above n. 50, para 31.
[60] See FAO Expert Consultation on Legal Issues (2001), above n. 50, para. 25.
[61] FAO, *Report No. 673* (above n. 51), para. 16; *FAO Fisheries Report No. 741* (2004) Report of the Expert Consultation on Implementation Issues Associated with Listing Commercially-Exploited Aquatic Species on CITES Appendices, paras. 79–80.
[62] FAO Expert Consultation on Legal Issues (2001), above n. 50.

the high seas.⁶³ They argue that a conflict could occur where an Appendix I species could be lawfully harvested on the high seas but not allowed to be introduced into the port of a CITES party.⁶⁴ A further concern has focused on the freedom within UNCLOS for coastal states to set allowable catch limits within their EEZ,⁶⁵ which some states consider to be potentially restricted by a CITES listing under Appendix II. That Appendix, it will be recalled, covers species 'in order to avoid utilization incompatible with their survival'.⁶⁶ Trade in these species is conditional on a finding that the trade will not be detrimental to the wild population. But even if a non-detriment finding is made, CITES preserves the right of parties to implement stricter domestic measures for Appendix II species, including complete prohibition of imports. Such trade measures, it is claimed, could conflict with the freedom of coastal states to set catch limits.⁶⁷

A further concern of FAO members relates to the applicability of the CITES 'look-alike' clause to the fisheries context. When species are listed in Appendix II of CITES, the Convention extends this listing to 'look-alike' species.⁶⁸ Listing of species that resemble the listed species is necessary for effectiveness and allows customs officials to check specimens without having to distinguish between closely resembling species. When all thirty-two species of seahorses were listed in Appendix II at COP 12, twenty-six were included under the look-alike provision.⁶⁹

Some FAO members have expressed concern that the application of the 'look-alike' clause will have negative impacts on the fishing industry and fishing communities, presumably because more species than necessary will be listed on this basis. They have also expressed concern that the regime will require large levels of monitoring and control.⁷⁰ A similar concern relates to the alleged incompatibility between the CITES regime and aquaculture. Aquaculture is a sector of growing economic importance in fisheries, and the query has been voiced within the FAO about whether CITES can cope with the distinction between wild and farmed fisheries.⁷¹

---

⁶³ Ibid., para. 38.   ⁶⁴ Ibid.   ⁶⁵ UNCLOS, Arts. 61, 297.
⁶⁶ CITES, Art. II(2).   ⁶⁷ CITES, Art. XIV(1).
⁶⁸ CITES, Art. II(2)(b) applies to 'other species which must be subject to regulation in order that trade in specimens of certain species referred to in sub-paragraph (a) of this paragraph may be brought under effective control'. See further CITES listing criteria (2007) (noted in Chapter 2, n. 192) Annex 2b.
⁶⁹ *FAO Fisheries Report No. 741* (above n. 61), para. 50.
⁷⁰ Ibid., para. 49.   ⁷¹ Ibid., para. 85(9).

An additional concern has been raised by some states about the listing of species in more than one Appendix of CITES ('split-listing').[72] Split-listing aims to accommodate the fact that species may be threatened in one region but not another. For example, when the minke whale was added to the CITES Appendices, it was listed in Appendix I except for the population of West Greenland, which was listed in Appendix II. FAO members have expressed concern that split-listing practices will require increased resources and will lead to the listing of some fish populations in Appendix II when they would otherwise not be included.[73]

3 Responses

The opposition to CITES' role may be met with a number of responses, which are supported by the text of the Convention, but also require a detailed and contextual analysis of state practices and adaptability.

(a) Textual mandate

As described above, some states have expressed concern about the mandate of CITES and uncertainties surrounding marine species that are 'introduced from the sea'. CITES parties sought to resolve this complexity when, in 2007, they agreed on a resolution affirming the applicability of UNCLOS provisions determining jurisdiction of marine areas.[74] The CITES reference to 'the marine environment not under the jurisdiction of any State' was agreed to mean 'marine areas beyond the areas subject to the sovereignty or sovereign rights of a State consistent with international law, as reflected in [UNCLOS]'.[75] On this basis, CITES regulation of trade extends to listed marine species that are 'introduced' into a state from a national vessel fishing in the high seas. Read together with the general definition of trade, CITES provisions thus apply to the introduction for *domestic* consumption of species from the high seas, in addition to the general import, re-export and import of *all* listed marine species (whether taken from within the EEZ or the high seas).

Given this definition, CITES will apply to both the international trade in fish and fishing for domestic consumption from the high seas. It will

---

[72] CITES, Art. I allows species to be identified according to geographically separate populations.
[73] FAO Expert Consultation on Implementation (2004), above n. 61, para. 61.
[74] CITES Resolution Conf. 14.6 (2007).   [75] Ibid.

not generally be used, however, for species that are harvested within a state's EEZ and consumed within that state; somewhat paradoxically, if there is no 'market' for the fish beyond this domestic consumption, the species may become overfished with no available CITES protection.

Notwithstanding apparent agreement over the synergies in maritime jurisdictional provisions of CITES and UNCLOS, controversy about CITES' role in fisheries has remained. States opposing the resolution in 2007 pointed out that UNCLOS did not enjoy universal jurisdiction and claimed that it represented a 'creeping jurisdiction' or 'sovereignization' of the sea.[76] They remained unhappy about the fact that the 'enclosure' of the EEZ in UNCLOS was being replicated by CITES – presumably giving an advantage to coastal states that could fish for listed CITES species within their EEZ and, provided they do not export the species, escape the CITES requirements. As such, the reconciliation of the CITES regime and the EEZ regime through a conference resolution agreed by CITES parties was unfavourable for some.

(b) Policy adaptation

Apart from concerns about CITES' textual mandate, some states alleged underlying legal conflicts between UNCLOS and CITES. Yet such claims are overstated, at least in legal terms. For example, the claim that a CITES listing would abrogate an UNCLOS-sanctioned freedom to fish on the high seas fails to acknowledge that the freedom to fish is heavily qualified.[77] In addition, the concern of coastal states that their right to set allowable catch limits within their EEZ would be affected by an Appendix II listing is overblown given that CITES does not interfere with the coastal state's harvesting, for domestic consumption, of species within its own EEZ.[78]

The concerns reveal not just misunderstandings about UNCLOS freedoms, but also about trade freedoms. For example, experts consulted by the FAO considered that a potential conflict would arise if a CITES party challenged a non-detriment finding made by other CITES parties and instituted stricter domestic measures that were inconsistent with the rights of coastal states to establish allowable catch limits within the EEZ.[79] This claim essentialises the UNCLOS, CITES and (implicitly) WTO

---

[76] See CITES SC54 Doc. 19, p. 4.
[77] See further Chapter 2, esp. p. 39. This was also the conclusion of the FAO Expert Consultation on Legal Issues (2001), above n. 50.
[78] FAO Expert Consultation on Legal Issues (2001), above n. 50, para. 34.   [79] Ibid.

regimes. The UNCLOS provision does not guarantee a right of coastal states to export. Nor does CITES 'allow' for stricter domestic trade measures, as suggested by the experts consulted by the FAO. Instead, CITES preserves any existing right of states to adopt such measures, a right that is also preserved by the WTO regime.[80]

The concerns about the CITES 'look-alike' clause and the 'split-listing' of species are also based on rather impressionistic understandings of the CITES and FAO fisheries management regimes. The practice of CITES in listing 'look-alike' species can adapt to the special situations arising in fisheries. Such adaptation is not new to the FAO, which is well-versed in the challenges of monitoring and controlling traded fish specimens. The FAO Sub-Committee on Fish Trade continues to consider initiatives such as traceability and labelling schemes. Indeed, when the FAO consulted experts in 2004 about CITES implementation issues, the experts noted that CITES and the FAO could collaborate on these initiatives as alternatives to the look-alike clause.[81] There is also sufficient flexibility with the CITES permit procedures to adequately accommodate fish stocks reared using aquaculture methods.[82] As such, the look-alike clause is not fundamentally inappropriate for fisheries but merely requires increased understanding of fisheries management methods.

Moreover, the notion of 'split-listing' species between Appendices is not unknown in fisheries management regimes. Addressing different degrees of endangerment for species according to different locations is a significant and ongoing challenge. Methods include the marking of fish stocks to identify and differentiate specimens.[83] Adapting these methods for the CITES licensing system may well require collaboration and additional resources, but is not a basis for rejection if there are gains for fisheries conservation and management. The concerns raised about achieving a policy fit between the CITES and FAO fisheries management regime thus point to questions about how the regimes can interact for the betterment of the conservation and management of endangered marine species.

(c) Institutional bias and insensitivities

The concerns about alleged insensitivities and biases within the CITES regime have suggested an irreconcilable bias regarding core concepts of science and economics. At first glance, the concerns about science

---

[80] See e.g. discussion of GATT, Art. XX, in Chapter 2, pp. 75-6.
[81] FAO Expert Consultation on Implementation (2004), above n. 61, para. 85(6).
[82] Ibid., para. 85(9).   [83] Ibid., para. 59.

appear to demonstrate a direct conflict between the rationalities of science and the principle of precaution. This conflict is apparent in many examples of regime interaction between the WTO and environmental law. For example, the Biosafety Protocol to the CBD enshrines a precautionary approach in order that states importing GM products exercise precaution in the face of scientific uncertainty, yet the application of the precautionary principle in the WTO is far from certain.[84] This conflict between attitudes to science has been central to disputes involving import restrictions on a range of products such as hormone treated beef and genetically modified food.[85]

There is a danger, however, that in conflating concerns about the use of science into conflicts between regimes, the complexities within regimes themselves are ignored. Within the regime of the law of the sea, for example, the precautionary approach is endorsed in the Fish Stocks Agreement[86] and the FAO Code of Conduct.[87] In addition, fisheries laws and institutions recognise the difficulties of producing and interpreting scientific data. For example, the FAO's role in providing information and recommending research with respect to fisheries and conservation is dependent on the provision of information by states and RFMOs.[88] Yet, while states have obligations to report fisheries statistics under the FAO Constitution,[89] the FAO acknowledges the disincentives for fishers to report on sensitive commercial information about the location and numbers of fish.[90] These examples suggest that the precautionary approach is often followed in the law of the sea. As such, the suggestion that CITES is inappropriate because it applies a precautionary approach is incorrect.

The second set of concerns about bias focused on the CITES regime insensitivities about the need for economic development. Yet the

---

[84] Panel Report, *EC – Biotech*, para. 7.89; AB Report, *EC – Hormones*, paras. 123–124.
[85] Panel Report, *EC – Biotech*; AB Report, *EC – Hormones*; see further Joanne Scott, *The WTO Agreement on Sanitary and Phytosanitary Measures: A Commentary* (2007).
[86] Fish Stocks Agreement, Art. 6.   [87] FAO Code of Conduct, Art. 6.5.
[88] The responsibilities of states and RFMOs for collecting and exchanging data for stock assessments are set out in the Fish Stocks Agreement (Art. 12), the FAO Code of Conduct (Art. 6.4) and the Compliance Agreement.
[89] FAO Constitution Art. XI.
[90] FAO, *Strategy for Improving Information on Status and Trends of Capture Fisheries* (Strategy-STF) Rome, 2003, approved by consensus at COFI on 28 February 2003. See also FAO, *The State of World Fisheries and Aquaculture 2002* (SOFIA) 61 ('Deliberate misreporting or non-reporting by legal and illegal fishers and other participants (processors, traders) is cited by most managers as a key problem').

presentation of concerns about economic biases as a conflict between institutions misstates the issue of resource utilisation within both CITES and the fisheries regime. Within CITES, for example, the issue of 'sustainable utilisation' has been debated for many years. Parties' growing awareness of the lack of resources for conservation within developing countries led to a move away from an ethos of strict preservation to one of sustainable utilisation. Revisions to the CITES listing criteria reflect this change,[91] and the tension between preservation and utilisation continues to inform proposals for listing and lobbying between states. In the law of the sea, the notion of 'sustainable utilisation' is also not uncontested. For example, there is increasing recognition of an ecosystem approach to utilisation in such instruments as the FAO Code of Conduct.[92] This approach recognises that the allocation of 'maximum sustainable yields' for different species may still lead to degradation if the rest of the ecosystem is insufficiently protected. The ecosystem approach is adopted by the CBD.[93] As such, the suggestion that there is a conflict between the norms of 'utilisation' and 'conservation' in the meeting of fisheries and environmental regimes is an oversimplification.

It is also a misstatement to suggest that the law of the sea and CITES are in conflict due to their treatment of the needs of developing countries. UNCLOS, the FAO Code of Conduct, the CBD and CITES all formally endorse the need for special attention to the interests of developing countries.[94] The suggestion that the needs of developing countries are insufficiently met in CITES vis-à-vis fisheries regimes also fails to acknowledge the complexity of interests of developing countries. Developing-country interests are plural and complex. While it is true that the fisheries sector is of major economic importance to developing

---

[91] CITES listing criteria (2007), noted in Chapter 2, n. 192. Cooney discusses how initial prohibitions on the trade of species such as elephants had counterproductive effects in poor areas of developing countries because there were minimal incentives for local communities to protect species that could not be legally traded: Rosie Cooney, 'CITES and the CBD: Tensions and Synergies' (2001) 10 RECIEL 259; see also Timothy Swanson, 'The Evolving Trade Mechanisms in CITES' (1992) 1 RECIEL 57 (arguing that CITES failed to recognise that habitat loss and population pressures were the main factors in species extinction).

[92] FAO Code of Conduct, Art 6.2; FAO Reykjavik Declaration (2001); World Summit Plan of Implementation (2002) para. 30(d). For background, see Stuart Kaye, *International Fisheries Management* (2001), 267–86.

[93] See Chapter 2, n. 237.

[94] See, e.g., UNCLOS, Preamble, Art. 62; Fish Stocks Agreement, Part VII (Arts. 25, 26 and 27); FAO Code of Conduct, Art. 5.2; CBD, Preamble, Art. 1. For the approach within CITES, see above n. 91.

countries,⁹⁵ such interests are not uniformly directed towards resource exploitation and trade. For example, the CITES proposal to list the whale shark by India and the Philippines may well be attributed to national interests in ecotourism.⁹⁶

It may be that some developing countries will indeed wish to avoid the listing of marine species, either because they have fishing industries that exploit the species or because they wish to provide lucrative licences to distant-water fishing vessels through access agreements. Other developing countries, however, will welcome listing under CITES but require technical and financial assistance for any resulting administrative burdens.

(d) Overlap, not normative conflict

My consideration of the concerns about CITES' mandate, expertise and institutional bias suggests that, for the protection of endangered marine species, the relationship between the regimes of the law of the sea and CITES is one of overlap rather than normative conflict. This finding is in contrast to other legal analyses. For example, Franckx considers the central issue of analysing the listing of commercially exploited marine species in CITES to be the 'application of successive treaties relating to the same subject-matter in general international law'.⁹⁷ His point of reference is a marine species that is covered by two or more instruments, and he gives the example of Atlantic bluefin tuna on the high seas, the regulation of which is addressed in UNCLOS, the FAO Compliance Agreement, the Fish Stocks Agreement and, potentially, CITES.⁹⁸

The central question for Franckx is to assess the priority of CITES vis-à-vis the other instruments and ask which instruments should take precedence. In answering this question, Franckx endorses the steps of first looking to conflict clauses, then using treaty interpretation to determine priority and, if these steps are inconclusive, applying the *lex posterior* rule.⁹⁹ These steps are consistent with the ILC's tool-box for

---

⁹⁵ See chapter 1 n. 105 and accompanying text.
⁹⁶ See above n. 33 and accompanying text. Note also the 'global' nature of migratory fish resources, which contrast in practical terms with land-based species such as the elephants, whose listing on CITES was perceived by some as 'stealing opportunities' from African nations.
⁹⁷ Franckx, above n. 38, 216.   ⁹⁸ See above n. 38 and surrounding text.
⁹⁹ Ibid., 228 (citing J B Mus, 'Conflicts Between Treaties in International Law' (1998) 45 *Netherlands International Law Review* 208, 231).

dealing with problems of fragmentation. Franckx notes that CITES predates many fisheries instruments, and applies VCLT Article 30 and the conflict clauses in UNCLOS, CITES, the Fish Stocks Agreement and other instruments to determine the relationship between them. Rather than stating which treaty has priority, Franckx concludes by noting the difficulties of applying the conflict clauses in practice and stating that much will depend upon the intentions of the parties.[100]

Franckx's conclusions are unsurprising given the difficulties of applying principles to determine the precedence between conflicting treaties. They support my suggestion that the ILC's tool-box may be inutile for addressing situations of ongoing diversity and overlapping mandates in fisheries governance.[101] More importantly, however, his starting assumption should be tested. Franckx assumes that there is a conflict between the relevant treaty provisions. As described above, however, the concerns about CITES' role in fisheries management reflect concerns about institutional expertise and bias but do not demonstrate conflicting norms. His approach also varies from the presumption against conflict, which requires an assumption that norms are in harmony unless disproved by evidence of real conflict.[102] A presumption both *for* or *against* conflict has consequences for continuity and change in international law.[103] Such presumptions therefore need to be dealt with openly, and with an awareness of the likely impact for dominant regimes within fisheries governance.

Instead of determining the priority of norms, the more pressing concern is to ascertain how CITES, UNCLOS, the Fish Stocks Agreement and the FAO instruments coexist in an effective way. Indeed, Franckx concedes this when he concludes his analysis of the priority of norms in fisheries management by observing that increased cooperation by regimes is 'the most urgent and overarching need'.[104] This leads to an

---

[100] Franckx, above n. 38, 231-2.
[101] Note the presentiment of the ILC itself on establishing the priority of norms from different regimes: see Chapter 1, n. 54 and accompanying text.
[102] See Wilfred Jenks, 'Conflict of Law-Making Treaties' (1953) 30 *BYIL* 401, 427-9.
[103] Joost Pauwelyn, *Conflict of Norms in Public International Law: How WTO Law Relates to Other Rules of International Law* (2003), 242-4 (warning of the limits of the presumption against conflict, because if one presumes an absence of conflict, one might be biased towards continuity and away from change).
[104] Franckx, above n 38, 232, paraphrasing Birnie and Boyle's analysis of CITES and land-based species, Patricia Birnie and Alan Boyle *International Law and the Environment* (2002) 634-5.

THE RESTRICTION OF TRADE IN ENDANGERED MARINE SPECIES   153

inquiry into the interaction between the regimes, and particularly the relevant stakeholders of the FAO and CITES.

(e) From forum shopping to regime interaction

My analysis of the opposition to CITES' role in fisheries management leads from questions of the correct allocation of forum to questions of regime interaction. Interaction between stakeholders from the relevant regimes can remedy uncertainties about CITES' competence to deal with marine species. For example, staff members from the FAO participated in discussions leading to the CITES resolution on 'introduction from the sea', in which CITES parties confirmed the relevance of UNCLOS in determining jurisdictional limits.[105] The perceived problems with CITES' practices in listing look-alike species, splitting species between Appendices and listing farmed fish can be addressed by enhancing expertise within the CITES regime. So, too, can perceptions of institutional bias. In this regard, open and accessible interaction between regimes can promote the examination of alleged institutional insensitivities such as differences in the use of scientific evidence of fluctuating fish stocks.

The CITES regime already envisages inter-regime collaboration with fisheries organisations. There is a duty for the CITES Secretariat to consult with CITES parties, the FAO and relevant RFMOs on proposals by CITES parties that are related to marine species.[106] This consultation is aimed at obtaining scientific data and 'ensuring coordination with any conservation measures enforced by such bodies'.[107] The CITES Secretariat is required to 'communicate the views expressed and data provided by these bodies and its own findings and recommendations' to the CITES parties before they vote on the proposals.[108] Within the FAO, the Code of Conduct encourages states to cooperate in complying with relevant international agreements regulating the trade in endangered species.[109]

Revisions to the listing criteria agreed by CITES parties are suggestive of a developing attitude among states to regime interaction.[110] The listing criteria's former acknowledgement of the competence of certain IGOs in marine species management was deleted by CITES parties in 2002.[111] In its place, the parties recall the inter-agency consultation

---

[105] CITES CoP14 Doc. 18.1 (detailing CITES workshop).   [106] CITES, Art. XV.   [107] Ibid.
[108] Ibid.   [109] FAO Code of Conduct, Art. 11.2.9.   [110] CITES listing criteria (2007).
[111] See further CoP12 Doc. 58, Annex 3; CoP13 Doc. 57.

envisaged by CITES Article XV.[112] As such, the CITES parties have chosen to promote regime interaction rather than an *ex ante* assessment of competences.

Similarly, the listing criteria formerly required proposing parties to 'consult in advance with the relevant competent intergovernmental organisations responsible for the conservation and management of the species, and take their views fully into account', as well as provide detailed information on current trade and conservation efforts.[113] This requirement was amended by the COP in 2004. Parties are now required to provide details of relevant international instruments and provide an assessment of their effectiveness in ensuring conservation, protection and management.[114]

These amendments to the listing criteria suggest that CITES parties are reluctant to follow a predetermined allocation of competence vis-à-vis CITES and fisheries IGOs but prefer to assess the capability and adaptability of regimes. This finding resonates with my analysis of the treatment of fisheries subsidies in Chapter 3, which demonstrated how opposition to the role of the WTO gave way to inter-regime learning and the entrenchment of ongoing interaction between the WTO, FAO and other organisations and participants. As such, my emphasis moves to the adaptability of the CITES regime to the challenges of listing marine species and the cooperative relationship between the relevant institutions. Efforts for collaboration and cooperation between the two regimes have coalesced around the negotiation of a single and highly contested instrument: the Memorandum of Understanding (MOU) between the FAO and the CITES Secretariat.

C The Memorandum of Understanding between CITES and the FAO

The increased demand for CITES to conserve and manage marine species has led to proposals to formalise the interaction between the CITES regime and fisheries regimes. While these attempts are not intended to

---

[112] CITES listing criteria (2007), Preamble ('RECOGNIZING further that the Secretariat, in accordance with [Art. XV], shall consult intergovernmental bodies having a function in relation to marine species').
[113] CITES Conf. Resolution 9.24 at ninth Conference of the Parties (1994) Annex 6, para. 4.1.2.
[114] See CITES CoP12 Doc. 58, Annex 3; CITES CoP13 Doc. 57. See further Chapter 2, n. 193 and accompanying text.

make new law, as was the case with the fisheries subsidies rules considered in Chapter 3, they have aimed to produce an institutional agreement in the form of an MOU between the CITES Secretariat and the FAO. This section first describes the growing trend for such agreements between international organisations. In examining their legal status, I refer to principles of international institutional law to assess the express or implied powers of the FAO and CITES Secretariats to enter into an MOU. Next, I investigate how the MOU was negotiated. While the MOU is an apparent attempt to strengthen institutional collaboration, other strategic aims, such as the allocation of primacy of one regime over another in fisheries matters, may have animated drafting proposals; such strategies become more clear with a detailed examination of the evolution of the MOU. In the light of the jostling for legal primacy between regimes, I next examine the substantive and procedural constraints on the MOU. Finally, I consider the national policy coordination that allowed states to agree on a negotiating position that reconciled interests between disparate domestic agencies such as fisheries and conservation. Such coordination was important for finalising the MOU, but is also integral to wider interaction between the FAO fisheries management and CITES regimes.

1 Legal status

Agreements setting out means of cooperation and collaboration are increasingly used by IGOs. They are seen to be of particular value in ocean matters. The UN Consultative Process, which identifies areas where coordination and cooperation in ocean matters should be enhanced,[115] encourages organisations 'to conclude memorandums of understandings with a view to avoiding duplication of work and to designate focal points'.[116] MOUs between organisations are part of a wider corpus of agreements that involve IGOs, member states and non-member states.[117]

The capacity of IGOs to enter into MOUs is generally accepted,[118] although much depends on questions of legal personality, express or implied powers and the resulting character of the agreement itself. In contrast to the general view that an IGO must have international legal

---

[115] See pp. 53-5. [116] See UN Consultative Process, UN Doc A/63/174 (2008) para. 133.
[117] See Henry Schermers and Niels Blokker, *International Institutional Law* (2003), 1114–17 (paras. 1748–50).
[118] Ibid., and sources referred to therein.

personality before entering into agreements with other organisations, some commentators have suggested MOUs are contracted without international legal personality.[119] Many IGOs have express powers to enter into international agreements, but for others the powers will be implied based on their functions and competences.[120] On this basis, MOUs between international organisations must be confined to whatever is necessary for the organisations to perform their functions, as expressed in their constitutions. However, according to others, MOUs are not binding rules and are voluntary collaborations of a 'purely technical and financial' nature.[121]

The FAO has full legal status,[122] and its constituting documents expressly envisage that it will enter into relations with other international organisations.[123] Members have approved 'guiding lines' (sic) regarding relationship agreements between the FAO and IGOs,[124] which note the desirability of concluding such formal agreements. The guidelines contain criteria to be applied by the FAO for recognising the intergovernmental character of the partner IGO, which provide that the relevant IGO should be set up by a treaty between states, should be composed of members designated by governments and should receive income mainly, if not exclusively, from governments. The guidelines state that cooperative agreements should include methods of liaison and cooperation, such as information-sharing, reporting, a mutual right of agenda-setting and arrangements for reciprocal representation at meetings.[125]

The powers of the CITES Secretariat to enter into collaborative arrangements with other IGOs are less settled. The CITES treaty text authorises the Secretariat 'to arrange for and service meetings of the Parties' and 'to perform any other function as may be entrusted to it' by the parties.[126] By contrast to the FAO constitutive documents, and other

---

[119] Pieter Jan Kuijper, 'Some Institutional Issues Presently Before the WTO' in Daniel Kennedy and James Southwick (eds.) *The Political Economy of International Trade Law: Essays in Honour of Robert E Hudec* (2002) 81, 108 (referring to agreements between the WTO Secretariat and other international Secretariats).

[120] On implied powers, see *UN Reparations* (1949) ICJ Rep 174.

[121] Kuijper, above n. 119. [122] FAO Constitution, Art. XVI (1).

[123] Ibid., Art. XIII(1). See also *Vienna Convention on the Law of Treaties between States and International Organizations or between International Organizations* (1986) (not yet in force). The FAO became a signatory to the agreement on 29 June 1987.

[124] FAO Guiding Lines Regarding Relationship Agreements Between FAO and IGOs (Resolution of tenth session of FAO Conference).

[125] Ibid., para. C. [126] CITES, Art. XII(2)(a), (i).

MEAs such as the Convention on Biological Diversity,[127] the CITES treaty text does not make express reference to international personality. The COP is authorised to 'make such provision as may be necessary to enable the Secretariat to carry out its duties'.[128] To date, the COP has not made express reference to legal personality.[129]

Although the CITES treaty text does not provide express legal personality, the CITES Secretariat operates under the assumption that it has international personality and the legal capacity to enter into MOUs.[130] It has regularly entered into MOUs with other Secretariats.[131] However, there is uncertainty about this practice. Even if it is accepted that the autonomous institutional arrangements of the CITES regime give rise to implied powers, it is not clear, for example, whether the legal personality resides in the CITES Secretariat, the COP itself or even the international organisation hosting the CITES Secretariat, namely UNEP.[132] In 2006, the CITES Standing Committee proposed a draft resolution for the fourteenth COP which would recognise the independent international legal personality of the CITES Secretariat and entrust the Secretariat to 'coordinate with other relevant international bodies and, in particular to enter into such administrative and contractual arrangements as may be required for the effective performance of its functions'.[133] This draft resolution has not appeared on the agenda of subsequent Conferences of the Parties. Uncertainty remains about the legal basis for the CITES regime to enter into formal institutional arrangements of coordination with other regimes.

If it is accepted that the CITES and FAO Secretariats are empowered to enter into the MOU, the limitations of their express or implied powers should be noted because such limitations may constrain the substance of the MOU. There is a presumption that international organisations act

---

[127] CBD, Art. 24(d).   [128] CITES, Art. XI(3)(a).
[129] CITES SC54 Doc. 8 para. 7.   [130] Ibid., para. 6.
[131] See, e.g. Memorandum of Cooperation between the CBD and CITES noted in Chapter 2, n. 239.
[132] The COP model utilised by many MEAs has been assessed by commentators as having the character of an IGO: Robin Churchill and Geir Ulfstein, 'Autonomous Institutional Arrangements in Multilateral Environmental Agreements: A Little-Noticed Phenomenon in International Law' (2000) 94 *AJIL* 623, 632–3. The authors further note that an interpretation that the international organisation hosting the MEA Secretariat does not have international legal personality to enter into collaborative arrangements is not tenable: 655. See also Philippe Sands, *Principles of International Environmental Law* (2003) 109.
[133] CITES SC54 Doc. 8.

within their powers,[134] but if the FAO and CITES Secretariats were to conclude cooperative arrangements that exceeded their mandates, they would be acting ultra vires. With this constraint in mind, the following section examines the evolution of the MOU.

2 Evolution of the MOU

There were seven drafts of the MOU agreed over the course of three years of meetings:

   (i) CITES Standing Committee Chair-led draft (2003);[135]
  (ii) FAO COFI-led draft (2003) (also known as the Friends of the COFI-chair text);[136]
 (iii) CITES draft (2004);[137]
  (iv) FAO COFI-FT draft (2004) (also known as the Japan–USA compromise text);[138]
   (v) FAO–CITES Secretariat negotiated draft (2004);[139]
  (vi) CITES Standing Committee negotiated draft (2004) (also known as the Norway–Australia compromise text);[140] and
 (vii) Final text (2006).[141]

The drafts reveal differing perspectives and strategies about CITES' role in fisheries conservation and management, as I describe below.

(a) Proposals to limit CITES' role

In February 2002, a meeting of the FAO's Sub-Committee on Fish Trade provided a forum for some states to express their concern about the impact of CITES on the areas of competence of the FAO and RFMOs. After discussing the increased inclusion of marine species in the CITES

---

[134] *UN Expenses* (1962) ICJ Rep 151, 168.   [135] Reproduced at CITES SC49 Doc. 6.3.
[136] FAO Committee on Fisheries, Rome, 2003. Consensus was not reached on this text: see *FAO, Fisheries Report No. 702* (2003) para. 48. Reproduced at Appendix G of that Report.
[137] As finalised by the Chair of the CITES Standing Committee, CITES, January 2004 and circulated to Ninth FAO Sub-Committee on Fish Trade (copy on file with author).
[138] As endorsed by the Ninth Session of the FAO Sub-Committee on Fish Trade, Bremen, Germany, 10–14 February 2004 and reproduced in *Fisheries Report No. 736* (2004), Annex E.
[139] As presented to fifty-first CITES Standing Committee, Bangkok, October 2004 (CITES SC51 Doc. 8, Annex).
[140] As presented to fifty-third CITES Standing Committee Geneva, 2005, (SC53 Doc. 10, Annex).
[141] Agreed by CITES and FAO in September–October 2006; Reproduced at CITES SC 53 Doc. 10.1 and FAO *Fisheries Report No 807* (2006) Appendix F. I reproduce the full text at Appendix B below.

appendices, the meeting noted its understanding of institutional competences in fisheries management:

> 18. Several countries reiterated their reservations about the role of CITES in relation to resources exploited by fisheries. The Sub-Committee held the view that FAO and the mandated regional fisheries management organizations (RFMOs) were the appropriate international bodies on fisheries and fisheries management. The Sub-Committee also underlined the importance of CITES Article 14 regarding the relationship between CITES and UNCLOS and its implementation agreement. The view was expressed that CITES should be seen as a complementary instrument in the protection of such resources, e.g. in cases where management regimes are not in place, and that a CITES listing should be limited to exceptional cases only and when all relevant bodies associated with the management of the species in question agreed that a listing would be advantageous. Some countries expressed support for the role of CITES in fisheries management, stating that it could not replace traditional fisheries management.[142]

This paragraph became important to drafts of the MOU, as I describe below, and as such its language warrants close attention. It affirms the role of the FAO and RFMOs in fisheries management. It then points to the CITES conflicts clause that recognises that CITES is without prejudice to the development of UNCLOS or present and future claims concerning the law of the sea.[143] The third sentence, beginning with the passive use of '[t]he view was expressed...', is the extreme position that the FAO and RFMOs can overrule attempts to list marine species in CITES. The final sentence suggests that some members were not convinced of this extreme view.

The meeting resulted in a request for the Secretariats of the FAO and CITES to coordinate the drafting of an MOU 'to facilitate dialogue and exchange of information', and urged the Secretariats to proceed in a manner so that the MOU could be considered and possibly approved at the 2003 meetings of COFI and the CITES Standing Committee.[144] The Sub-Committee recommended that the MOU describe a process by which the FAO and CITES would coordinate scientific and technical advice on proposals to list commercially exploited aquatic species.

In November 2002, the CITES COP met in Santiago, Chile and adopted proposals to list a number of commercially exploited marine species.[145] The United States proposed that the MOU issue be considered by the

---

[142] FAO *Report No 673* above n. 51 para. 18.
[143] See especially CITES, Art. XIV (4), (6).   [144] FAO *Report No 673* above n. 51, Appendix F.
[145] See above n. 31 and accompanying text.

COP,[146] and the CITES Secretariat sought direction on whether collaborative mechanisms with the FAO should address both fish and timber sectors.[147] Japan proposed a resolution to progress the development of an MOU.[148] Under the heading of 'synergy and cooperation', this draft resolution proposed important changes to CITES competence, as demonstrated by excerpts from the draft resolution, which:

AFFIRMS that FAO and the mandated [RFMOs] are appropriate inter-governmental bodies responsible for fisheries and fisheries management;
  AGREES that, in cases where there is no responsible [RFMO] and where trade is having a significant negative impact on conservation, the listing of commercially-exploited fish species in the Appendices may temporarily serve a useful conservation purpose.

The attempt to limit CITES' role to marine areas without RFMOs, and even then for only temporary periods, was not adopted by the COP. Instead, a decision was made to direct the CITES Standing Committee to work with the FAO in the drafting of the MOU.[149] The decision recognised the 'primary role of the FAO and regional fisheries management organizations in the fisheries management' and the 'role of CITES in regulating international trade' in endangered species.[150]

As the Standing Committee was not due to meet until after the next FAO forum in February 2003, the Chair of the Standing Committee instructed the Secretariat to develop, in consultation with him, a draft MOU ('CITES Standing Committee Chair-led draft').[151]

In its short preamble, this CITES Secretariat-led draft recognised 'that the aims and purposes of CITES and FAO are related and in conformity with each other and that strengthened cooperation between CITES and FAO would better ensure the achievement of those aims and purposes'.[152] The draft identified (i) the need for a procedure for the scientific evaluation of proposals to list commercially exploited aquatic species in the CITES Appendices; (ii) the need for capacity building for natural resource management; (iii) the need to identify technical and legal issues of common interest; (iv) the coordination of work through annual meetings and shared reporting;

---

[146] CITES CoP12 Doc. 16.2.2.
[147] Ibid. (noting the competence of the FAO in both fish trade and timber issues).
[148] CITES CoP12 Doc. 16.2.1.   [149] CITES CoP12 Decision 12.7 (2002).
[150] See further Wold, above n. 31, 396.
[151] CITES SC49 Doc. 6.3 (reproducing Notification to the Parties).   [152] Ibid.

THE RESTRICTION OF TRADE IN ENDANGERED MARINE SPECIES   161

and (v) general provisions on termination and the maintenance of separate budgetary responsibility.

This CITES Secretariat-led draft was provided by a CITES representative to COFI at its meeting in Rome in February 2003. Instead of discussing the MOU in the Plenary session, COFI established a 'Friends of the Chair Group'. After some members objected to the participation of the CITES Secretariat in the Group, representatives from the CITES Secretariat and all other IGOs and NGOs were expelled from its meetings.[153] After working separately from the Plenary for a number of days, the Group developed its own draft MOU, which differed substantially, both in substance and tone, from the CITES Secretariat-led draft.

The FAO COFI-led MOU draft[154] contained a long preamble that recognised the 'primary role of sovereign states, FAO and regional fisheries management organizations in fisheries conservation and management' and noted that CITES could not replace traditional fisheries management. The COFI-led MOU draft then sought to narrow the jurisdiction of CITES in two ways. First, it expressed the view that CITES listing would only occur if agreement from relevant RFMOs had been obtained. Secondly, it provided that the CITES listing of commercially exploited aquatic resources should be limited to exceptional cases. Given there was no consensus between Group members on this narrowing of CITES' competences, these views were reproduced as square-bracketed text in both the preamble[155] and the draft articles.[156] The draft also included a preambular reference to the 'three medium-term strategic objectives for fisheries', none of which included CITES listing of endangered marine species.

When the COFI-led MOU draft was presented to the COFI Plenary, some members expressed further disagreement with its attempts to limit CITES' competence.[157] The Committee agreed that an informal group should continue to work at the MOU 'at opportune times including at the Ninth Session of COFI-FT in 2004'.[158]

---

[153] Reported by CITES: Doc. SC49 Summary Report (Rev.1) at 3.
[154] Published in FAO, *Fisheries Report No. 702* (above n. 136) Appendix G.
[155] Ibid. (reproducing as square-bracketed text the view expressed within the Sub-Committee on Fish Trade (noted above n. 142) that CITES listing 'should be limited to exceptional cases only').
[156] Ibid., draft Art. 6.
[157] FAO, *Fisheries Report No. 702* (above n. 136), para. 47 ('Some Members also expressed the view that FAO should produce a draft MOU containing a process for increased cooperation without policy pronouncements').
[158] Ibid., para. 48.

(b) Alternative views

In April 2003, the CITES Standing Committee met in Geneva. It considered the CITES Secretariat-led draft of the MOU described above, as well as a proposal by Japan to amend this draft.[159] Aiming to reach agreement on the text during the meeting, the Standing Committee established a working group which comprised Japan, Norway, Saint Lucia, the United Kingdom, the United States, Australia, Chile, Egypt, the Secretariat and two NGOs.[160] This working group produced a draft but did not reach consensus on its contents.[161]

The Standing Committee considered further steps. Following a tight vote, it agreed that the CITES Secretariat-led MOU was the draft to work from. It decided that the Secretariat would invite comments from parties on this draft and make the comments available on the CITES website, and that the Chair would negotiate directly with the FAO and provide a resulting document to the next meeting of the Standing Committee for its consideration. An alternative proposal by Norway, who preferred that parties' comments on the draft MOU be provided to the Standing Committee to review before further involvement by the Chair, was not adopted.[162]

On the basis of this agreed process, the CITES Secretariat invited comments from the parties,[163] which were received up to June 2003 and posted on the website. The CITES Secretariat made some changes, although the result did not differ markedly from the original CITES Secretariat-led draft. On 5 January 2004, the Secretary-General of CITES wrote to the Director-General of the FAO attaching the revised CITES draft ('CITES draft (2004)') for the consideration of the FAO at the Sub-Committee on Fish Trade the following month.[164]

(c) Development of the FAO and CITES drafts

The FAO Sub-Committee on Fish Trade met in February 2004 in Bremen. Papers distributed to the parties included a copy of the COFI-led draft

---

[159] See CITES SC49 Doc. Inf 3.

[160] The two NGOs were the International Fund for Animal Welfare (IFAW) and the IWMC (formerly International Wildlife Management Consortium): see further below n. 233 and accompanying text.

[161] CITES Doc. SC49 Summary Report (Rev.1) ('There were difficulties over semantics as well as fundamental principles').

[162] The conduct of the vote on Norway's proposal was contentious, with Norway alleging that rules of procedure were inappropriately applied in relation to a tied outcome: ibid.

[163] Notification to the Parties No. 2003/030 (6 May 2003).

[164] See letter from Willem Wijnstekers, Secretary-General of CITES to Dr Jacques Diouf, Director-General of the FAO dated 5 January 2004.

MOU on which consensus had not been reached at COFI, with square brackets around the contentious text. The revised CITES draft, which had been sent on 5 January, was not included in the conference papers,[165] a somewhat surprising omission given that other documents finalised at the end of January were included as informal papers for participants.[166]

The CITES draft was not circulated until the conclusion of the conference. The Chair instead decided to establish a working group that would work in parallel to the Plenary.[167] It was to be chaired by Japan and the United States, and was to operate within a narrow mandate of considering only the COFI-led draft. It met for several hours in two separate sessions.

The working group produced its own draft MOU (reported as the 'Japan–USA compromise text'[168] and noted here as the FAO COFI-FT draft (2004)). This was based on the COFI-led draft but modified some of that draft's attempts at limiting CITES jurisdiction. It agreed that the contentious square-bracketed text in the preamble and body of the MOU (which sought to make listings contingent on RFMO and FAO approval, and only in exceptional cases)[169] could be deleted from the draft. However, the preamble continued to assert this controversial position, at least implicitly, by incorporating the 'results' of the 2002 COFI-FT meeting at which this view was expressed.[170] A new preambular paragraph also affirmed the 'rights and duties of all States pertaining to fishing activities outlined in [UNCLOS]' and emphasised the goal of sustainable utilisation.[171]

The working group reported back to the Plenary with this compromise text, and recommended that it be sent to the Chair of the CITES Standing Committee with a request that it be considered at its next meeting. It also recommended that FAO staff be mandated to conclude the draft with CITES, but according to a strict process by which they would consult with certain members if substantive changes were involved:[172]

---

[165] See FAO Doc. COFI:FT/IX/2004/3 (stating that papers are current to December 2003).
[166] See, e.g., COFI:FT/IX/2004/Inf.8 'Report of the Expert Consultation on Fish Trade and Food Security' (Casablanca, Morocco, 27–30 January 2003).
[167] See FAO, *Fisheries Report No. 736* (2004) 6–8.   [168] Ibid., Annex E.
[169] Ibid. On the earlier square-bracketed text, see above nn. 155 and 156.
[170] Ibid. The preamble continued to refer to the contentious para. 18 noted above n. 142.
[171] Ibid.   [172] See FAO *Fisheries Report No. 736* (2004) para. 22.

If, in discussions, CITES requested changes that the FAO Secretariat thought might be unacceptable to some Members, the FAO Secretariat would informally consult with *those* Members. If those Members did not raise any objections, the FAO Secretariat could sign the MOU. [Emphasis added]

The working group did not identify the FAO members that might have found CITES' requested changes unacceptable, but it presumably intended this paragraph to relate at least to Japan and Norway. When these recommendations were presented to the Plenary, Argentina challenged the working group's suggested process. It suggested that the word 'those' be deleted from the first line of the working group's recommendation, as reproduced above.

This small amendment would have changed the procedure from one of a potentially private discussion between FAO staff and interested parties such as Japan to a provision for multilateral consultation with the membership as a whole. However, although the Plenary appeared to accept Argentina's amendment, it was not reproduced in the final report.[173]

Argentina's concern about the suggested procedure is important. On one level, it is understandable that the FAO works closely with members and is sufficiently aware of which members would find aspects of an MOU with CITES unacceptable. Yet this type of interaction may be indicative of the 'capture' of the FAO by particular states. The suggested procedure may also be inconsistent with the FAO's own rules on the procedure to be followed by the FAO in entering into collaborative arrangements, which state that 'proper consultation with governments' must be secured.[174] These rules are directed at collaborative arrangements between the Organization and national institutions or private persons, rather than MOUs with other IGOs. However, the emphasis in those rules on proper consultation is similarly persuasive for collaborative agreements between IGOs.

After the working group had presented the Japan–USA compromise text, it reported on the CITES revised draft MOU. It stated that there were 'no substantive differences between that and the FAO proposal and there was nothing contradictory in the two texts',[175] a statement that is in direct contrast to the working group's concern that future

---

[173] The apparent acceptance by Plenary of Argentina's amendment was recorded by the author, who attended the meeting as a non-government delegate with the Australian delegation.
[174] FAO Constitution Art. XIII(4).
[175] FAO *Fisheries Report No. 736* (above n. 138) para. 24.

negotiations with CITES would lead to unacceptable changes to the FAO proposal.

(d) Further consideration of the FAO draft

Using the Japan–USA compromise text agreed at the FAO Sub-Committee on Fish Trade as the basis for their negotiations, the FAO and the CITES Secretariat attempted to finalise the MOU in the course of 2004. The 'limited negotiating authority' of the FAO was later noted by the Chair of the CITES Standing Committee.[176] By the time of the next meeting of the CITES Standing Committee, in April 2004, there was still no agreed MOU text.[177]

By October 2004, a draft FAO–CITES negotiated MOU was ready for presentation to the CITES Standing Committee. It met twice in October in Bangkok during the thirteenth CITES COP. On 1 October 2004, the Chair presented the draft FAO–CITES negotiated MOU.[178] This text used the Japan–USA compromise text agreed at COFI-FT as a base but included amendments aimed at bringing 'more balance to the text with regard to CITES'.[179] The preamble included square-bracketed text on which agreement had not been reached, which reflected disagreement over whether CITES 'cannot replace, but rather has sought to complement' or 'has not sought to replace but rather complement'.[180] In addition, the preamble recognised that 'peoples and States are and should be the best to conserve their own wild fauna and flora'.[181]

Some provisions of the MOU implied that CITES should be deferential to the FAO in proposals to amend its Appendices. For example, the draft reproduced the suggestion made by the FAO Committee of Fisheries about the use of FAO's reviews of proposals to amend CITES Appendices, as follows:

In order to ensure maximum coordination of conservation measures, the CITES Secretariat will consider, to the greatest extent possible, the results of the FAO scientific and technical review of proposals to amend the Appendices, and technical and legal issues of common interest and the responses from all the relevant bodies associated with management of the species in question [as well as the substance of the preambular paragraphs of this MOU] in its advice and recommendations to the CITES Parties.[182]

---

[176] CITES SC51 Doc. 8.   [177] CITES Doc SC50 Summary Report (15–19 March 2004).
[178] CITES SC51 Doc. 8, Annex.   [179] CITES SC51 Doc. 8.   [180] Ibid., Annex (para. 10).
[181] Ibid. (para. 1).   [182] Ibid. (para. 17).

In providing that CITES would 'consider, to the greatest extent possible', the results of the FAO's scientific and technical review of proposals to amend the CITES Appendices, the MOU retreated from the FAO Committee of Fisheries' earlier language that CITES would 'incorporate to the greatest extent possible' such results.[183]

The CITES Standing Committee revised this draft but failed to agree to it, and instead the Standing Committee agreed that Australia and Norway should have bilateral discussions on the revised text. Interventions on the topic were made by four countries, two observers and two NGOs – the Humane Society International[184] and Defenders of Wildlife.[185]

At the fifty-second meeting of the CITES Standing Committee on 14 October 2004, Australia reported that it had produced a text with Norway. This text had been circulated to parties by the Chair on 7 October 2004.[186] Two Committee members objected to substantive discussion on it.[187] The Standing Committee agreed that discussion be postponed until the fifty-third meeting.[188]

In March 2005, the FAO Committee on Fisheries met in Rome. The Secretariat described how it had agreed on a compromise text with the Chairperson of the CITES Standing Committee. This was the draft FAO–CITES negotiated MOU presented to the Standing Committee a few months before.[189] Some COFI members disputed that the Secretariats should be able to conclude such a text, and asserted that the only 'official FAO text' was the one agreed by a majority of members of the Sub-Committee on Fish Trade in Bremen.[190]

Again, the Committee considered it necessary to establish a separate working group. Within that group, many members considered that the

---

[183] On the earlier language, see FAO COFI-led draft (2003), para. 6 (reproduced in FAO, *Fisheries Report No. 702*, Appendix G); FAO COFI-FT draft (2004), para. 6 (reproduced in FAO, *Fisheries Report No. 736* (2004) Annex E).

[184] The HSI 'works with national and jurisdictional governments, humane organizations, and individual animal protectionists in countries around the world to find practical, culturally sensitive, and long-term solutions to common animal issues, and to share an ethic of respect and compassion for all life'. See further www.hsus.org/hsi/about_us/.

[185] 'Defenders of Wildlife is a national, nonprofit membership organization dedicated to the protection of all native animals and plants in their natural communities.': see www.defenders.org/about_us/index.php. Its International office 'has a long history of advocacy on behalf of threatened marine species. This advocacy has focused on conserving dolphins, sharks, and sea turtles.'): see www.defenders.org/programs_and_policy/international_conservation/index.php.

[186] CITES SC53 Doc. 10, Annex ('CITES Standing Committee negotiated draft (2004)').

[187] CITES SC 53 Doc. 10, para. 7.   [188] Ibid.   [189] See above n. 178.

[190] FAO *Fisheries Report No. 780* (2005) para. 58, referring to the FAO COFI-FT draft (2004).

FAO–CITES negotiated MOU[191] was the appropriate draft to work from, but there was no consensus on this view. The Committee resolved this impasse in the following way. First, it agreed that the text produced by the Secretariats was unofficial, and that the only draft MOU approved by the FAO body was the text agreed in Bremen.[192] Secondly, it recognised that the CITES Standing Committee would be free to consider any text for an MOU.[193] The Committee then said the matter would be reconsidered again in Bremen by the Sub-Committee on Fish Trade at its tenth session after feedback from CITES.[194] The Committee encouraged members to coordinate with their domestic fisheries and environmental agencies to achieve a consistent position on the MOU.[195]

(e) Finalising the MOU

The CITES Standing Committee met for its fifty-third session in June 2005 in Geneva. The MOU text negotiated between Australia and Norway in Bangkok, which updated the FAO–CITES negotiated draft, was included in the meeting materials.[196] The text was different from the draft FAO–CITES-negotiated MOU in several ways. For example, the preamble was rephrased to emphasise the jurisdiction of CITES. Instead of CITES having 'a role in regulating international trade' in threatened species, CITES had 'a primary role'.[197] In addition, the preamble no longer included the aim of 'maintaining the goal of sustainable utilization as stated in the FAO Code of Conduct for Responsible Fisheries'.

In the main body, the provision that stated that the FAO would continue to provide advice to CITES on, and be involved in future revision of, the CITES listing criteria was now a square-bracketed item.[198] Moreover, notifications of proposals to amend appendices no longer had to be provided to FAO 'as soon as possible'.

The provision that set out a procedure for the CITES Secretariat's consideration of the results of the FAO scientific and technical review of proposals was also changed.[199] On the one hand, instead of merely '*considering*, to the greatest extent possible' these results, the CITES

---

[191] See above n. 178. [192] FAO *Fisheries Report No. 780* (above n. 190) para. 60.
[193] Ibid. [194] Ibid., para. 61.
[195] Ibid., para. 62 ('The Committee agreed on the importance of ensuring that there was consistency in the positions of Members at meetings of FAO and CITES on the issue of a MOU').
[196] CITES Standing Committee negotiated draft (2004), reproduced in CITES SC53 Doc 10, Annex.
[197] Ibid. [198] Ibid., para 14. [199] Ibid., para 17.

168   TRADING FISH, SAVING FISH

Secretariat was now required to '*respect* [them], to the greatest extent possible'.[200] Weakening this provision, however, was the fact that the CITES Secretariat's 'respect' for the FAO's scientific and technical review was no longer to be expressly reflected in its recommendations to the CITES parties on whether a marine species should be listed. In addition, the CITES Secretariat was also no longer required to consider the substance of the preamble during this listing procedure.

Some members continued to express dissatisfaction with this draft, so the Chair suspended discussions and suggested that he conduct informal consultations.[201] On the basis of the consultation, the Chair then produced a revised document.[202]

The revised draft deleted the entire preamble. In addition, the square brackets around the following phrase had been removed: 'FAO will continue to provide advice to CITES on, and be involved in any future revision of, the CITES listing criteria.'[203] The text providing that CITES would 'respect, to the greatest extent possible' the results of the FAO scientific and technical review of listing proposals was retained.[204]

Mexico complained that it had been excluded from the informal consultations that led to the revised text.[205] It contested the agreed procedure for the CITES Secretariat's consideration of FAO scientific and technical reviews of specific proposals to list marine species. It stated that the Spanish version contained the phrase 'take into account', rather than 'respect' and this version should be reflected in the English text. In response, Iceland expressed its regret for not involving other members in coming to the consensus text but stated that it refused to replace the word 'respect' with 'take into account' or 'consider'. Japan agreed that the word 'respect' should be maintained.[206] The FAO and the Humane Society also made interventions, although the substance of their comments is not on record.[207] The Standing Committee approved this draft and agreed that it be forwarded to the FAO for consideration.

The FAO Sub-Committee on Fish Trade met from 30 May to 2 June 2006. The Sub-Committee agreed to this version,[208] although many members thought its wording was 'not perfect'.[209] The procedure for the CITES Secretariat's consideration of the FAO's scientific and

[200] Ibid. (emphasis added).   [201] CITES SC53 Summary Record (Rev.1).
[202] CITES SC53 Doc. 10.1. The full text of the MOU appears in Appendix B below.
[203] Ibid., para. 3.   [204] Ibid., para. 6.   [205] CITES SC53 Summary Record (Rev.1), 6.
[206] Ibid., 5-6.   [207] Ibid., 5.   [208] Reproduced at Appendix B below.
[209] FAO, *Fisheries Report No. 807* (above n. 141), para. 24.

technical reviews of proposals to amend the Appendices continued to attract controversy. Brazil expressed its concern that the phrase 'respect to the greatest extent possible' was used and would have preferred 'take into account'.[210] The Philippines was dissatisfied with the termination arrangements of the MOU and would have preferred to allow only for its modification.[211]

3 Substantive and procedural constraints

Before assessing the MOU's impact on regime interaction between the FAO and CITES, I conclude this section by highlighting some institutional and procedural problems with the development of the MOU. In particular, I place in legal context the attempt by key states to shape regime interaction by reinforcing the 'primary' role of the FAO and RFMOs in the MOU. This requires a revisit to principles of international institutional law, and an investigation into treaty law and legal process. My findings have important consequences for the use of cooperative agreements between Secretariats as tools of regime interaction.

(a) International institutional law

Japan's proposal to insert into the MOU a requirement that the prior approval of the FAO, or of its subsidiary bodies or RFMOs, be granted before commercially exploited marine species could be listed in the CITES Appendices, can be assessed according to the implied powers of the CITES and FAO Secretariats described above. The CITES treaty text contains no requirement that the views of the FAO should prevail on issues relating to trade restrictions for marine species. Japan's attempted revisions to the role of the FAO are thus inconsistent with the CITES treaty text. Japan's proposal was not accepted in the final MOU, but if it had been, the resulting provision may well have exceeded the express and implied powers of the FAO and the CITES Secretariat.

The consequences of the FAO and CITES Secretariats acting ultra vires in this way are unclear. On one view, an MOU is void if it exceeds the Secretariats' powers. Yet a key question is whether there is an organ that can rule upon a claim that a Secretariat has acted ultra vires. Establishing such an organ requires agreement, and the international procedures are unwieldy and undeveloped as compared to the European

---

[210] Ibid., para. 25.   [211] Ibid.

legal context.[212] Agreement to challenge the acts of international organisations as ultra vires could in theory be forthcoming in a dispute where states consented to the compulsory jurisdiction of the ICJ. Alternatively, the ICJ could exercise its advisory jurisdiction.[213] These avenues will differ depending on the constituting documents within the regimes themselves; in the WTO regime, for example, there is no procedure to challenge as ultra vires the acts of the organisation.[214]

Related to the question of whether the FAO and the CITES Secretariat could be found to be acting ultra vires is the question of whether the MOU is binding. This is a matter of some uncertainty. Memoranda of Understanding between *states* are generally considered to be non-binding instruments, unless it can be shown that the parties intended otherwise.[215] For such agreements, intention is to be gauged from the wording of the text.[216]

The decision to include termination provisions in the MOU has bearing on its legal status. Amongst some opposition, the MOU provides for both termination and amendment by the FAO and CITES Secretariats.[217] Such a provision is usually indicative of the binding nature of the agreement.[218] However, the MOU also contains the cooperative verb 'will' rather than 'shall', which has been considered in other contexts to indicate a non-binding arrangement.[219] The level of agreement or opposition to the MOU is also relevant,[220] and the contentiousness with which

---

[212] See further Dan Sarooshi, *International Organizations and their Exercise of Sovereign Powers* (2005) 108, 119–20 (considering measures a state can take against an organisation on which it has conferred powers and comparing legal remedies for ultra vires acts in the context of the EC and the ECJ); see also José Alvarez, 'The New Treaty Makers' (2002) 25 *Boston College International and Comparative Law Journal* 213.

[213] *UN expenses* (1962) ICJ Rep 151. See further Schermers and Blokker, above n. 117, 592 (para. 912).

[214] Pauwelyn, above n. 103, 44–7, 144–5, 285–6.

[215] Anthony Aust, 'The Theory and Practice of Informal International Instruments' (1986) 35 *ICLQ* 787, 800–4 (writing on international instruments between states).

[216] On agreements between states that are equivocal on whether they are binding, see ibid.; see also Kelvin Widdows, 'What is an Agreement in International Law' (1979) 50 *BYIL* 117; Oscar Schachter, 'The Twilight Existence of Nonbinding International Agreements' (1977) 71 *AJIL* 296.

[217] MOU, Art. 8; note opposition by the Philippines, above n. 211 and accompanying text.

[218] See e.g. Churchill and Ulfstein, above n. 132, 651.

[219] Ibid., 650–1 (considering the MOU between the COP of the UNFCCC and the Global Environment Facility).

[220] See e.g. in relation to declarations of states within international forums: *Military and Paramilitary Activities in Nicaragua (Merits)* (1986) ICJ Rep 16 (paras. 203–5), where the Court considered the general consensus accompanying a relevant UN Declaration in its reliance on it.

the issue of cooperation has been treated in meetings of the FAO might indicate, on the one hand, a failure of the MOU to bind or, on the other, that the MOU represents an important and binding compromise about a difficult topic.

(b) Treaty amendment and modification

Aside from issues of institutional law, the negotiation of the MOU gives rise to important issues of treaty law. There are a number of constraints on the ability of states to amend or modify treaties. These constraints become relevant because the wrangling between states during the negotiation of the MOU could lead to a particular characterisation of the MOU. Rather than an agreement between international organisations, the MOU could be seen as an agreement to amend the core treaty text of CITES.

As I demonstrate above, CITES expressly recognises its potential role in the listing of marine species. The strategy of Japan and others, in attempting to limit the role of CITES in marine listings, could be considered an amendment of the Convention.

Treaty amendment is permitted but must conform to the VCLT. This provides for the *inter se* modification of a treaty where 'two or more parties to a multilateral treaty ... conclude an agreement to modify the treaty as between themselves alone'.[221] *Inter se* modification is permissible if the possibility of such a modification is provided for by the treaty, and provided it does not affect the rights or obligations of the other parties.

Reservations and denunciation are available within CITES.[222] However, any amendment requires a two-thirds majority of an extraordinary meeting of the CITES COP.[223] As such, the attempts to limit CITES' role in the protection of endangered marine species by the Japanese-led MOU drafts could not be considered as lawful *inter se* modification unless they were announced as such and were voted upon.

This finding accords with the need for treaty law to provide consistency and stability. Admittedly, this need must always be balanced with law's receptiveness to changing conditions,[224] and a finding that MOUs could never modify the constitutive treaties of IGOs would be overly formalist. The acknowledgment that international law is a process does

---

[221] VCLT, Art. 41. [222] CITES, Arts. XXIII, XXIV. [223] Ibid., Art. XVII.
[224] This is demonstrated most clearly by the use of custom in international law: see Rosalyn Higgins, *Problems and Process: International Law and How We Use It* (1994) 18–22.

not preclude the finding, however, that a party to CITES may not limit its jurisdiction unless it has followed set procedures for open and participatory amendment.

(c) Uses and abuses of rules of procedure

A further aspect of international law of relevance to the development of the MOU is the nature and use of rules of procedure in international forums. The constitutive instruments of the CITES Secretariat[225] and the FAO[226] contain rules for access, representation and participation by international bodies, including IGOs and NGOs. These instruments also allow for the development and application of other rules of procedure, such as the provision for committees described as 'Friends of the Chair' or 'working groups' which operate in secret. These rules of procedure had significant impact upon the development of the MOU.

In the 2003 meeting of COFI, the representative of the CITES Secretariat had been invited to assist in the drafting of the MOU. A working group, comprised of Japan, the United States and others, was established to consider the issue of the MOU, and it was to report back to the Plenary. The working group wished to conduct its deliberations in secret, excluding the CITES representative and any other representatives from IGOs or NGOs.[227] This action seems to be in line with a rule of procedure which provides that the Committee may decide to restrict attendance at private meetings to representatives of member states 'in exceptional circumstances'.[228] That the Committee considered the drafting of an MOU to facilitate cooperation with another IGO amounted to 'exceptional circumstances' reflects the contentiousness with which certain states viewed CITES' role. Such contentiousness will often be present in regime interaction, but it is doubtful that closed meetings are the most useful response.

The tension between the need for efficiency and free debate was apparent in the conduct of the working groups themselves. During the working group at the 2004 Sub-Committee on Fish Trade, participating

---

[225] CITES, Art. XI (7); see further pp. 57–61. See also CITES Standing Committee Rules of Procedure (2009).
[226] See FAO General Rules, Rule XVII. See also FAO 'Cooperation with International Governmental Organizations'; FAO, 'Guiding Lines Regarding Relationship Agreements Between FAO and IGOs'; FAO, 'Cooperation with NGOs'; FAO, 'Granting of Observer Status'.
[227] See above n. 153 and accompanying text.
[228] Rules of Procedure of the Committee of Fisheries, Rule III(3)(c).

members were told by the United States and Japan that debate would not be opened on the substantive sections of the draft MOU that had already been agreed at COFI. This effectively removed the ability of the parties to debate any language used in the draft CITES MOU, and any other aspect of the FAO draft. Instead, debate was to concentrate on the square-bracketed texts. Whether an appropriate balance between efficiency and openness was reached is questionable.

Of interest, too, is the decision to delay the distribution of the CITES draft MOU at the 2004 Sub-Committee on Fish Trade. During the initial stages of the Plenary, staff from the FAO intimated that the revised CITES draft had been received. After rumblings from the floor (but no formal interventions), in which a number of parties suggested that their views had not been taken into account in the CITES process, the Chair decided to defer its distribution. This action is in apparent breach of the FAO's rules of procedure, which provide that the views of intergovernmental organisations are to be circulated freely and without abridgement.[229]

These examples suggest that certain states used, and sometimes extended, FAO rules of procedure to constrain open debate and collaboration in the development of the MOU. Such practices present challenges for regime interaction. This issue is revisited in Part III of the book.

Within the CITES regime, too, my examples demonstrate that rules of procedure impacted substantively on the MOU text, although the CITES procedures gave rise to different challenges in the development of the MOU. The CITES procedures provide somewhat more fulsome rights to participate. The Chair of the CITES Standing Committee may invite any person, including a representative of any body or agency, to attend as an observer, provided that 'any such person, body or agency is technically qualified in protection, conservation or management of wild fauna and flora'.[230] Such bodies have the right to participate on certain agenda items as agreed by the Committee, but not to vote.[231] The Committee can withdraw the right of any observer to participate.[232]

Challenges arose in the negotiations leading to the MOU due to the disparate and sometimes contradictory interests of the non-state participants. There is a strong tension between the two NGOs that participated in the CITES Standing Committee working group to negotiate the MOU. One group, the IFAW, campaigns on animal welfare and

---

[229] FAO General Rules of Procedure, Rule XVII.
[230] CITES Standing Committee Rules of Procedure (2005), Rule 6.1.   [231] Ibid.   [232] Ibid.

protection.²³³ The other, the IWMC, is a pro-utilisation group that seeks to 'protect the sovereign rights of independent states in their conservation efforts'.²³⁴ The draft preamble of the CITES Secretariat-led MOU drafts reproduced the view 'that peoples and States are and should be the best to conserve their own wild fauna and flora', suggesting that the IWMC successfully influenced the Standing Committee. However, most advocates of fisheries conservation, including most NGOs, would probably not share this wholesale endorsement of the ability of states to conserve fisheries. This example demonstrates the difficulties of finding the 'voice' for marginal interests, such as interests in fish sustainability, in regime interaction, an issue that I address in Part III.

4 National policy coordination

At various times during the fraught negotiations over the MOU, it was clear that states were unable to reconcile competing interests within their domestic agencies. Fisheries policy and environment protection policy are often developed by different domestic government portfolios and implemented by different agencies. The domestic agencies do not always coordinate or collaborate on marine issues. This problem is exacerbated when domestic agencies have not implemented the core requirements for even one regime. For the CITES regime, for example, approximately half of all CITES parties lack appropriate legislation to implement their obligations.²³⁵

The lack of national policy coordination impacted upon the development of the MOU. In response to problems in finalising the MOU, the FAO Committee on Fisheries recommended that fisheries delegates work with environmental delegates to achieve a consistency of position.²³⁶ Collaboration between relevant domestic agencies within states was therefore seen as a precursor to collaboration between CITES and the FAO.

The need for national policy coordination is not limited to the wrangling over the MOU. Collaboration between domestic agencies is also important for subsequent implementation of CITES obligations. National departments responsible for CITES permits and licences are commonly different from fisheries departments. In Hong Kong, for

---

[233] The IFAW seeks to 'achieve lasting solutions to pressing animal welfare and conservation challenges': www.ifaw.org.
[234] The mission of the IWMC is to 'promote sustainable use of wild resources' whilst protecting the independence of states: www.iwmc.org/iwmcinfo/statement.htm.
[235] See Chapter 2, n. 201.   [236] See above n. 195.

example, the domestic management authority responsible for administering the CITES licensing scheme is the Agriculture, Fisheries and Conservation Department specific to the Special Administrative Region of Hong Kong. The designated scientific authority is the People's Republic of China's Endangered Species Advisory Committee.[237] These functions are separate from Hong Kong's Trade and Industry Department. In Australia, the management authority and the scientific authority are located in the Department of Environment and Water Resources. This is a separate department from Australia's Department of Agriculture, Fisheries and Forestry and Department of Foreign Affairs and Trade. By contrast, the United States houses the CITES management authority and scientific authority in the Department of the Interior, which includes the United States Fish and Wildlife Service. Its Department of Agriculture's Animal and Plant Health Inspection Service and the Department of Homeland Security take on enforcement responsibilities for plants. The United States has reported that coordination of these agencies takes place through a CITES Coordinating Committee, in which the USTR participates, and in subcommittees, including a subcommittee exclusively devoted to marine issues.[238]

The issue of national policy coordination has gained attention in other areas of fisheries governance. In 2006, delegates to the UN Consultative Process commented on the 'serious coordination and cooperation gap among the [many intergovernmental] organizations and agencies themselves and between national governmental bodies'.[239] Some delegates concluded that the responsibility for remedying this situation lay with the countries themselves:

It was noted by some that States are responsible for ensuring the necessary cooperation and coordination among the various agencies and that their delegates should better cooperate and coordinate their work through international organizations. Such cooperation and coordination posed a challenge at the national level, where many departments might have oceans-related mandates but did not always work in a coordinated manner.[240]

An improvement to national policy coordination is increasingly discussed as a means of resolving uncertainties about conflicting treaty

---

[237] This information was provided by Hong Kong to the WTO: WTO Doc. TN/TE/W/28. This information is also available via links to domestic contact points on the CITES international website: www.cites.org/common/directy/e_directy.html.
[238] See submission by United States to the WTO: WTO Doc. TN/T/W/40, paras. 15–18.
[239] UN Doc A/61/156 para. 106.   [240] Ibid.

obligations and other issues associated with the fragmentation of international law. For example, in negotiations over the relationship between WTO agreements and specific trade obligations in MEAs, Australia and the United States have submitted that the solution to potential conflicts between treaties lies within domestic coordinating bodies.[241] This has led, in particular, to the provision about information on domestic policy coordination of CITES commitments by WTO members.[242] National policy coordination has major impacts on regime interaction, and is considered in greater depth in Part III.

D Entrenching interaction through the MOU

The MOU between the FAO and the CITES Secretariat entrenches regime interaction. It requires ongoing exchange of information and sets out a degree of responsibility of each organisation to have regard to that information. Other than this, there is no delimitation of organisational roles. In contrast to the draft proposals described in the previous section, the MOU makes no reference to any relative competences between the FAO and CITES Secretariat. Instead, it provides for cooperation over marine species conservation in five major areas: information-sharing and observation; capacity building; joint involvement in the CITES listing criteria; consultation and review by the FAO of CITES listing proposals; and resource allocation and reporting.

1 Information-sharing and observership

The MOU provides for communication and regular information-exchange between the FAO and CITES Secretariats, as follows:

> The signatories will communicate and exchange information regularly and bring to each other's attention general information of common interest and areas of concern where there is a role for the other to play. The signatories will be invited as observers to meetings under their respective auspices where subjects that are of common interest will be discussed.[243]

This provision for observership first appeared in the MOU drafts discussed within the FAO; the original CITES draft did not refer to it. Unusually, the MOU drafts did not reflect the possibility of routine observership, as is

---

[241] See e.g. WTO Docs. TN/TE/W/40; TN/TE/W/45.
[242] Papers were submitted by Hong Kong (TN/TE/W/28), the United States (TN/T/W/40), Australia, (TN/TE/W/45), the EC (TN/TE/W/53) and Switzerland (TN/TE/W/58).
[243] MOU, Art. 1.

regularly practised, for example, in some WTO Committees such as the WTO Committee on Trade and Environment. Instead, the MOU provides for ad hoc observership 'where subjects that are of common interest will be discussed'.

The issue of 'common interest' may prove contentious in future applications of the MOU. Its characterisation remains within the discretion of each Secretariat. A number of factors could facilitate the awareness of issues of 'common interest' and information-exchange outside of the provisions of the MOU. For example, Schermers and Blokker suggest that 'the prestige of the international civil service' may stimulate collaboration between IGOs.[244] On the other hand, they note an opposite effect caused by rivalry between secretariats.[245] The social and professional norms of the FAO and CITES Secretariats may change through practices that develop outside of the context of the MOU, such as staff exchanges,[246] workshops in other forums[247] and even through the increasing public awareness of fisheries depletion.

The MOU does not address RFMOs and it is unclear whether the information-exchange between CITES and the FAO will further facilitate cooperation between CITES and RFMOs. Such cooperation could improve the conservation measures of both CITES and RFMOs, but currently faces much resistance, as the example of the Patagonian toothfish – commonly known in North America as Chilean Sea Bass – makes clear.

In 2002, attempts were made to improve cooperation between CITES and the regional fisheries management organisation of the Antarctic region, CCAMLR. Australia sought to address IUU fishing of Patagonian toothfish by linking CITES certification requirements with the catch documentation scheme of CCAMLR.[248] The COP resolved that CITES parties should adopt the relevant catch document used by CCAMLR and implement requirements for verification.[249] The resolution also provided for information-sharing and a promotion of the CCAMLR

---

[244] Schermers and Blokker, above n. 117, 1087 (para. 1706).   [245] Ibid.
[246] See, e.g., ibid., 1099 (para. 1723) (approving exchanges between Secretariats to enhance collaboration).
[247] See e.g. multistakeholder workshops by UNEP and others discussed in Chapter 3.
[248] See CITES CoP12 Doc. 44 (draft resolution that a 'Catch Document issued in accordance with the requirements of the CCAMLR Catch Documentation Scheme for *Dissostichus* species is equivalent to and an acceptable substitute for a certificate of introduction from the sea or export permit issued under CITES').
[249] CITES Resolution Conf. 12.4.

conservation regime among CITES parties. CITES parties were directed to report to the CITES Secretariat on the use and verification requirements of the CCAMLR catch documentation.[250] The CITES Secretariat was directed to compile information on parties' practices in this regard and invite CCAMLR to consider, at its meetings, how further cooperation between CITES and CCAMLR could be progressed.[251]

Attempts by Australia to update these requirements were not accepted at the following COP in 2004.[252] The CITES Secretariat noted that the CITES treaty text could not serve as a basis for an ongoing reporting mechanism within CITES for the Patagonian toothfish, and that CITES parties should report on their use and verification requirements of the CCAMLR catch documentation scheme directly to CCAMLR because of 'insufficient resources [for the CITES Secretariat] to serve as an intermediary for ongoing reporting related to non-CITES species'.[253]

The Patagonian toothfish example is important in two major respects. First, it suggests that a failure to achieve lasting cooperation between CITES and RFMOS such as CCAMLR may be attributed to the background political controversies over the species at issue. The Patagonian Toothfish is imminently endangered. Indeed, Australia had originally sought its listing in CITES Appendix II at the COP in 2002, but withdrew its proposal before the matter went to vote due to opposition from CCAMLR members.[254] Such controversies will often accompany marine species, as the discussion on sharks, seahorses and bluefin tuna made clear. Perhaps if the MOU had called for cooperation between CITES and RFMOs (or indeed a new MOU between CITES and CCAMLR) the cooperation between Secretariats would have been less dependent on the opposing wills of states. Secondly, the fact that Patagonian toothfish has not been accepted as warranting listing in CITES, at least for the present, may be attributed to the problems over institutional collaboration. Commentators have noted that cooperation over documentation and permit schemes is dependent on the prior CITES listing of a species.[255] The implication is that cooperation will only occur *after* a species is listed, or at least when a relevant proposal for listing has been

---

[250] CITES CoP12 Decision 12.57 (trade in toothfish). See also CITES Notification to Parties No. 2003/081 (December 2003).
[251] CITES CoP12 Decisions 12.58, 12.59 (trade in toothfish).
[252] See Australia's proposed resolution in CITES CoP13 Doc. 12.3.
[253] CITES CoP13 Doc.12.3, p. 2.   [254] Little and Orellana, above n. 53, 24.
[255] Murphy, above n. 55, 556.

lodged with CITES. This constraint is present in a further area of cooperation set out in the MOU: capacity building.

2 Capacity building

The MOU provides for cooperation between the Secretariats to facilitate capacity building:

> The signatories will cooperate as appropriate to facilitate capacity building in developing countries and countries with economies in transition on issues relating to commercially-exploited aquatic species listed on the CITES Appendices.[256]

The CITES-led drafts had sought cooperation on capacity building for a wider range of issues. First, cooperation was sought on the building of capacity for a broad range of issues relating to commercially exploited aquatic species,[257] regardless of whether such species had been listed in the CITES Appendices. The final MOU limits the ambit of cooperation to species that have already entered CITES' jurisdiction by dint of inclusion in the Appendices. Secondly, the CITES-led drafts had sought cooperation to build capacity for law-enforcement.[258] The final MOU did not reproduce this provision, although the need for cooperation to build the capacity of states to enforce compliance with CITES may be implied by the reference to cooperation on 'issues relating to commercially-exploited aquatic species listed on the CITES Appendices'.[259] The lack of express recognition of the benefits of cooperation in enforcement has implications not only for CITES obligations but also for the compliance of related FAO instruments, as I discuss further when I consider the dispute-settlement aspects of restraining trade in endangered marine species below.

3 Involvement in CITES listing criteria

The CITES listing criteria, which set out biological, trade and other factors for parties to consider when they vote on listing proposals, have been highly controversial in the marine species context.[260] The MOU provides that the FAO and the CITES Secretariat will continue to cooperate on the CITES listing criteria and that the FAO will be involved in future revisions.[261] As described above with reference to the

---

[256] MOU, Art. 2.
[257] CITES Standing Committee Chair draft (2003), draft Art. 2(1); CITES draft (2004), draft Art. 2.
[258] Ibid.   [259] MOU, Art. 2.   [260] See above n. 57.   [261] MOU, Art. 3.

CITES-led drafts, CITES had not originally endorsed the need for the FAO to provide advice or be involved in the revision of CITES listing criteria. In lawyer slang, the matter appeared to be a 'deal-breaker' in the final discussions leading to the conclusion of the MOU.

The FAO drafts also implied that the listing criteria were not the only basis for the FAO and the CITES Secretariat's evaluation of proposals for amendments to the CITES Appendices. The following additional sentence questioned the sole basis of the listing criteria for the evaluation of proposals to amend the Appendices:

> These criteria will be the primary basis for the evaluation of proposals for amendment of the CITES Appendices by the FAO and the CITES Secretariat and for subsequent actions of all CITES Parties.[262]

Other states opposed the notion that there were additional sources from which to evaluate proposals to amend the CITES Appendices and it was not reproduced in the final MOU. Instead, the MOU provides that the FAO and CITES will work together in their evaluation of proposals to amend the CITES Appendices 'based on the criteria agreed by the Parties to CITES'.[263]

The controversies over the CITES listing criteria demonstrate a large disparity between the views of states and the relevant Secretariats. The status of, and method of revising, the CITES listing criteria was clearly controversial for some FAO members. Yet for the Secretariats themselves, collaboration was ongoing and successful in this area. The 'epistemic communities' within the FAO and CITES had already embraced the need for shared understanding about specific scientific and technical issues relevant to aquatic species.[264]

For example, the CITES listing criteria were revised at the thirteenth CITES COP in Bangkok in 2004. The FAO contributed to these revisions. It had convened a technical consultation on the 'suitability of the CITES Criteria for listing commercially-exploited aquatic species' in 2001.[265] This technical consultation was endorsed by the FAO Sub-Committee on Fish Trade, which conveyed it to the CITES Secretariat as the formal input to the CITES review process.[266] The CITES Secretariat appreciated 'the fresh perspectives that the involvement of FAO had brought to the

---

[262] FAO COFI-FT draft (2004), Art. 4; noted above n. 168.   [263] MOU, Art. 4.
[264] On 'epistemic communities', see Peter Haas, 'Introduction: Epistemic Communities and International Policy Coordination' (1992) 46 *International Organization* 1.
[265] Above n. 57.   [266] FAO *Fisheries Report No. 673* (above n. 51) para. 16.

THE RESTRICTION OF TRADE IN ENDANGERED MARINE SPECIES   181

process'.[267] Revisions to the listing criteria may be attributed to the FAO's input. For example, the Preamble now states:

NOTING the objective to ensure that decisions to amend the Convention's Appendices are founded on sound and relevant scientific information, take into account socio-economic factors, and meet agreed biological and trade criteria for such amendments;[268]

This inclusion may be attributed to the FAO's emphasis on the 'best scientific information available',[269] although it may well be attributed to CITES' own documents.[270] At any rate, viewed in the light of ongoing collaboration on revisions to the CITES listing criteria, the final MOU reflects a fully established practice between the two IGOs, rather than a novel activity that required agreement by states.

A final point of note about the CITES listing criteria is the relevance of national policy coordination, as described above. When the results of the FAO's technical consultation on the suitability of CITES listing criteria to commercially exploited aquatic species was conveyed to FAO members in 2002, FAO members were reminded that decisions made to list species were made by CITES parties, and that national delegations at CITES meetings should be 'properly briefed ... to take into account the views of the fisheries authorities if fishery matters were to be promoted'.[271] This view was reiterated by the representative of CITES present at the meeting. He stated that countries should 'resolve internally any differing views amongst their relevant agencies and departments on the role of CITES and the listing criteria if effective progress was to be made'.[272] Secretariat collaboration and national policy coordination were thus both clearly important in effective regime interaction in the endangered marine species context. The implications of this demonstrated importance are developed further in Part III of the book.

4 Consultation and review of listing proposals

The text of CITES requires the CITES Secretariat to consult with the FAO on listing proposals, 'especially with a view to obtaining scientific data these bodies may be able to provide and to ensuring co-ordination with

---

[267] Ibid., para. 19.   [268] CITES Listing Criteria (2007).
[269] FAO *Fisheries Report No. 673* (above n. 51) para. 16.
[270] See CoP12 Doc. 58, noting CITES Objective 2.2 of its Strategic Vision.
[271] FAO *Fisheries Report No 673* (above n. 51) para. 15.   [272] Ibid., para. 19.

any conservation measures enforced by such bodies'.²⁷³ This had led to useful practices, such as scientific and technical evaluations by the FAO on all of the marine species that were proposed for inclusion in the Appendices at the thirteenth COP. This accompanied other collaboration, such as the heavy involvement of the FAO in the review by the Animal Committee of the significant trade in queen conch. Moreover, the collaboration has worked both ways: the CITES Animal Committee, for example, participated in the development of the FAO IPOA-Sharks²⁷⁴ and FAO representatives participated in a CITES workshop on legal issues relating to 'introduction from the sea'.²⁷⁵

The MOU builds on these efforts and sets out a process for the FAO to consult with CITES when it evaluates proposals to include, transfer or delete commercially exploited aquatic species in the CITES Appendices.²⁷⁶ It is helpful to break the process of consultation down into five steps:

(i) The CITES Secretariat will inform the FAO of 'relevant' proposals;
(ii) The FAO will carry out a scientific and technical review of the proposals (usually done by an ad hoc expert panel);
(iii) The FAO will transmit the result to the CITES Secretariat;
(iv) The CITES Secretariat will make its own findings and recommendations on the proposals. In doing so, it will take 'due account' of the FAO review. In order to ensure 'maximum coordination of conservation measures', it will also 'respect, to the greatest extent possible' (A) the FAO review of proposals, (B) the technical and legal issues of common interest; and (C) the responses from all the relevant bodies associated with the management of the species in question; and
(v) The CITES Secretariat will communicate the FAO views and data, in addition to its own findings and recommendations, to the parties.

The decision to include a process for consultation over CITES listing is in variance to previous drafts of the MOU. The CITES drafts did not endorse a process for consultations over CITES listings. The FAO drafts, however, had originally indicated that CITES listings should be limited to exceptional cases only and should be contingent on agreement from all relevant management organisations.²⁷⁷ The FAO drafts also provided that the CITES Secretariat 'will incorporate to the greatest extent

---

²⁷³ CITES, Art. XV. ²⁷⁴ CITES Resolution Conf. 12.6; CITES Decision 13.42.
²⁷⁵ The workshop was conducted in 2005 and the FAO provided comments on the draft report: CITES CoP14 Doc.18.1, 4. This led to the resolution noted above n. 74.
²⁷⁶ MOU, Arts. 4–6.
²⁷⁷ See square-bracketed text of COFI-led draft (2003) noted above nn. 155, 156.

possible' the position of the FAO and RFMOs when it gave advice and recommendations to CITES parties on species listing proposals.[278]

The final MOU is a rather confused compromise that appears to provide two standards for the CITES Secretariat's deference to the FAO review. On the one hand, the CITES Secretariat is to take 'due account' of the FAO review in its evaluation of proposals to amend Appendices I or II.[279] In addition, there is a general reference to the need for CITES to 'respect, to the greatest extent possible' the position of the FAO and RFMOs in conservation measures.[280] This terminology remains controversial for several states.[281] The final phrasing means that it remains unclear whether the standard of deference to the FAO and RFMOs that the CITES Secretariat must follow in its recommendations on proposals for listings of marine species is one of 'due account' or one of the greatest possible 'respect'.

Recent practice has demonstrated the difficulties of implementing this arrangement. In 2007, Germany proposed the Appendix II listing of two shark species.[282] The FAO convened an ad hoc expert panel which reviewed the proposals and concluded that the available evidence did not support the proposed listing in Appendix II.[283] Germany did not agree with the FAO conclusions on the grounds of disputed methodology.[284] The FAO responded with a detailed defence of its methodology by reference to the CITES listing criteria.[285] When the proposals came to be voted upon, they each failed to meet the required two-thirds majority and were rejected.[286]

The difference in interpretation (or, at the least, application) of the CITES listing criteria between the FAO and the CITES Secretariat has

---

[278] See FAO COFI-led draft (2003), above n. 154, Art. 6 and FAO COFI-FT draft (2004), above n. 168, Art. 6.
[279] MOU, Art. 5.   [280] Ibid., Art. 6.
[281] See e.g. Mexico, above n. 205 and accompanying text, and Brazil, above n. 210 and accompanying text.
[282] The *Lamna nasus* (porbeagle) and the *Squalus acanthias* (spiny dogfish): see CITES CoP14 Prop. 15 and Prop. 16.
[283] The ad hoc panel met from 26 to 30 March 2007: see FAO, *Fisheries Report No. 833* (2007) submitted to CITES as CITES Doc. CoP14 Inf. 38.
[284] CITES CoP14 Inf. 48, 'Comments on the FAO Assessment of the CITES Amendment' (7 June 2007).
[285] CITES CoP14 Inf. 64 (14 June 2007).
[286] The porbeagle shark proposal was rejected with 54 votes in favour and 39 against. The spiny dogfish proposal was rejected with 57 votes in favour and 36 against. The EU sought successfully to reopen debate at the Plenary, but the proposal was rejected in a secret ballot, with 55 votes in favour and 58 against: see IISD, Vol. 21 No. 61 *Earth Negotiations Bulletin* (Monday, 18 June 2007).

caused some tension between the organisations.[287] It is clear that the MOU's provision for the CITES Secretariat to 'respect' and take 'due account' of FAO recommendations will remain as controversial as the underlying political issues relating to the species themselves. This is also clear from the reporting practices of the Secretariats, which forms a separate area of potential cooperation under the MOU.

5 Reporting and resource allocation

The MOU provides that the FAO and the CITES Secretariat will allocate specific resources for activities carried out jointly under the MOU.[288] It further provides that the secretariats to CITES and FAO will periodically report on work completed under the MOU to the CITES COP and the FAO Committee on Fish Trade, respectively.[289] This provision reflects some of the suggestions for review made by the CITES drafts. Left out of the final MOU, however, was a process by which the secretariats would meet annually to discuss implementation and prepare joint work plans, which was suggested by the CITES drafts. FAO states appeared to have preferred a unilateral review of the CITES COP. An example is the FAO's review on the assessment of proposals to amend the CITES Appendices, which are conducted by ad hoc FAO Expert Advisory panels.[290] The FAO Sub-Committee on Fish Trade has resolved that after each CITES COP, the FAO 'should undertake an evaluation of whether the recommendations of the ad hoc Expert Advisory Panel had been taken into account and, if not, why they had not been'.[291]

E Settling disputes

The settlement of disputes relating to the restriction of trade in endangered marine species could arise in a number of forums, including the WTO, a law of the sea tribunal, the Permanent Court of Arbitration and the ICJ. Further detail on these avenues is provided in Chapter 2. This section concentrates instead on interaction between the CITES and FAO regimes to assess the impact of the MOU on resolving disputes. As such,

---

[287] Interview by author with staff members of the Food and Agriculture Organization (Rome, 2 July 2008). Tensions may also be arising over the funding of the ad hoc panel's work, with Japan seeking to be a primary supporter and other members preferring funding from the regular programme: see FAO *Fisheries Report No 830* (2007), para. 35 and FAO *Fisheries and Aquaculture Report No. 902* (2009) para. 32.
[288] MOU, Art. 9.   [289] Ibid., Art. 7.   [290] See e.g. CITES CoP14 Doc.18.1.
[291] FAO, *Fisheries Report No 807* (above n. 141) para. 35.

I give less emphasis on formal dispute settlement procedures and more emphasis on general issues of compliance. This change of focus resonates with the approach of Chayes and Chayes and others, who argue that adherence to international law is based on active management by parties and other actors rather than coercion.[292]

Compliance with CITES obligations is currently rather weak. CITES relies on individual parties to enforce international trade restrictions within their borders, yet national implementation rates are poor.[293] Similarly, the FAO instruments such as the IPOA-Sharks suffer from poor implementation, and the measures such as the Code of Conduct are voluntary and unenforceable. That is not to say that these instruments have no effect on conservation, but rather that there is real scope for improvement of compliance within both regimes.

A lack of capacity is often a core reason for non-compliance.[294] The MOU addresses capacity building.[295] Yet, as I mention above, it does not include suggested provisions that would have provided for cooperation in enforcement.[296] The lack of recognition of the benefits of cooperation in enforcement might be a missed opportunity, not only for enforcement of CITES obligations but also for related compliance issues for FAO instruments and other obligations arising from the law of the sea. Observers of overlapping fisheries regimes have suggested that the effectiveness of regimes is enhanced when 'capabilities under one regime are used to induce compliance under another'.[297] I made similar observations with respect to the interaction between the WTO and fisheries regimes in Chapter 3.

Monitoring is considered to be one of the most important measures for effective international policies for transboundary and common problems.[298] Monitoring could be enhanced by the use of disparate techniques such as CITES' certification requirements and the FAO's

---

[292] Abram Chayes and Antonia Handler Chayes, *The New Sovereignty: Compliance with International Regulatory Agreements* (1995).
[293] See Chapter 2, n. 201.   [294] Chayes and Chayes, above n. 292, 13–15.
[295] MOU, Art. 2.
[296] See above n. 258 and accompanying text.
[297] Olav Schram Stokke, 'Conclusions' in Olav Schram Stokke (ed.) *Governing High Seas Fisheries: The Interplay of Global and Regional Regimes* (2001) 303, 349.
[298] See Robert Keohane, Peter Haas and Marc Levy, 'The Effectiveness of International Environmental Institutions' in Peter Haas, Robert Keohane and Marc Levy (eds.) *Institutions for the Earth: Sources of Effective International Environmental Protection* (1993) 3, 16; see also Elinor Ostrom, *Governing the Commons: The Evolution of Institutions for Collective Action* (1990).

monitoring of stocks. The regular information-sharing and observership entrenched in the MOU may well facilitate such ideas. Yet the MOU does not entrench periodic review of its *own* provisions. Ongoing monitoring of the collaborative arrangements between the FAO and the CITES Secretariat is instead subject to unilateral initiatives. That said, the high political and economic stakes in the listing of marine species probably means that attention and interest in the MOU will not wane.

The MOU's role in facilitating compliance is related to questions about compliance with the MOU itself. As described above, uncertainties persist about the legal status of the MOU, although its form and character suggest it is a binding document. Failure to comply with the MOU provisions will presumably lead to the same legal issues I have noted with respect to organs acting ultra vires.

F Conclusions

This chapter examined the restriction of trade in marine species according to the CITES regime. Such restrictions are increasingly sought for commercially lucrative species such as sharks, seahorses and bluefin tuna. Moves to restrict trade will affect significant political and economic interests, and many states prefer alternative conservation and management tools that are used in regimes such as the FAO. Opposition to CITES' role has therefore been expressed against the background of a fragmented international legal order. Several states have sought to restrict CITES' role by drafting provisions in a Memorandum of Understanding between the FAO and the CITES Secretariat that would give the former a power of veto in proposals to list marine species. This is an example of the complexities of regime interaction and the chapter sought to investigate the development and application of the MOU in this context.

Although the opposition to CITES' listing of marine species has been presented by some states as a response to potentially 'conflicting norms', the law of the sea and CITES are mutually compatible. Instead, allegations of conflicting norms are often forum-shifting strategies by major fishing states that oppose CITES listing of marine species or are concerned about the perceived lack of expertise and institutional bias within CITES and increased costs. This suggests that international lawyers should critically consider the interaction between relevant regimes, and the motives of those seeking 'coherence', before they deploy techniques to resolve conflicting norms. Indeed, in the

CITES–FAO context, attempts to resolve conflicting norms, such as according priority to successive treaties,[299] are inconclusive. More understanding is needed about regimes and the interaction between them if problems of fragmentation and the diversification of international law are to be addressed,[300] and the chapter sought to contribute to such an understanding.

Regime interaction itself should enhance understanding and scrutiny of regimes. For example, in opposing the role of CITES, some states implied that the regime of CITES embraced a precautionary approach to regulation while fisheries law required more rigorous scientific results. Contrary to these claims, however, the regimes of the law and CITES both adopt precautionary approaches in different contexts. Concepts that are represented by some states as 'agreed' within regimes should be subject to ongoing scrutiny. As a further example, some states misrepresented the discourse of 'development' to claim within the FAO that the CITES regime disadvantaged developing countries by promoting conservation. I demonstrated that such arguments failed to acknowledge the pluralist interests of developing countries such as in ecotourism. There is a need to avoid essentialist and simplistic assumptions about regimes in order to promote effective regime interaction.

It is useful to place the motives of the states that contested CITES' role in a wider context. Japan, for example, has been a party to CITES since 1980, UNCLOS since 1996, and the Fish Stocks Agreement since 2006. It has been a member of the FAO since 1951. Its opposition to CITES is not based on a lack of consent to international fisheries law, but rather on its preference for a form of fisheries management that does not involve restricting trade through CITES listings. Like its opposition to the development of fisheries subsidies disciplines in the WTO,[301] Japan's motivation to question the competence of CITES is based on a desire to control the form of international fisheries management, and perhaps even the forum. This has implications for the use of international norms from disparate regimes in treaty interpretation, and in particular the view that parallel membership of regimes is required before norms can be regarded as interrelated.[302]

The chapter described the negotiations to develop the MOU and analysed attempts made over the course of three years to restrict

---

[299] See, for example, the analysis by Franckx, above n. 38.
[300] ILC Analytical Study, 249 (para. 493(b)).   [301] See Chapter 3.
[302] This issue is discussed in Chapter 5 with particular reference to VCLT, Art. 31(3)(c).

CITES' role in listing marine species. If, for example, the attempts to restrict CITES' role are conceived as a form of treaty amendment, the MOU as proposed by Japan would have been subject to additional procedural requirements for CITES parties. However, the MOU is more correctly assessed as an agreement between the FAO and the CITES Secretariat, rather than between their respective members. As such, the FAO and the CITES Secretariat must act within their powers in agreeing an MOU, and any amendment to CITES' role in listing marine species would have been ultra vires.

The chapter revealed that rules of procedure within the FAO and CITES regime served to facilitate or constrain discussion on the MOU. Observers and even participating states were excluded from certain working groups. A further impediment to the MOU's development was a lack of national policy coordination on fisheries issues. By contrast, several procedures for cooperation between the FAO and the CITES Secretariat were being followed well before, and perhaps *despite*, the development of the MOU.

The MOU itself provides promising indications for cooperation. Indeed, states have begun to call for a similar model for relations between the FAO and the WTO, especially on the issue of subsidies.[303] There are, however, a number of constraints with the MOU as agreed between the CITES Secretariat and the FAO. One is the ad hoc nature of mutual observership, which is tied to a rather disputable concept of 'common interest'. Although capacity building is a welcome development, it could have been extended further to cover species that are not yet listed in the Appendices. In addition, the requirement that the CITES Secretariat must 'respect, to the greatest extent possible' and take 'due account' of the FAO review of listing proposals is rather fraught, as the recent attempt to list two shark species demonstrates. Furthermore, when considered in the context of the need to enhance compliance with CITES and FAO instruments, the MOU could have generated more proactive management, especially of enforcement. Together, these issues are relevant for the development of a legal framework of regime interaction, which I offer in Part III.

---

[303] FAO, *Fisheries and Aquaculture Report No. 902* (2009) para. 33.

5 Adjudicating a fisheries import ban at the WTO

> [T]he Appellate Body has, on occasion, been forced to confront strong arguments that its interpretation should not be totally confined to embellishing trade policies ... This is an area that the jurisprudence will need to develop further.[1]

This chapter investigates the interaction that occurs between the WTO and other regimes in the settlement of trade disputes. I take as my case study the dispute brought at the WTO against a United States import ban on shrimp from certain countries. This ban was part of a series of US measures designed to prevent the drowning of sea turtles in shrimp trawling nets; the United States claimed that the ban was justified for environmental reasons. The WTO panel and Appellate Body (AB) applied WTO law to resolve the dispute, but drew also on fisheries law and international environmental law in establishing relevant facts and in interpreting WTO treaty norms. This chapter considers these methods and compares regime interaction in other relevant WTO cases.

In doing so, the chapter uses the term 'non-WTO sources'[2] to describe treaties, other rules of international law, standards, guidelines and other materials produced by international bodies outside of the WTO. Some of these sources, such as CITES, are binding on at least some of the WTO members. Others have a non-binding character but still serve to guide the behaviour of states, such as the Resolution on Assistance to Developing Countries adopted in conjunction with the Convention on

---

[1] Consultative Board (to Director-General Supachai of the WTO), *The Future of the WTO: Addressing Institutional Challenges in the New Millennium* (2004) para. 236.
[2] See further Chapter 1, above n. 93 and accompanying text.

189

the Conservation of Migratory Species of Wild Animals (CMS), on which the AB partly relied in *US – Shrimp*.

A Shrimp fisheries and marine by-catch

While the previous chapter dealt with direct threats to endangered marine species, this chapter derives from less direct but equally important threats: the loss of marine life through by-catch. The unintended killing of marine species and seabirds is a major problem in contemporary fisheries. Modern fishing techniques are often indiscriminate and cover extensive marine areas, with the result that many non-targeted marine species are caught in wide nets and other fishing gear. According to some estimates, 38 million tonnes, or at least 40 per cent of global marine production, is caught as by-catch every year.[3]

The harvesting of shrimp, one of the most important globally traded fishery commodities,[4] contributes to large-scale drowning or killing of non-target species. Threatened and high-profile species that are caught as bycatch by shrimp trawling techniques include dolphins, seahorses, dugongs, albatrosses, penguins and – prominently – sea turtles.[5] Sea turtles, which are listed as endangered in Appendix I of CITES, are particularly threatened by shrimp fisheries. Their nesting sites and feeding grounds are often located near the most intensely trawled waters, and their entanglement and capture in shrimp trawls has been identified as causing more sea turtle mortality than all other human activities combined.[6]

B The forum shop: regimes of relevance

The UN Fish Stocks Agreement and the FAO Code of Conduct both address the issue of marine by-catch and shrimp fisheries. The Fish Stocks Agreement provides that signatories shall:

> minimize ... discards, catch by lost or abandoned gear, catch of non-target species, ... and impacts on associated or dependent species, in particular endangered species, through measures including, to the extent practicable,

---

[3] WWF 'Fact Sheet, Bycatch' (April 2009) (defining by-catch as anything caught that is either unmanaged or unused).

[4] Around six million tonnes of shrimp are produced each year, most of which enters the world market. Annual exports of shrimp are currently worth more than US$10 billion: *FAO Fisheries Technical Paper No. 475*, 'Global Study of Shrimp Fisheries' (2008) 37.

[5] Ibid., 47.   [6] Ibid., 50, and citations therein.

the development and use of selective, environmentally safe and cost-effective fishing gear and techniques.[7]

The FAO Code of Conduct for Responsible Fisheries addresses a similar provision at both states and RFMOs,[8] and calls for accurate and reliable data on by-catch to be compiled.[9] Both sets of provisions are worded in a non-binding way and, as described in Chapter 2, face difficulties in compliance and enforcement.

Other regimes recognise the threatened status of species affected by by-catch. Some regional conventions seek to regulate fishing activities so as to reduce the incidental capture or mortality of sea turtles.[10] Less directly, because their legal mechanisms are designed to restrain different threats, multilateral environment agreements such as CITES and CMS list sea turtles as endangered.[11] Perhaps due to the limitations of these international instruments in mitigating the effects of shrimp trawling methods on sea turtles, the United States acted unilaterally, giving rise to complaints within a different international regime: the WTO.

1 The US ban on shrimp products

At the end of the 1980s, the United States instituted a series of measures to conserve sea turtles. Sea turtles had been subject to protection in domestic environmental legislation since 1973. This legislation generally prohibited the taking of endangered sea turtles within the US, within the US territorial sea and the high seas.[12] In addition, the international trade in sea turtles was restricted by CITES; sea turtles were listed in CITES Appendix I.

These mechanisms all targeted the direct taking and trading of sea turtles. However, research from marine scientists indicated that the main cause of sea turtle mortality was not the direct taking of

---

[7] Fish Stocks Agreement, Art. 5(f).   [8] FAO Code of Conduct, Art. 7.6.9.
[9] Ibid., Art. 12.4. See also the global FAO/GEF/UNEP project 'Reduction of the Environmental Impact of Tropical Shrimp Trawling through the Introduction of Bycatch Reduction Technologies and Change of Management' (Project EP/GLO/201/GEF) (www.fao.org/fishery/gefshrimp/en).
[10] See e.g. Inter-American Convention for the Protection and Conservation of Sea Turtles 1996, Art. IV.2(h).
[11] See further Chapter 2; these regimes still monitor the effects of by-catch: see, e.g., CMS Press Release, 'Whales, dolphins and porpoises suffer dramatic declines from by-catch in fishing nets' (4 February 2010).
[12] See Endangered Species Act 1973 (US) ('ESA').

specimens but the incidental catching of sea turtles in trawling nets used in shrimp fisheries.[13]

In response, the US National Marine Fisheries Service (NMRF) developed and promoted the use of turtle excluder devices (TEDs) for use by shrimp trawlers. These devices allowed sea turtles to escape from trawling nets. After voluntary programmes failed to enhance the use of TEDs, the United States issued regulations to make them compulsory for all domestic shrimp trawl fishing where sea turtles were present.[14]

Recognising that sea turtles are migratory, the United States also enacted legislation in 1989 that promoted a multilateral strategy for the protection and conservation of sea turtles. Section 609 of Public Law 101-102 called for the initiation of negotiations with foreign governments for the development of relevant bilateral or multilateral agreements.[15] It also prohibited the import into the United States of shrimp harvested with technology that could adversely affect sea turtles,[16] unless the President certified to Congress that (i) the relevant harvesting nation had a regulatory programme and incidental take rate comparable to that of the United States; or (ii) that the fishing environment of the relevant harvesting nation did not pose a threat to sea turtles. The main tool for harvesting nations to achieve this certification was to demonstrate that shrimp were harvested 'by commercial shrimp trawl vessels using TEDs comparable in effectiveness to those required in the United States'.[17] Other exceptions to the ban existed, including for nations that harvested shrimp exclusively by artisanal means.[18]

The ban and exemptions had significant effects in countries that wished to access the United States' significant market for shrimp products. In just over a decade, the United States had certified nineteen countries as having adopted programmes to reduce the incidental capture of sea turtles in shrimp fisheries comparable to the US programme,

---

[13] See, e.g., National Research Council, National Academy of Sciences (1990) *Decline of the Sea Turtles: Causes and Prevention*, Washington DC, cited in Panel Report, *US – Shrimp*, para. 2.5.
[14] These regulations were pursuant to the ESA and became fully effective in 1990.
[15] Section 609 of Public Law 101-102, codified at 16 United States Code (U.S.C.) § 1537 ('Section 609').
[16] Section 609(b)(1).
[17] See 'Revised Notice of Guidelines for Determining Comparability of Foreign Programs for the Protection of Turtles in Shrimp Trawl Fishing Operations' 61 Fed. Reg. 17342 (19 April 1996) ('1996 Guidelines'). These had been elaborated and revised in United States guidelines in 1991, 1993 and 1996, and subject to litigation in the International Trade Court.
[18] Ibid.

sixteen countries as having shrimp fisheries in only cold waters where there was essentially no risk of taking sea turtles and eight countries on grounds that their fishermen only harvested shrimp using manual rather than mechanical means to retrieve nets.[19]

2 Complaint at the WTO

India, Malaysia, Pakistan and Thailand were not certified by the US programme and their shrimp products were therefore banned from the United States. They considered that this embargo was contrary to their rights under the WTO Agreements. After consultations with the United States failed to resolve the matter, the Dispute Settlement Body (DSB) established a panel in February 1997.

The complainants' main claim was based on the prohibition of quantitative restrictions in GATT Article XI.[20] This provision had been grounds for earlier complaints against the United States for fisheries import bans. In 1990, after the US imposed an embargo on yellowfin tuna caught with purse-seine nets because they threatened dolphins, a GATT Panel upheld a claim by Mexico.[21] In 1992, the EEC and the Netherlands were successful in their claim against the same embargo as it applied to countries that acted as 'intermediaries' in importing yellowfin tuna and then exporting it on to the United States.[22] In both cases, the GATT Panels found that the United States had violated Article XI.[23]

The United States did not dispute that 'with respect to countries not certified under Section 609, Section 609 amounted to a restriction on the importation of shrimp within the meaning of Article XI:1'.[24] It claimed, however, that the ban was justified under the permitted exceptions listed in GATT Article XX.

GATT Article XX provides a list of general exceptions to members' GATT obligations. The relevant provisions, together with the *chapeau*, read:[25]

Subject to the requirement that such measures are not applied in a manner which would constitute a means of arbitrary or unjustifiable discrimination between countries where the same conditions prevail, or a disguised restriction

---

[19] See further Panel Report, *US – Shrimp*, para. 2.16.
[20] The complainants also included GATT Art. I and Art. XIII in their claim.
[21] GATT Panel Report, *US – Tuna I*.   [22] GATT Panel Report, *US – Tuna II*.
[23] GATT Panel Report, *US – Tuna I*, paras. 5.17 – 5.19; GATT Panel Report, *US – Tuna II*, para. 5.10.
[24] Panel Report, *US – Shrimp*, para. 7.13.
[25] GATT Art. XX(b),(g). See further Chapter 2 esp. pp. 75–77.

on international trade, nothing in this Agreement shall be construed to prevent the adoption or enforcement by an [Member] of measures:

...

(b) necessary to protect human, animal or plant life or health;

...

(g) relating to the conservation of exhaustible natural resources if such measures are made effective in conjunction with restrictions on domestic production or consumption...

A key task for the WTO Panel was therefore to interpret the meaning of GATT Article XX. There was much scope here for the Panel to draw on relevant rules of international law and other non-WTO sources. In its ruling, however, the Panel did not consider whether the trade measures fell within Article XX(b) or (g), focussing instead on the *chapeau* of Article XX and finding that the measures were unjustifiably discriminatory.[26]

The ruling that the United States had violated the GATT was upheld by the AB. The AB was, however, very critical of the Panel's approach and reversed its interpretative analysis of GATT Article XX.[27] Drawing heavily on non-WTO sources, the AB decided that the US measure met the conditions of Article XX(g),[28] but that the US authorities had failed to meet the requirements of the *chapeau* in applying the measures to other countries.[29]

After this ruling, there was a further round of litigation at the WTO. Malaysia claimed that the United States compliance measures were inadequate and filed a dispute pursuant to DSU Article 21.5.[30] In implementing the AB's ruling, the United States had attempted to negotiate with other countries to protect sea turtles. It did not amend Section 609 but instead an interpretative note from the Department of State changed the application of the provision to the extent that it would not apply to states that had a 'comparatively effective framework' of sea turtle protection. This interpretative note effectively removed the requirement that states demonstrate the use of TEDs in their fishing sector in order to get access to the US market. Malaysia claimed that these measures did not amount to compliance.[31]

Given that the issue in the Article 21.5 dispute centred on the United States' efforts to conclude an agreement on sea turtles, there was much scope for the Panel to draw on non-WTO sources in its inquiries about

[26] Panel Report, *US - Shrimp*, para. 7.44; see further below n. 51 and accompanying text.
[27] AB Report, *US - Shrimp*, paras. 117-122; see further below n. 55 and accompanying text.
[28] Ibid., paras. 125-145.   [29] Ibid., paras. 176, 184.
[30] Panel Report, *US - Shrimp (Art 21.5)*, para. 1.4.   [31] Ibid.

the state of the international negotiations.³² The United States and Malaysia invoked a range of treaties to demonstrate efforts to engage in negotiations.³³ The Panel differentiated between a requirement to reach agreement and a duty to negotiate in good faith and found that the latter was the relevant standard against which to judge the US.³⁴ It found that the measures adopted by the United States were justified under Article XX as long as there were 'ongoing serious good faith efforts to reach a multilateral agreement' on turtle conservation and the use of TEDs.³⁵ The Appellate Body upheld this decision.³⁶

C Settling the dispute: scope for regime interaction

This section discusses how the adjudicating bodies in *US – Shrimp* drew on non-WTO sources in three ways:

(i) as applicable law between the disputing parties;
(ii) as interpretative tools for the understanding of relevant WTO treaty terms in accordance with customary rules of interpretation; and
(iii) in establishing facts in support of, or against, a claim of violation of the WTO law.

1 Applicable law

The issue of applicable law is not straightforward at the WTO. On one reading of the DSU, WTO panels may apply all law applicable between the parties, including from sources outside the WTO.³⁷ As such, if the disputing parties are also parties to an MEA, that MEA can be raised as a

---

[32] Ibid., especially para. 5.77.
[33] Ibid., especially paras 3.76, 3.81–3.82, 3.96–3.97, 3.104; see para. 5.76 for Panel's conclusions.
[34] Ibid., paras. 5.63, 5.67. [35] Ibid., para. 6.1.
[36] AB Report, *US – Shrimp (Art 21.5)*, para. 153.
[37] See Panel Report, *Korea – Procurement*, para. 7.96 with respect to customary international law: 'Such international law applies to the extent that the WTO treaty agreements do not "contract out" from it. To put it another way, to the extent there is no conflict or inconsistency, or an expression in a covered WTO agreement that implies differently, we are of the view that the customary rules of international law apply to the WTO treaties and to the process of treaty formation under the WTO.' See further David Palmeter and Petros Mavroidis, 'The WTO Legal System: Sources of Law' (1998) 92 *AJIL* 398, 409; Lorand Bartels, 'Applicable Law in WTO Dispute Settlement Proceedings' (2001) 35 *JWT* 499; Joost Pauwelyn, *Conflict of Norms in Public International Law: How WTO Law Relates to Other Rules of International Law* (2003), 460 and the ILC Analytical Study, para. 169. For the use of 'applicable law' in selected international disputes, see Campbell McLachlan, 'The Principle of Systemic Integration and Article 31(3)(c) of the Vienna Convention' (2005) 54 *ICLQ* 279 and references therein.

defence to a claim of WTO violation.[38] On the other view, WTO panels are restricted to applying WTO law.[39] This view is largely based on the panels' terms of reference as set out in DSU Article 7.1, which provides that panels are to examine the matter at issue 'in light of the relevant provisions ... in the covered agreements cited by the parties to the dispute'.[40] The situation can be compared to non-violation complaints under GATT Article XXIII, which can be heard by an adjudicating body if a WTO member considers that the benefits that accrue to it under the GATT are nullified or impaired by the conduct of another WTO member even if there is no breach of WTO obligations.[41] If the impugned conduct relates to non-WTO legal obligations, such as a breach of a country's obligations under the Fish Stocks Agreement that has led to trade effects, the panel or AB may be required to assess the conduct of the respondent member according to another regime.[42]

There was no need for the Panel in *US – Shrimp* to directly confront this issue. The grounds of the claims of the complaining parties and the United States' defence were all based on the GATT and there was no suggestion that the applicable law was anything other or more than the GATT. However, both the United States and the complainants referred to CITES at the Panel stage, and the panel noted that all the disputing parties were parties to CITES.

Even though there was no need for the Panel to consider any non-WTO rules as applicable law, it did indirectly apply CITES in the course of its reasoning. After noting the disputing parties' membership of CITES, the Panel found that the turtle species covered by the United States' trade measures were listed in CITES Appendix I. It then noted that while CITES concerned trade in 'endangered species', the import prohibition at issue restricted the trade in shrimp rather than the listed turtles. The Panel then ruled that CITES did not require its parties to adopt specific methods of conservation such as TEDs.[43] As such, the

---

[38] If the WTO and relevant MEA rules were found to conflict, the panel would apply a rule of recognition such as *lex specialis* to determine which rule prevails. For further discussion of the *lex specialis* rule, see ILC Analytical Study, paras. 46–422.
[39] See, e.g., Gabrielle Marceau, 'Conflict of Norms and Conflicts of Jurisdictions: The Relationship between the WTO Agreement and MEAs and other Treaties' (2001) 35 *JWT* 1081, 1116.
[40] DSU Art. 7.1   [41] GATT Art. XXIII; DSU Art. 26.
[42] See further Bagwell, Mavroidis and Staiger, noted in Chapter 2, n. 309 and accompanying text.
[43] Panel Report, *US – Shrimp*, para. 7.58.

Panel implied that CITES did not provide a relevant defence for the US trade measure at issue in the case.

The issue of applicable law arose indirectly in *US - Shrimp (Art 21.5)*. The Panel held that the efforts of the United States to negotiate with Malaysia could be measured against the efforts it had made in concluding with countries of the Caribbean and Western Atlantic region the Inter-American Convention for the Protection and Conservation of Sea Turtles.[44] The Panel concluded that the 'Inter-American Convention can reasonably be considered as a benchmark of what can be achieved through multilateral negotiations in the field of protection and conservation'.[45] On appeal, Malaysia claimed that the word 'benchmark' had the connotation of a 'legal standard' that was unsupportable.[46] The AB rejected that the Panel had used the Inter-American Convention as a legal standard and stated that it had been merely a 'basis for comparison'.[47] The AB thus clarified that the non-WTO source was applied by the Panel to establish facts rather than as law.

2 Treaty interpretation

A WTO adjudicating body may draw on non-WTO rules of international law in interpreting WTO treaty terms. According to the DSU, WTO panels must follow customary norms of treaty interpretation.[48] These norms are codified, at least in part, in the VCLT.[49] VCLT Articles 31 and 32 provide:

Article 31 General rule of interpretation

1. A treaty shall be interpreted in good faith in accordance with the ordinary meaning to be given to the terms of the treaty in their context and in the light of its object and purpose.

...

3. There shall be taken into account, together with the context:
   (a) any subsequent agreement between the parties regarding the interpretation of the treaty or the application of its provisions;
   (b) any subsequent practice in the application of the treaty which establishes the agreement of the parties regarding its interpretation;
   (c) any relevant rules of international law applicable in the relations between the parties.

---

[44] Panel Report, *US - Shrimp (Art 21.5)* para. 5.71.   [45] Ibid.
[46] AB Report, *US - Shrimp (Art 21.5)* para. 3.13.   [47] Ibid., para. 130.   [48] DSU Art. 3.2.
[49] The AB has considered VCLT Articles 31 and 32 to have each attained the status of rules of customary or general international law: see, e.g., respectively, AB Report, *US - Gasoline*, pp. 15–16; AB Report, *Japan - Alcohol II*, p. 9.

4. A special meaning shall be given to a term if it is established that the parties so intended.

Article 32 Supplementary means of interpretation

Recourse may be had to supplementary means of interpretation, including the preparatory work of the treaty and the circumstances of its conclusion, in order to confirm the meaning resulting from the application of article 31, or to determine the meaning when the interpretation according to article 31:
  (a) leaves the meaning ambiguous or obscure; or
  (b) leads to a result which is manifestly absurd or unreasonable.

The *US – Shrimp* dispute hinged on whether the United States' measures were justified by GATT Article XX. Consequently, the interpretation of the terms of GATT Article XX was key to the case. The Panel and the AB differed significantly in their method of interpreting Article XX and in their use of non-WTO sources for this purpose.

The Panel interpreted Article XX without recourse to rules of international law or other non-WTO sources.[50] Instead, the Panel focused on the object and purpose of the whole of the GATT 1994 and the Marrakesh Agreement. It concluded that measures which 'undermine the WTO multilateral trading system'[51] must be regarded as 'not within the scope of measures permitted under the chapeau of Article XX'.[52] It ruled that Article XX required parties to adopt multilateral solutions to trade issues and, by acting unilaterally, the US measures constituted unjustifiable discrimination because it found no reason to ascertain any provisional justification for a measure if the *chapeau* was not satisfied.[53] The Panel did not go on to consider Article XX(b) or (g).[54]

The AB overturned the Panel's interpretative approach to Article XX.[55] It ruled that the Panel erred in only referring to the *chapeau* of Article XX and instead began its analysis with reference to Article XX(g). In doing so, the AB made repeated references to non-WTO sources.

(a) Appellate Body's interpretation of Article XX(g)

The main opportunity for regime interaction occurred when the AB interpreted GATT Article XX(g) by reference to a number of relevant rules of international environmental law. There were two major aspects of Art XX(g) that required interpretation: whether the measure related

---

[50] Panel Report, *US – Shrimp*, paras. 7.33 – 7.46.   [51] Ibid., para. 7.44.
[52] Ibid., para. 7.62.   [53] Ibid., para. 7.61.   [54] Ibid., para. 7.63.
[55] AB Report, *US – Shrimp*, para. 122.

to 'exhaustible natural resources'; and whether it was a measure 'relating to conservation'.

In interpreting the meaning of 'exhaustible natural resources', the AB had to rule on whether it included sea turtles. India, Pakistan and Thailand had contended that 'exhaustible' meant 'finite resources such as minerals, rather than biological or renewable resources'.[56] Malaysia, too, construed Article XX(g) as applying to 'nonliving exhaustible natural resources'.[57]

The AB disagreed. It first interpreted 'exhaustible' by reference to scientific consensus about the risks of extinction of species that are otherwise capable of reproduction.[58] This scientific consensus was to be found in the work of international bodies such as the World Commission on Environment and Development.[59] On whether the term 'natural resources' was restricted to non-living resources, the AB found that it was 'pertinent to note' that international Conventions and declarations such as UNCLOS, the CBD and Agenda 21 made frequent references to natural resources as embracing both living and non-living.[60]

In referring to these non-WTO sources, the AB did not identify whether the disputing parties or the WTO membership as a whole had ratified them. An exception was the Resolution on Assistance to Developing Countries adopted in conjunction with the CMS, on which the AB partly relied in its interpretation of 'natural resources'. The AB noted that India and Pakistan had ratified the Convention but that Malaysia, Thailand and the United States were not parties.[61]

The AB did not refer to specific paragraphs of the VCLT rule of interpretation in its interpretation of 'exhaustible natural resources',[62] although it is likely that it was drawing on VCLT Article 31(3)(c) or VCLT Article 31(1).[63] It then followed the rule on supplementary means of interpretation in VCLT Article 32 to conclude its interpretation of 'exhaustible natural resources'. It considered the drafting history of Article XX, the GATT-adopted reports of *US – Tuna (Canada)* and *Canada – Herring and Salmon* and the principle of effectiveness in treaty interpretation.[64] After this interpretative exercise, the AB ruled that

---

[56] Panel Report *US – Shrimp*, para. 3.237.   [57] Ibid., para. 3.240.
[58] AB Report, *US – Shrimp*, para. 128.   [59] Ibid.   [60] Ibid., para. 130.   [61] Ibid.
[62] cf The AB's interpretation of the *chapeau*, for which it relied upon general principles of law according to VCLT Art. 31(3)(c): see ibid., para. 158, noted below n. 77 and accompanying text.
[63] See Pauwelyn, above n. 37, 256.   [64] AB Report, *US – Shrimp*, para. 131.

'exhaustible natural resources' could include living or non-living species such as the sea turtles at issue in the case.[65]

The second major part of Article XX(g) requiring interpretation was the issue of whether the US measure was one 'relating to conservation'.[66] Again, the AB drew on non-WTO rules to determine whether there was a relationship between the measures at stake and the policy goal of conserving exhaustible natural resources.

In determining whether the measures 'related to' conservation, the AB considered whether the means were reasonably related to the ends.[67] In ascertaining the importance of the policy goal (the 'ends'), the AB referred to CITES and found that the policy of protecting sea turtles was shared by all disputing parties and third parties,[68] and the 'vast majority of the nations of the world'.[69] In this regard, the AB noted that there were 144 states parties to CITES at the time of the appeal and that these included all parties and third parties.[70]

In considering whether the measures designed to achieve this policy goal, as contained in Section 609, constituted a 'means' to this 'end', the AB considered the science behind the use of TEDs and drew on the Panel's consultation with scientific experts.[71] The AB held that Section 609 was 'not disproportionately wide in its scope and reach in relation to the policy objective of protection and conservation of sea turtle species'[72] and concluded that the measures 'related to' conservation.[73]

(b) Appellate Body's interpretation of Article XX *chapeau*

The other major provision that fell for interpretation by the AB was the *chapeau* of Article XX. As noted above, the Panel had interpreted the *chapeau* in accordance with the broad object of the WTO Agreements, and without reference to other rules of international law, in finding that the US measure amounted to unjustifiable discrimination. The AB departed from the Panel's approach and again drew heavily on relevant rules of international law.

The AB articulated the interrelationship between international trade law and international environmental law by noting that the WTO

---

[65] Ibid., para. 134. [66] Ibid., paras. 135–142. [67] Ibid., para. 141.
[68] I note that the countries that appear as third parties to appeals are known as 'third participants' according to the AB Working Procedures; I have retained the language of 'third parties' throughout.
[69] AB Report, *US – Shrimp*, para. 135. [70] Ibid. [71] Ibid., para. 140.
[72] Ibid., para. 141. [73] Ibid., para. 142.

negotiators had added 'sustainable development' to the Preamble to the Marrakesh Agreement.[74] The AB recalled, too, that WTO members had established a permanent CTE and, at the same time, had taken note of the Rio Declaration and Agenda 21.[75] These Uruguay Round initiatives demonstrated for the AB that its interpretation of the *chapeau* should take into account the notion of sustainable development.[76] Moreover, general principles of law were relevant in guiding the content of the 'good faith' obligations of the *chapeau*.[77]

The AB also drew on relevant rules of international law in assessing whether the United States met the requirements of the *chapeau*. As mentioned above, the AB had noted in general that the *chapeau* to Article XX required the AB to keep in mind the objective of sustainable development. On the specific question of whether there had been 'unjustifiable discrimination' in the application of the trade measure, the AB considered whether there had been a failure to engage in multilateral negotiations as required by 'environmental protection policy'.[78]

This was not merely a factual inquiry, and the AB continued to build on its conception of the requirements of the *chapeau*. The AB recalled that both WTO and non-WTO rules recognised the need for concerted and cooperative efforts to conserve highly migratory species such as sea turtles.[79] The AB excerpted relevant parts of the Rio Declaration, Agenda 21, the CBD and the CMS.[80] The AB noted that WTO members had endorsed multilateral solutions when they referred to the mutually supportive relationship of MEAs and WTO Agreements.[81]

(c) The need for parallel membership of relevant rules

In citing this body of laws and declarations from international environmental law, the AB did not investigate whether they had been ratified or endorsed by all the disputing parties, except for its reference to CITES to demonstrate the genuineness of the policy of sea turtle conservation.[82] Similarly, the AB did not investigate whether all the WTO members had ratified and endorsed the MEAs it cited. The only overture about any need for parallel membership between these bodies of law and the WTO membership was an implication that WTO members had taken note of

---

[74] Ibid., para. 152. [75] Ibid., para. 154. [76] Ibid., para. 155. [77] Ibid., para. 158.
[78] Ibid., para. 167. [79] Ibid., para. 168. [80] Ibid.
[81] Ibid (citing the WTO Report of the CTE (1996)).
[82] AB Report, *US – Shrimp*, para. 135 noted above n. 70 and accompanying text.

the Rio Declaration and Agenda 21 and had incorporated them into the work of the CTE.[83]

In *US – Shrimp (Art 21.5)*, by contrast, the Panel recorded that Malaysia and the United States, the two disputing parties, were bound by the rules of international law that had been cited by the AB in *US – Shrimp*.[84] In doing so, the Panel implied that VCLT Article 31(3)(c) incorporated as valid tools of interpretation relevant rules of international law that were binding on the parties to the dispute.[85] The Panel used this as evidence that reinforced the need for the United States to achieve multilateral solutions and use non-trade restrictive measures in sea turtle conservation.[86]

The implication that 'relevant rules of international law' that are interpretative tools pursuant to VCLT Article 31(3)(c) are rules that are binding on the 'disputing parties' was not followed in subsequent WTO litigation. In *EC – Biotech*, the United States, Canada and Argentina challenged the EC's importation measures surrounding genetically modified organisms. While not relating specifically to fisheries, the relevance of genetically modified organisms to environmental issues more generally means the decision is particularly worthy of consideration here. In *EC – Biotech*, the Panel declined to take account of 'relevant rules of international law' pursuant to VCLT Article 31(3)(c). The Panel understood that the requirement in Article 31(3)(c) that such rules be 'applicable in the relations between the parties' meant that the rules had to be binding on all the WTO members,[87] rather than binding on the disputing parties, an interpretation advanced by the disputing parties themselves.[88] The Panel then established that the CBD and the Biosafety Protocol, which had been invoked by the respondent as tools of interpretation, did *not* have the same coverage of members as the WTO covered agreements (by noting, in particular, the fact that the United States had not ratified either).[89] The Panel concluded that it

---

[83] Ibid. (citing the WTO Decision on Trade and Environment (1994) Preamble and para. 2(b)).
[84] Panel Report, *US – Shrimp (Art 21.5)*, para. 5.57 (referring to the rules cited by the AB Report, *US – Shrimp*, para. 168, namely the Rio Declaration, Agenda 21, the CBD, the CMS, and the WTO Report of the CTE (1996)).
[85] Ibid. [86] Ibid., paras. 5.58–5.59. [87] Panel Report, *EC – Biotech* para. 7.68.
[88] Panel Report, *Biotech* para. 4.543 (US); para. 4.600 (Canada); para. 4.688 (Argentina). Canada subsequently amended its approach: see para. 7.60. This understanding of 'the parties' as parties 'to the dispute' has also been advanced by several commentators: see Palmeter and Mavroidis, above n. 37, 411. This is also implicit in Marceau, above n. 39, 1087.
[89] Panel Report, *EC – Biotech*, paras. 7.74–7.75.

could not take account of these treaties, or the precautionary principle that the EC had claimed to be a customary norm of international law, pursuant to VCLT Article 31(3)(c).[90]

The *EC – Biotech* Panel attempted to reconcile its approach with the AB in *US – Shrimp* by invoking VCLT Article 31(1) to draw on non-WTO sources in finding the 'ordinary meaning' of treaty terms.[91] The Panel considered that this provision allowed for the use of rules of international law that were not binding on the parties where those rules provided evidence of the 'ordinary meaning' of the treaty terms and were thus 'informative'.[92] Extending the well-known reliance by WTO panels on language dictionaries in finding the 'ordinary meaning' of terms,[93] the Panel thus incorporated international law instruments as sources of linguistic guidance. The Panel considered that this approach would not 'mandate' a consideration of relevant rules of international law, as compared with Article 31(3)(c).[94] However, if a rule of international law could 'shed light on the meaning and scope of a treaty term to be interpreted', a panel could have regard to it.[95] The Panel found its approach to be consistent with the AB's use of relevant rules of international law that were not binding on all the parties in *US – Shrimp* and declared that 'the mere fact that one or more disputing parties are not parties to a convention does not necessarily mean that a convention cannot shed light on the meaning and scope of a treaty term to be interpreted'.[96] The Panel stated that it had given careful consideration to various provisions of the CBD and the Biosafety Protocol on this basis. It concluded that it did not find it 'necessary or appropriate to rely on these particular provisions in interpreting the WTO agreements at issue in this dispute'.[97] It did, however, take account of a large number of reference materials provided to it by several international organisations, namely Codex, FAO, the IPPC Secretariat, WHO, OIE, the CBD

---

[90] Ibid., para. 7.89.   [91] Ibid., paras. 7.90–7.96.   [92] Ibid., para. 7.92.
[93] The use by WTO dispute settlement bodies of dictionaries has been criticised as an over-textual approach: see Henrik Horn and Joseph Weiler, 'European Communities – Trade Description of Sardines: Textualism and its Discontent' in Henrik Horn and Petros Mavroidis (eds.) *The WTO Case Law of 2002* (2003) 248, 248. The AB has nodded at the limitations of dictionaries: see, e.g., AB Report *Canada – Aircraft* para. 153.
[94] Panel Report, *EC – Biotech*, para. 7.92 cf para. 7.69.   [95] Ibid., para. 7.95.
[96] Ibid., para. 7.92. The Panel continued in a footnote: 'Equally, in a case where all disputing parties are parties to a convention, this fact would not necessarily render reliance on that convention appropriate.' This may indicate reticence by the Panel to refer to non-WTO sources as applicable law.
[97] Ibid., para. 7.95.

Secretariat and UNEP, to assist its search for ordinary meaning.[98] These materials included Conventions, standards and guidelines of these international organisations, in addition to glossaries and reference works.[99]

According to the method of treaty interpretation adopted by the Panel in *EC – Biotech*, WTO adjudicating bodies will never be *required* to take non-WTO sources into account, except for the rare event that a rule of international law is binding on the WTO membership as a whole.[100] Instead, non-WTO sources may be 'informative' in understanding the terms of the WTO covered agreements. I assess the correctness of this ruling below,[101] but it is important to note here that this classification of non-WTO sources blurs the line between treaty interpretation and the use of such sources in ascertaining facts.

3 Relevant facts

Under the DSU, a WTO panel is required 'to make an objective assessment of the matter before it, including an objective assessment of the facts of the case'.[102] As such, a WTO panel has a duty to have regard to all relevant evidence, and should not conduct a *de novo* review nor defer to national authorities.[103] It can review evidence that post-dates the establishment of the panel.[104] As well as having an obligation to consider all evidence presented to it,[105] the panel is empowered to seek relevant information.[106]

It was this factual inquiry, rather than a need to interpret WTO treaty terms, that formed the basis of the Panel's use of non-WTO sources in *US – Shrimp*. The Panel drew on CITES, the CMS and the IUCN when determining the status of sea turtles.[107] As mentioned above, the Panel did not go on to consider the relevance of their status to the text of GATT Article XX.

Like the Panel, the AB also used non-WTO sources to ascertain the status of sea turtles. After interpreting the terms of Article XX(g), it

---

[98] Ibid., para. 7.96. [99] Ibid.
[100] I note, however, that treaties will never be binding on all WTO members because the 'parties to the WTO Agreement' include customs territories that are simply unable to be parties to treaties like the CBD.
[101] See below n. 287 and accompanying text. [102] DSU Art. 11.
[103] AB Report, *EC – Hormones*, para. 117.
[104] AB Report, *EC – Selected Customs Matters*, para. 188; see also AB Report, *Brazil – Retreaded Tyres*, para. 193.
[105] See AB Report, *EC – Hormones*, para. 133. [106] DSU Art. 13; see further nn. 129ff.
[107] Panel Report, *US – Shrimp*, para. 2.3.

established their threatened condition by first noting that all the parties at this stage of the dispute had appeared to concede that sea turtles were exhaustible. It then reinforced this finding by referring to the CITES listing[108] (rather than the IUCN or the CMS) and the scientific expertise considered by the Panel.[109] This scientific expertise was itself dependent on non-WTO sources. For example, one of the experts referred the Panel to the FAO Code of Conduct. The Panel cited this as evidence of the types of conservation methods generally endorsed by the international community, which recognised the need for states to endorse environmentally safe fishing gear (such as TEDs) but in the context of mutual consultation and cooperation.[110]

The AB also referred to international environmental law in order to answer the specific factual question of whether the United States met its obligations to cooperate. Here, whether the United States was a member of the relevant MEAs was a key factor in the AB's reasoning. The AB noted that the United States had attempted to engage diplomatically by negotiating the Inter-American Convention, which it had concluded with five other countries.[111] The AB found, however, that the United States had failed to use such multilateral procedures in other diplomatic areas, and recalled that the United States had not signed the CMS or UNCLOS, and had not ratified the CBD.[112]

The Panel and the AB used non-WTO sources to establish facts in the compliance dispute brought under DSU Article 21.5, as described above.[113] This example demonstrates that, like the distinction between treaty interpretation and fact-finding, the line between law and fact is blurred. For example, if the United States subsequently entered into an MEA with Malaysia, this MEA would arguably be a valid defence to a future WTO claim made in these terms.[114] The AB assessment of alleged violations of the Enabling Clause in *EC – Tariff Preferences* suggests a similar relationship between law and fact. International norms were recognised by the AB as providing a benchmark for behaviour for GSP

---

[108] AB Report, *US – Shrimp*, para. 132.   [109] Ibid., para. 133.
[110] Panel Report, *US – Shrimp*, para. 7.59; see AB Report, *US – Shrimp*, para. 77.
[111] AB Report, *US – Shrimp*, para. 169.   [112] Ibid., para. 171.
[113] I note that the Panel's analysis of the Inter-American Convention as providing a 'benchmark' was criticised by the AB in *US – Shrimp (Art 21.5)*, para. 130: see further above n. 47 and accompanying text.
[114] Pauwelyn, above n. 37, 465; see further Joost Pauwelyn, 'How to Win a World Trade Organization Dispute Based on Non-World Trade Organization Law?' (2003) 37 *JWT* 997.

donor states, so that an adjudicating body could objectively verify whether state behaviour accords with WTO obligations by inquiring into the standards set down in exogenous sources.[115]

In summary, there was scope for regime interaction in the US – Shrimp dispute in the use of sources from international environmental and fisheries law in the applicable law of the dispute, the interpretation of the relevant exceptions to the GATT and in the establishment of facts. Equally relevant to a discussion of regime interaction is the question of 'how' the Panel and the AB obtained these non-WTO sources.

D Settling the dispute: methods of regime interaction

The Panel in US – Shrimp was provided with information about other regimes from several different individuals and bodies, including the parties themselves and NGOs. On appeal, the AB made some important rulings about the accessibility of WTO dispute settlement proceedings. Although the hearings of the disputes are generally closed,[116] individuals and international organisations may provide information about other regimes through *amicus* briefs or through solicited consultation with adjudicating bodies. In this section, I distinguish between five available methods of regime interaction and analyse the approach of US – Shrimp and other relevant WTO disputes.

1 Panellists, AB members and the Secretariat

The access to non-WTO sources in WTO dispute settlement proceedings is facilitated if the adjudicators themselves have an understanding of and expertise in exogenous regimes. The ability of panellists, AB members and the Secretariat to act as conduits enhances regime interaction, even if it is not strictly interaction itself. This section assesses the relevant background and qualifications of these individuals.

There is no permanent panel at the WTO. Instead, ad hoc panellists are appointed for every dispute. According to the DSU, panellists are nominated by the Secretariat and appointed with the agreement of the

---

[115] AB Report, *EC – Tariff Preferences*, para 163, as discussed in Chapter 3 n. 246 and accompanying text.
[116] DSU Art. 14. But note that the parties may choose to open the proceedings to the public, which first occurred in 2005 in the ongoing *Hormones* litigation; see also *Canada – Continued Suspension* and *US – Continued Suspension*.

disputing parties.[117] If disputing parties fail to agree on the selection of panellists, the DSU provides that the Director-General shall appoint them. This commonly occurs.[118] The DSU also provides that panellists are to be selected 'with a view to ensuring [their] independence ..., a sufficiently diverse background and a wide spectrum of experience'.[119] They are considered to be 'well-qualified' if they have presented a case to a panel, served as a representative or senior trade policy official of a WTO member, worked in the WTO Secretariat or 'taught or published on international trade law or policy'.[120] The Secretariat keeps an indicative list of governmental and non-governmental individuals to assist in the selection of panellists, which is updated by the members every two years.[121] WTO members can obtain the *curricula vitae* of individuals on this list from the Secretariat.[122]

The AB, on the other hand, is a permanent institution. The WTO membership as a whole appoints seven members to serve four-year terms. These appointees are 'persons of recognized authority, with demonstrated expertise in law, international trade and the subject matter of the covered agreements generally'.[123] The DSU also provides that the AB members must be unaffiliated with any government, and the membership of the AB as a whole 'shall be broadly representative of membership in the WTO'.[124] Unlike the panel roster, the biographies of AB members are freely available.[125]

These DSU provisions do not require panellists or AB members to have experience in other international regimes, although the requirement that AB members have demonstrated expertise 'in law' has led to the selection of experts in public international law. In *US – Shrimp*, the three Panel members were government officials from Hong Kong, Brazil and Germany.[126] The Panel record does not indicate whether they possessed any additional expertise in species protection or fisheries and international environmental law.[127] In the appeal, the AB

---

[117] DSU Art. 8.
[118] William Davey, 'The Case for a WTO Permanent Panel Body' (2003) 6 *JIEL* 177.
[119] DSU Art. 8.2.   [120] Ibid., Art. 8.1.   [121] WTO Doc. WT/DSB/19.
[122] Ibid.   [123] DSU Art. 17.3.   [124] Ibid.
[125] See www.wto.org/english/tratop_e/dispu_e/ab_members_descrp_e.htm.
[126] WTO Doc. WT/DS58/9. The three Panel members were Michael Cartland (Hong Kong's first Permanent Representative to the GATT), Carlos Cozendey (an official with the Brazilian Ministry of External Relations, Trade Policy Division) and Kilian Delbrück (Germany).
[127] I note that the WTO Secretariat's provision to WTO members of panellists' *curricula vitae* (noted above n. 122) does not extend to requests from the public.

members had experience in international trade and public international law.[128]

Instead of requiring adjudicators to be experienced in other regimes, the DSU envisages that panellists can learn from relevant non-WTO sources through their ability to seek information from and consult external sources. The Panel in *US – Shrimp* made a significant ruling about its right to 'seek information ... from any individual or body which it deems appropriate'.[129] The Panel considered that this right did not extend to information from NGOs that were in the form of unsolicited *amicus* briefs.[130] According to this interpretation of DSU Article 13, the resort to non-WTO sources in a WTO dispute would depend heavily on the initiative of the panel members. Such initiative would in turn depend on the experience and background of panel members, who would have to know about relevant individuals and organisations in order to approach them. On appeal, the AB overturned the Panel's interpretation,[131] as I discuss in greater detail in my separate analysis of NGO *amicus* briefs as a method of regime collaboration.

Aside from the background of panellists and AB members, it is important to note the degree of assistance provided by the WTO Secretariat to the adjudicating bodies. As well as providing 'secretarial and technical support', the Secretariat has responsibility for 'assisting panels ... on the legal, historical and procedural aspects of the matters dealt with'.[132] These aspects may include secretariat assistance on historical and legal aspects of disputes which may draw on knowledge and experience of other regimes[133] as well as awareness of institutional methods of ongoing regime interaction within the WTO, such as the CTE, of which the Secretariat may be more cognisant than the disputing parties.[134] Although there is limited information about whether Secretariat assistance extends to the drafting of reports, it is clear that the awareness

---

[128] The AB members were Florentino Feliciano from the Philippines, James Bacchus from the United States and Julio Lacarte-Muró from Uruguay. Their biographies are available on the WTO website.
[129] DSU Art. 13.   [130] Panel Report, *US – Shrimp*, para. 7.8.
[131] AB Report, *US – Shrimp*, para. 110.   [132] DSU Art. 27.
[133] For a contrast between the influence of the Secretariat in GATT days, which had a narrow economic outlook, and the post-Uruguay Round Secretariat, which is more open to other perspectives, see Robert Howse, 'From Politics to Technocracy – and Back Again: The Fate of the Multilateral Trading Regime' (2002) 96 *AJIL* 94, 108, 117.
[134] On the importance of the CTE to the AB's interpretation of GATT Art. XX in *US – Shrimp*, see above n. 81 and accompanying text.

2 The parties' submissions

The second most important conduits for non-WTO sources in WTO disputes are the parties themselves. WTO members have wide standing rights to bring claims to the WTO[136] or to appear as third parties.[137] No other individuals or bodies have standing. Moreover, only members who are parties or third parties to a dispute have a *legal right* to make submissions to a WTO panel.[138] As such, 'a panel is *obliged* in law to accept and give due consideration only to submissions made by the parties and the third parties in a panel proceeding'.[139] In general, panels must base their decision only on the claims put to them.[140]

In *US – Shrimp*, the United States relied on international environmental laws and declarations such as UNCLOS and Agenda 21 as support for its submission that the use of TEDs had 'become a recognized multilateral environmental standard' and therefore was justified under GATT Article XX.[141] In *US – Shrimp (Art 21.5)*, the United States and Malaysia invoked a range of treaties to show that they had tried to negotiate on particular issues.[142]

The invocation of non-WTO sources by the parties themselves has featured in a number of WTO disputes. Most prominently, the policy issues that underlie the *EC – Biotech* case have been considered in many international forums. These issues are discussed under various terms of 'GM', 'biotechnology' and 'biosafety'.[143] For example, the Biosafety Protocol specifically addresses the 'transport, handling and use of living modified organisms resulting from modern biotechnology that may

---

[135] On the number and geographical representation of WTO Secretariat staff, see www.wto.org/english/thewto_e/secre_e/intro_e.htm; see also the internship programme: www.wto.org/English/thewto_e/vacan_e/intern_e.htm.
[136] GATT Art. XXIII; DSU Art. 3.7; see further AB Report, *EC – Bananas III*, para. 135.
[137] DSU Art. 10. [138] Ibid., See also DSU Art. 12 and Appendix 3.
[139] AB Report, *US – Shrimp*, para. 101.
[140] DSU Art. 7; see also DSU Arts. 11, 13. But see *Belgian Family Allowances*, where the GATT Panel famously extended the application of MFN to the Belgian scheme notwithstanding that the complaining parties had not impugned the scheme as a whole: see further Steve Charnovitz 'Belgian Family Allowances and the Challenge of Origin-Based Discrimination' (2005) 4 *World Trade Review* 7, 10.
[141] See Panel Report, *US – Shrimp*, para. 7.57. [142] See above n. 33 and accompanying text.
[143] The *Biotech* panel used the terms biotech products, GMOs, GM plants, GM crops or GM products interchangeably: see paras. 7.1–7.2.

have adverse effects on ... biological diversity'.[144] The use by the Panel of this international legal context was dictated largely by the claims and submissions of the disputing parties.[145]

The complaining parties in *EC – Biotech* based their claims on three WTO covered agreements: the SPS Agreement, the TBT Agreement and the GATT. In its defence, the EC claimed that three rules of international environmental law were relevant to the dispute and should be used by the Panel as interpretative tools according to DSU Article 3.2. First, the precautionary principle was said by the EC to be a general principle of law. Secondly, the EC invoked the Convention on Biological Diversity (CBD), which recognises the precautionary principle in its Preamble. Of the disputing parties, the EC, Argentina and Canada were bound by the CBD, while the United States had signed but not ratified it. Thirdly, the EC referred to the Biosafety Protocol, which lays down requirements for the transboundary movement of 'living modified organisms'.[146] It had been ratified by the EC but Argentina and Canada had only signed it and the United States had no involvement with it except for participation in the 'Biosafety Clearing-House' information-sharing mechanism. After a request by the Panel, the EC provided a list of provisions from these MEAs that it considered to be necessary for the Panel to take into account.[147]

It is important to note the influence of the disputing parties in framing the Panel's use of non-WTO sources in *EC – Biotech*. In its submissions, the EC did not claim that the rules of international law enshrined in the precautionary principle, the CBD and the Biosafety Protocol should be directly applied by the Panel.[148] It claimed instead that

---

[144] Biosafety Protocol, Art. 1. In its submissions, the EC emphasised that there were 103 signatories to the Biosafety Protocol: see *Biotech* para. 4.340. There are currently 157 parties: see www.cbd.int/biosafety/signinglist.shtml. On the number of regimes relevant to biotechnology, including the Biosafety Protocol and Codex, see Donald Buckingham and Peter Phillips, 'Hot Potato, Hot Potato: Regulating Products of Biotechnology by the International Community' (2001) 35 *JWT* 1.

[145] In addition to the multiple submissions of the four disputing parties, there were third-party submissions from Australia, Chile, China, New Zealand and Norway. The Panel also obtained written and oral evidence from international organisations and scientific experts and had access to three sets of *amicus* briefs: see below nn. 164, 193, 237 and accompanying text.

[146] See Biosafety Protocol, Art. 3(g).   [147] Panel Report, *EC – Biotech* para. 7.95.

[148] See, e.g., *EC – Biotech* Annex D, D-91, para. 18 with respect to the CBD and Biosafety Protocol: 'The European Communities is not inviting the panel to "apply" these instruments as such, but rather to ensure that the WTO rules are interpreted consistently with them.'

these rules were relevant to the interpretation of the WTO treaty terms. By framing the case in this manner, the EC missed an opportunity to invoke as a legal defence both the CBD, at least with respect to its obligations to the two complaining parties who had ratified it, and the precautionary principle, which as a general principle of law would have applied to all the disputing parties. As a consequence, the Panel in *EC – Biotech* did not need to address the issue of whether non-WTO law could be applied by a WTO adjudicating body as 'applicable law between the disputing parties' in defending an alleged WTO violation.[149] Nor did the Panel take account of these rules of international environmental law in its interpretation of the relevant WTO terms.[150] The Panel did, however, take into account a large number of treaties and soft law instruments as part of its quest to find the 'ordinary meaning' of SPS treaty terms, pursuant to VCLT Article 31(1).[151] The Panel obtained most of its information through consultation with scientific experts and IGOs, which I explain in the next section.

In addition to the disputing parties, submissions by third parties are crucial in elaborating norms for WTO panels. Panels are bound to give due consideration to these submissions as well as the disputing parties' submissions.[152] Third parties may file written or oral submissions in any proceedings for which they have notified their interest to the DSB.[153] These submissions may contain reference to non-WTO norms or clarify information about non-WTO norms that has been presented by others. For example, in *Brazil – Retreaded Tyres*, the United States clarified the basis of a factual submission that had been made by the Humane Society International, which had relied on a statement of the US Environment Protection Agency.[154] Unlike evidence from scientists, IGOs or NGOs considered below, the Panel is required to take these submissions into account.

3 Consultation with scientific experts

In certain disputes involving scientific or technical issues, the DSU provides that panels may consult with scientific experts or establish

---

[149] It may be said to have made an oblique reference to this issue: Panel Report, *Biotech* para. 7.72; see further Margaret A. Young, 'The WTO's Use of Relevant Rules of International Law: An Analysis of the Biotech Case' (2007) 56 *ICLQ* 907, 912–13.
[150] See above n. 90. [151] See above n. 98. [152] See above n. 139.
[153] DSU Art. 10; AB Working Procedures.
[154] Panel Report, *Brazil – Retreaded Tyres*, para. 5.158.

an advisory experts group.¹⁵⁵ Notwithstanding that this method of consultation is restricted to scientific and technical issues, such processes expose the panel to a range of non-WTO sources and often entail collaboration with IGOs.¹⁵⁶

In *US – Shrimp*, none of the disputing parties requested the Panel to consult scientific experts. The Panel instead decided to seek information on its own initiative.¹⁵⁷ The Panel met with these experts over a two-day period while the parties were present.¹⁵⁸ The Panel asked the experts about the habitat and migratory patterns of sea turtles and local conservation approaches.¹⁵⁹ The Panel then gave the parties an opportunity to comment in writing on the experts' replies. These comments, and the experts' answers, are reproduced in the Panel Report.¹⁶⁰

As well as basing their answers on a wide body of scientific and technical data, the experts referred to non-WTO sources. One expert, for example, referred to UNCLOS and the FAO Code of Conduct to emphasise his view of the importance of the precautionary principle.¹⁶¹ In addition, all the experts referred to the IUCN and CITES in describing the critical status of depletion of sea turtles species.¹⁶²

Similarly, in the *EC – Biotech* dispute the scientific experts consulted by the Panel referred to a number of rules of international law and other non-WTO sources. The EC had argued that these experts should be consulted on the meaning of certain terms in the SPS Agreement. The complaining parties opposed this request on the basis that the terms were to be assessed by applying the rules of treaty interpretation.¹⁶³ The Panel asked the experts about three categories of scientific and technical information surrounding the products at issue in the dispute.¹⁶⁴ Although this request for information appears to be quite limited, the

---

¹⁵⁵ DSU Art. 13; See also SPS Art. 11.2; TBT Agreement Arts. 14.2–14.4, Annex 2. Note also provisions in draft Annex VIII of SCM Agreement for fisheries subsidies disputes: see Chapter 3, n. 227 and accompanying text.

¹⁵⁶ I note that collaboration with other IGOs is often key to the selection of scientific experts: see panel reports in *Japan – Agricultural Products II* para. 6.2 (involving Secretariat of the IPPC); *EC – Hormones* paras. 8.7–8.9 (involving Codex and the IARC) and *EC – Biotech*, para. 7.21 (involving the CBD, Codex, FAO, IPPC, OIE and WHO). Cf *US – Shrimp*, where the Panel did not seek assistance from an IGO in selecting experts, but asked the disputing parties to provide suggested names: *US – Shrimp* para. 5.5.

¹⁵⁷ Panel Report, *US – Shrimp*, para. 7.9 (although it asked the disputing parties for suggestions: see para. 5.5).

¹⁵⁸ Ibid., para. 1.9.   ¹⁵⁹ Ibid., para. 5.1.   ¹⁶⁰ Ibid., Part V.   ¹⁶¹ Ibid., para. 5.12.

¹⁶² Ibid., para. 5.19 (Dr S Eckert), para. 5.42 (Dr J Frazier), para. 5.68 (Mr H-C Liew); para. 5.71 (Dr I Poiner); see also reference to IUCN in para. 5.60 (Mr M Guinea).

¹⁶³ Panel Report, *EC – Biotech* para. 7.19.   ¹⁶⁴ Ibid., para. 7.18.

Panel expected the experts to draw on rules and guidelines of international organisations in providing their advice.[165] For example, the Panel asked the experts to comment on how the relevant scientific documentation relied on by the EC member states in establishing their safeguard measures compared with documentation of several international organisations, including Annex III of the Biosafety Protocol. The Panel then gave the disputing parties an opportunity to comment on its representation of the experts' evidence at the interim review stage.[166]

Scientific expertise was also sought in the *EC – Hormones* dispute brought by Canada and the United States. Again non-WTO rules and sources were brought to the attention of the Panel by the appointed experts. The claim was brought against the EC ban of hormone-treated beef. This ban was based on the EC's scientific assessment that hormone-treated beef posed an unacceptable risk to human health. Such bans are consistent with the SPS Agreement, but only if they meet certain criteria. Canada and the United States claimed that the ban was not based on the relevant international standard produced by Codex or on a valid scientific risk assessment.

It was therefore necessary for the *EC – Hormones* Panel to consider a range of scientific evidence in order to rule on whether the EC had violated the SPS Agreement. This included, in addition to the Codex standard, scientific studies conducted by the International Agency for Research on Cancer (IARC), a separate governmental body of the WHO.[167] Studies from the IARC had suggested that added hormones could be carcinogenic, and the EC claimed that these studies constituted a relevant risk assessment on which it based its trade measure. As such, the Panel was required to mediate between works from different international organisations.

Contrary to the disputing parties' expectations that the Panel would constitute an advisory expert group, the *EC – Hormones* Panel instead

---

[165] Ibid., Annex H-170. The Panel referred to IPSM, Codex principles and Annex III of the Biosafety Protocol. Canada disputed that Annex III could be construed as an 'international standard' in these terms: Annex I-2, para. 119.

[166] See, e.g., the disagreement between the parties at the interim review stage over the representation of expert opinion on antibiotic-resistant marker genes: *EC – Biotech* paras. 6.36–6.41.

[167] The IARC was established in 1965 and its twenty members include Germany, France, Italy, the UK and the US (the founding members), in addition to Australia, Belgium, Canada, Denmark, Finland, India, Ireland, Japan, Norway, the Netherlands, Korea, Russian Federation, Spain, Sweden, Switzerland and Austria: see www.iarc.fr/en/about/membership.php.

selected five experts. This decision was apparently motivated by the desire of the Panel that the experts would act in their individual capacities and not attempt to arrive at a consensus on the issues raised.[168] The Panel selected these experts from a list of names provided by both Codex and the IARC.[169] In addition to these five experts, an expert from Codex was in attendance over the two-day consultation period. The parties also included scientists in their delegations.[170]

The experts contributed not only scientific expertise but also assisted the Panel in ascertaining the relevance of the work of different IGOs. The experts referred to the work of a number of international organisations and committees. In addition to scientific explanations about the safety of hormones, the experts described the organisation and processes of relevant international organisations,[171] including the work of the Joint FAO/WHO Expert Committee on Food Additives (JECFA), on which the Codex standard was based. Indeed, one of the experts selected by the Panel was a member of the JECFA, a decision that the EC unsuccessfully appealed.[172]

The Panel used this information in its ruling that the EC had violated the SPS Agreement, a conclusion that was upheld by the AB. The Panel's finding that the work of the IARC was insufficient to support the EC measures, in particular, was based on the evidence of the scientific experts.[173] As such, the Panel used its consultation with scientific experts to rank a number of non-WTO sources.

The evidence of experts promotes regime collaboration by exposing adjudicators to facts and norms from other regimes. Such evidence is different, however, from the evidence presented by parties, in the sense that adjudicators are not mandated to take it into account. For example, if facts are presented by appointed experts, a panel has 'a substantial margin of discretion as to which statements are useful to refer to

---

[168] See Panel Report, *EC – Hormones*, para. 8.7. The AB upheld the Panel's selection and use of experts in this way: AB Report, *EC – Hormones*, paras. 146–149.
[169] Panel Report, *EC – Hormones*, paras. 8.7–8.9. At the interim review stage, the EC asked that the Panel's process of selection be recorded in the final report.
[170] Ibid., para. 8.9.
[171] See e.g. explaining the relationship between FAO, WHO and Codex: ibid, Annex, para. 26 (Dr Randall).
[172] AB Report, *EC – Hormones* paras. 37, 146–149.
[173] The experts had stated that the JECFA had taken into account the conclusions of the IARC reports: see Panel Report, *EC – Hormones*, para. 8.129. In addition, the Panel noted that unlike the Codex standard, the IARC reports were not specific to the use of hormones in meat, a finding that was upheld by the AB: *EC – Hormones*, paras. 199–200.

explicitly'.[174] Even if a panel misrepresents the opinion of experts, such an error will 'not amount to the egregious disregarding or distorting of evidence' required to violate DSU Article 11.[175] Although panels and the AB can draw on the evidence of experts to better understand and apply non-WTO sources, their duty to take the evidence into account is different from their duty to take account of parties' submissions.

4 Consultation with IGO secretariats

WTO panels are empowered to consult with other IGOs over the use and relevance of non-WTO sources as part of the general power to seek information,[176] their general duty to provide an objective assessment of the facts,[177] and their specific powers with respect to particular IGOs such as the IMF.[178] In *US – Shrimp*, the Panel did not consult IGOs such as the CITES Secretariat. This may be surprising given the Panel's reliance on international environmental law in determining the conservation status of sea turtles, yet it is typical for many panels that rule on alleged violations of WTO agreements in the context of existing non-WTO norms and institutions.

Aside from seeking assistance in the nomination of experts,[179] WTO adjudicating bodies have consulted IGOs in only a small number of cases.[180] Two cases, in particular, demonstrate the tension for panels in allowing IGOs to participate in panel proceedings in the face of concerns about overlapping competencies between regimes. In *India – Quantitative Restrictions*, the Panel consulted with the IMF on issues relating to India's balance of payments, as provided in DSU Article 13 and GATT Article XV.2. India claimed on appeal that the Panel improperly delegated its judicial function to the IMF. The AB disagreed that there was a delegation or a violation of the Panel's duty under DSU Article 11, and found instead that the Panel had critically accepted the views of the IMF and considered other data and opinions.[181] In *EC – Chicken Cuts*, a dispute involving contested customs classifications of frozen chicken,

---

[174] AB Report, *EC – Hormones*, para. 138.
[175] Ibid., para. 144; see also AB Report, *EC – Asbestos*, para. 162.
[176] DSU Art. 13. See further SPS Agreement, Art. 11.2; TBT Agreement, Art. 14.2.
[177] DSU Art. 11.
[178] GATT Art. XV.2; see *India – Quantitative Restrictions*, Panel Report paras. 5.12–5.13; AB Report, para. 77; see also AB Report, *Argentina – Textiles and Apparel*, paras. 82–86.
[179] See above n. 156.
[180] See e.g. *India – Quantitative Restrictions*, *EC – Geographical Indications*, *EC – Chicken Cuts*, *EC – Biotech*.
[181] AB Report, *India – Quantitative Restrictions*, para. 149.

the panel sought information from the World Customs Organization (WCO). In providing information, the WCO suggested that its own settlement procedures should be followed by the parties before the Panel proceeded.[182] The Panel rejected this view, partly because it lacked the authority to refer the dispute to the WCO and partly because of its own duty to rule on the matter.[183] The tension suggested by these cases accounts for the restraint with which WTO adjudicating bodies consult other organisations when attempting to rule on matters involving other regimes.

The general reluctance of WTO adjudicating bodies to consult IGOs is manifest in a variety of other ways. First, in the small number of cases where panels have engaged in IGO consultation, panels have consulted IGOs in a restrictive way. Secondly, panels have often failed to consult in circumstances where the parties have deemed such consultation to be necessary. Thirdly, adjudicating bodies have failed to consult with IGOs when the disputes involve norms exogenous to the WTO and in which certain IGOs have particular competences.

The first example of the reluctance of WTO adjudicating bodies to consult IGOs is demonstrated by the restricted way in which such consultations have been conducted. Consultations were restrained even in early GATT disputes. For example, in *Thai Cigarettes* a GATT Panel was asked to rule on whether a Thai import ban on cigarettes could be justified on GATT Article XX grounds. On Thailand's request, and with the agreement of the parties, the Panel consulted a representative from the World Health Organization (WHO) over a one-day period.[184] The WHO representative referred to recent experience in developing countries that opened their markets for cigarettes, which suggested that less trade-restrictive alternatives to an import ban, such as restrictions on advertising and marketing techniques, were ineffective public health policies.[185] It also noted that foreign cigarettes would be likely to be more attractive to women and young people than local brands and thus constitute a threat to a vulnerable population.[186] In response, the United States argued that the WHO was not competent to address the 'health consequences of the opening of the market for cigarettes'.[187] The Panel agreed with the WHO expert that smoking

---

[182] Panel Report, *EC – Chicken Cuts*, para. 7.53.
[183] Ibid., para. 7.56. This part of the Panel ruling was not the subject of the appeal.
[184] GATT Panel Report, *Thai-Cigarettes*, paras. 5, 27, 50.   [185] Ibid., para. 27.
[186] Ibid., para. 52.   [187] Ibid., para. 58.

constituted a serious public health risk.¹⁸⁸ However, it rejected Thailand's argument that the import restriction was 'necessary'.¹⁸⁹ In its findings, the Panel did not refer to the evidence submitted to it by the WHO expert. It referred instead to WHO general resolutions urging member states to take measures addressed at the advertising, promotion and sponsorship of tobacco and other non-trade discriminatory health measures.¹⁹⁰ This move can be criticised as excluding relevant evidence from the WHO representative consulted.¹⁹¹

The Panel in *EC – Biotech* consulted with a number of IGOs but did so in a very confined way. The United States and other countries had considered the issue of import measures on GM food in a variety of international forums before the case was filed at the WTO.¹⁹² During the proceedings, the Panel consulted with a number of IGOs that had had involvement with the dispute. This consultation was restricted in two ways. First, the Panel asked a number of IGOs to provide reference documents and other materials to 'assist the Panel in ascertaining the meaning of certain terms and concepts'.¹⁹³ Thus, notwithstanding the articulation of its general power to consult, the Panel emphasised that its use of the relevant international rules and guidelines was empowered by the need to ascertain the 'ordinary meaning' of SPS terms.¹⁹⁴ As such, the only evidence considered by the Panel was that related to treaty interpretation, pursuant to VCLT Article 31(1), rather than the establishment of facts.

Secondly, the *EC – Biotech* Panel was careful to stress that in conducting these consultations, it had taken into account the views of the disputing parties.¹⁹⁵ A similar approach was taken by the Panel in *EC – Geographical Indications*, which closely involved the disputing parties in its requests for assistance from WIPO with respect to the interpretation of the Paris Convention for the Protection of Industrial Property.¹⁹⁶ While this approach is reasonable in an adversarial procedure, there was arguably no need to consult the disputing parties in this way given

---

¹⁸⁸ Ibid., para. 72.   ¹⁸⁹ Ibid., para. 81.   ¹⁹⁰ Ibid., paras. 77, 79–80.
¹⁹¹ Cf the unlikelihood that such an exclusion would be subject to WTO appellate review given the discretionary nature of DSU Art. 13: AB Report *EC – Sardines*, para. 302, see below n. 202 and surrounding text.
¹⁹² See Buckingham and Phillips, above n. 144.   ¹⁹³ Panel Report, *EC – Biotech* para. 7.31.
¹⁹⁴ Ibid., para. 7.96.
¹⁹⁵ Ibid., para. 7.31.
¹⁹⁶ Panel Report, *EC – Geographical Indications*, paras. 2.16–2.18. The views of the parties as to the request and WIPO's reply were published in the final report.

the Panel's wide powers to seek information. Instead, this partial deference to the disputing parties demonstrates the influence of the disputing parties in the conduct of the consultations. In the *EC – Biotech* case, in particular, this deference demonstrates the influence of the disputing parties in the interpretation of treaty terms according to VCLT Article 31 (1). It reveals that the disputing parties shape the interpretative context of the WTO treaty terms to a far greater degree than acknowledged by the Panel.[197]

A second example of the limited consultation by WTO panels is demonstrated by cases where a request from a party for IGO consultation has been denied. In *Argentina – Textiles and Apparel*, the Panel declined the request of both disputing parties to consult the IMF about non-WTO sources that related to an allegedly GATT-inconsistent statistical tax. The sources in question were undertakings between Argentina and the IMF. Argentina claimed that it was collecting a statistic tax for 'fiscal' purposes in the context of its undertakings with the IMF.[198] The Panel noted that Argentina did not argue that it was required to impose this specific tax in order to meet its commitments to the IMF, and accepted the United States' claim that the tax violated GATT Article VIII.[199] On appeal, Argentina claimed that the Panel had violated DSU Article 11 because it had not consulted with the IMF. The AB rejected this claim, and reiterated that a panel's power to seek information under DSU Article 13 is a grant of discretionary authority rather than a duty, and that the bounds of this authority had not been exceeded by the Panel.[200]

The AB made reference to this case in *EC – Sardines*, when it ruled on the Panel's decision to decline the request of one of the disputing parties to consult an IGO. The EC claimed that the Panel had breached its duty under DSU Article 11 to conduct 'an objective assessment of the facts' because it had not sought information from Codex. The *EC – Sardines* case rested on an alleged violation by the EC of its requirement to base its technical regulations on relevant international standards unless such standards were inappropriate or ineffective.[201] The EC's marketing standards for preserved sardines deviated from a Codex standard. In its defence, the EC claimed that the Codex standard was not relevant because it had a different product coverage and because it

---

[197] Contrast the Panel's restrictive conception of VCLT Art. 31(3)(c) noted above n. 87 and accompanying text; see further Young, above n. 149, 926.
[198] Panel Report, *Argentina – Textiles and Apparel*, para. 6.78.   [199] Ibid., paras. 6.79–6.80.
[200] AB Report *Argentina – Textiles and Apparel*, paras. 82, 84, 86.
[201] TBT Agreement Art. 2.4.

had not been adopted by consensus. In ruling on the status and validity of the Codex standard, the Panel did not find it necessary to consult Codex. The EC's claim that the Panel had erred in failing to consult was rejected by the AB.[202] It found there was no impropriety in the Panel's decision not to seek information because such a power is discretionary.[203]

The third example of the reticence to seek information from IGOs is demonstrated by cases where, in the face of disputed norms that have been agreed under the auspices of other IGOs, the panels have failed to consult those IGOs. The absence of an UNCTAD representative in *EC – Tariff Preferences* is notable. In that case, the Panel was asked to rule on the justiciability and legal effect of the Enabling Clause and whether the norm of non-discrimination applied to GSP policies. The Enabling Clause was decided by the GATT contracting parties after UNCTAD negotiations about the status of developing countries in world trade.[204] The parties and third parties made multiple submissions on the history and status of the Enabling Clause. UNCTAD, however, was not consulted.

In the *EC – Asbestos* case, too, the Panel preferred to seek relevant factual information from scientific experts rather than representatives from key IGOs. The Panel was asked to determine the WTO-consistency of a French import ban on certain fibres containing asbestos. A key issue of fact was whether the asbestos product posed a risk to human health. What is surprising is that the Panel did not seek information from the WHO and the IARC, which had both contributed to studies on the carcinogenic nature of asbestos, but instead asked four experts to give information about these studies. The Panel had consulted with the WHO, the IARC, the ILO, the IPCS and the ISO in identifying the experts.[205]

Largely absent, too, have been organisations involved in the classification of products, which could have very pertinent data on how to assess, for example, whether products are 'like' in claims relating to MFN or national treatment.[206] The International Coffee Organization, which classifies coffee products, was not consulted in the GATT dispute over Spanish classification of unroasted coffee.[207]

---

[202] AB Report *EC – Sardines*, para. 302.   [203] Ibid.
[204] Decision of the GATT Contracting Parties of 28 November 1979 (L/4903).
[205] Panel Report, *EC – Asbestos*, para. 5.20.
[206] GATT Arts. I, III; see further AB Reports in *Japan – Alcoholic Beverages II*, 21; *EC – Asbestos*, para. 101.
[207] GATT Panel Report, *Spain – Unroasted Coffee*. The GATT Panel did not rule on the ICO classifications in finding that the relevant products were like.

The restrictive approach of WTO adjudicating bodies to gaining evidence from IGOs contrasts with the interdependence set out in the WTO agreements. For example, the SPS and TBT Agreements depend on international bodies like Codex to harmonise non-tariff barriers (through standards)[208] by providing a 'multilateral scientific consensus'.[209] Relevant international bodies are expected to provide members with information, guidelines, risk assessment techniques and scientific and technical advice.[210] The generation by these institutions of important scientific and technical understandings sits uneasily with the *EC – Biotech* Panel's conception that they merely inform the 'ordinary meaning' of treaty terms. Yet while the Panel was aware of the broader status of international organisations in the SPS and TBT Agreements, it considered that Annex A of the SPS Agreement did not incorporate such co-existence.[211] The implications of the restrictive approach to IGO consultation will be discussed further below.

5 Amicus curiae *briefs from NGOs and others*

In accepting *amicus* briefs, WTO adjudicating bodies can access information about non-WTO sources that might be otherwise absent from the panel proceedings. The legal basis for accepting *amicus* briefs differs for panels and the AB; the former is part of panels' general power to consult[212] while the latter is based on the AB's broad authority to adopt procedures that do not conflict with the DSU or the covered agreements.[213] Moreover, while panels may take into account *amicus* submissions on both law and fact as part of their duty to make objective assessments of the matter,[214] the AB can only take into account legal submissions given it has no fact-finding powers.[215]

---

[208] See SPS Agreement Art. 3.1 and TBT Agreement Art. 2.4. See further *EC – Sardines*, esp. paras. 171–316. On the use by the WTO of definitions from standard-setting bodies in the TBT context, see TBT Agreement Art. 1.1.
[209] See Chapter 3; see further Doaa Abdel Motaal, 'The "Multilateral Scientific Consensus" and the World Trade Organization' (2004) 38 *JWT* 855.
[210] SPS Agreement Arts. 5.1, 5.7, Art. 6.1, Art. 9.1.
[211] Panel Report, *EC – Biotech*, para. 7.300.
[212] DSU Art. 13; AB Report, *US – Shrimp*, paras. 107–110.
[213] AB Report, *US – Lead and Bismuth II*, para. 39; see further on whether the AB's powers extend to unsolicited briefs, Petros Mavroidis, '*Amicus curiae* Briefs Before the WTO: Much Ado About Nothing' in Armin von Bogdandy, Petros Mavroidis and Yves Meny (eds.) *European Integration and International Co-ordination, Studies in Transnational Economic Law in Honour of Claus-Dieter Ehlermann* (2002) 317.
[214] AB Report, *US – Shrimp*, para. 106. [215] AB Report, *EC – Sardines*, para. 169.

The acceptance of *amicus* briefs was a controversial issue in *US - Shrimp*. Two groupings of NGOs submitted *amicus* briefs to the Panel without invitation.[216] India, Malaysia, Pakistan and Thailand requested the Panel not to consider the content of the documents.[217] The United States, in response, advocated the use by the Panel of information in the briefs, in line with its power to 'seek' information from any relevant source under DSU Article 13.[218] The Panel ruled that it did not have authority to accept the *amicus* submissions.[219] This ruling was reversed by the AB.[220]

The ruling by the AB in *US - Shrimp* that WTO panels may accept unsolicited *amicus* briefs was followed by a decision by the Appellate Body Secretariat, acting in consultation with AB members, to set down special procedures for the filing of *amicus curiae* briefs in the *EC - Asbestos* dispute.[221] Given the controversies about the *Asbestos* panel decision (which had ruled that certain asbestos and non-asbestos building materials were 'like products') the Secretariat expected a large number of briefs from public interest-led NGOs and provided that *amicus* parties were required to seek leave to file such briefs.

The production of these special procedures by the Secretariat gave rise to significant debate among WTO members, many of whom were concerned that, as a 'Member-driven organisation', access to WTO dispute settlement should be restricted to members. Convening a Special Meeting of the General Council, many members complained that they did not agree with the decision to accept *amicus curiae* briefs in *US - Shrimp* and that the Appellate Body Secretariat did not have the mandate to extend its procedures in this way.[222]

---

[216] (1) WWF and FIELD; and (2) the CMC and CIEL.
[217] Panel Report, *US - Shrimp*, para. 7.7.   [218] Ibid.
[219] Ibid., para 7.8; the Panel allowed the US to annex briefs from NGOs to its own submissions, and it annexed the document by the CMC and CIEL.
[220] AB Report, *US - Shrimp*, para. 110. This ruling led to several suggestions for the management of *amicus* briefs by the AB: Gabrielle Marceau and Matthew Stilwell, 'Practical Suggestions for *Amicus Curiae* Briefs Before WTO Adjudicating Bodies' (2001) *JIEL* 155 (written before the Appellate Body issued its controversial special procedures in the *Asbestos* dispute, see below n. 221); Georg Umbricht, 'An "Amicus Curiae Brief" on Amicus Curiae Briefs at the WTO' (2001) 4 *JIEL* 773.
[221] The additional procedure was adopted by the AB pursuant to Rule 16(1) of the AB Working Procedures.
[222] Minutes of the General Council Meeting of 22 November 2000, WTO Doc. WT/GC/M/60, dated 23 January 2001 (see esp. paras. 114–115 and 118). See further Robert Howse, 'Membership and its Privileges: The WTO, Civil Society, and the *Amicus* Brief Controversy' (2003) 9 *European Law Journal* 496, 496–7; Mavroidis, above n. 213.

This reticence among members to permit the filing of *amicus* briefs has often been linked to 'developing country concerns' but this is somewhat misleading. In *US – Shrimp* it was India, Malaysia, Pakistan and Thailand who were opposed to the *amicus* briefs. In the General Council, too, many developing countries have spoken out against the acceptance of *amicus* briefs.[223] Commentators have opined that the environmental and labour NGOs that are most likely to file briefs are likely to be based in the 'West' and to argue against the interests of developing countries to exploit their own resources.[224] There are many examples, however, where developing countries have relied on the mechanism of *amicus* briefs in support of their own submissions. For example, Brazil referred to an *amicus* filed by the Humane Society in its own submissions in the *Brazil – Retreaded Tyres* dispute.[225] In *EC – Sardines*, Peru relied on the submissions of a UK consumers' association.[226] At the appellate level, Morocco itself filed an *amicus* when it was out of time to act as a third participant.[227] It is a mistake to assume that developing countries are opposed to regime collaboration in dispute settlement through the use of *amicus* briefs.

In the fisheries context, *amicus* briefs have not focused solely on conservation issues. The regime interaction has extended to consumer, scientific and economic perspectives. For example, in *Australia – Salmon (Art 21.5)*, a group of 'Concerned Fishermen and Processors' from South Australia sent a letter to the Panel detailing the differences in treatment by Australia of imports of pilchard used as bait or fishfeed and imports of salmon.[228] In accepting and taking into account this information pursuant to its authority under DSU Article 13.1, the Panel noted that the information directly related to Canada's legal claim that Australia's differentiation of certain fisheries products was inconsistent with SPS Agreement Article 5.5.[229] The interest of these South Australians in

---

[223] e.g. Costa Rica noted that 'such a measure represented a risk for developing countries as it would put them in a situation where they would be short of possibilities of defence' (Minutes, above n. 222, para 70).

[224] See e.g. Jagdish Bhagwati, 'Afterword: The Question of Linkage' (2002) 96 *AJIL* 126, 126; see further Gregory Shaffer, 'The WTO Under Challenge: Democracy and the Law and Politics of the WTO's Treatment of Trade and Environment Matters' (2001) 25 *Harvard Environmental Law Review* 1.

[225] Panel Report, *Brazil – Retreaded Tyres*, para. 4.12.

[226] Panel Report, *EC – Sardines*, para. 7.132.   [227] AB Report, *EC – Sardines* para. 153.

[228] Panel Report, *Australia – Salmon (Art 21.5)*, paras. 7.8–7.9.   [229] Ibid., para 7.9.

supporting Canada's claim is unclear and further details are not provided in the Panel record.[230]

The other fisheries dispute that involved an *amicus* submission was *EC – Sardines*. At the panel stage, a UK consumers' association supported the challenge by Peru against the EC restriction of the product description of certain species of sardines.[231] The Panel drew on this material in determining European consumer product recognition,[232] which was challenged by the EC at both the interim review and appellate stages. On this issue, the AB ruled that the Panel 'did not exceed the bounds of [DSU Article 11] discretion' by taking account of the letter.[233] The AB also had to rule on whether it should accept additional *amicus* briefs that had been filed at the appellate stage. One was from Morocco, and provided both factual and legal information, including scientific arguments about the differences between the relevant sardine products and legal submissions about the relevant Codex standard.[234] The other brief was from a private individual and contained legal submissions on the interpretation of the TBT Agreement.[235] The AB found that it was entitled to consider these briefs, but did not find it necessary to take them into account.[236]

Although it seems settled that panels and the AB may accept *amicus* briefs, they have been very restrained in actually considering them. In the *EC – Biotech* case, for example, the Panel accepted briefs that had been compiled by more NGOs than in any other dispute, but it did not find it necessary to take them into account.[237] Even though the AB has accepted *amicus* briefs in a number of cases, there have been no cases where the AB took account of the *amicus* briefs in its reasoning. Even where such briefs do not feature in the decision-making by WTO adjudicating bodies, however, they still provide some scope for the

---

[230] I note that this group did not participate in the interviews of those who had sought to be a WTO *amicus curiae* by Leah Butler, 'Effects and Outcomes of *Amicus Curiae* Briefs at the WTO: An Assessment of NGO Experiences' (2006) (unpublished manuscript at http://nature.berkeley.edu/classes/es196/projects/2006final/butler.pdf).

[231] See Consumers' Association open letter to the Executive Director of the Advisory Centre on WTO Law, attached to Peru's submission: discussed in Gregory Shaffer and Victor Mosoti, 'The EC-Sardines Case: How North-South NGO-Government Links Benefited Peru' (2002) 6:7 BRIDGES.

[232] Panel Report, *EC – Sardines*, para. 7.132.   [233] AB Report, *EC – Sardines*, para. 300.

[234] Ibid., para. 169.

[235] Although the identity of the *amicus* was not provided in the record (see ibid., para. 153), the brief was from US law professor Robert Howse.

[236] AB Report, *EC – Sardines*, paras. 160, 170.   [237] Panel Report, *EC – Biotech*, para. 7.11.

dissemination of non-WTO norms. For example, some panels have reproduced the *amicus* briefs in their report annexes, thus enhancing transparency and accessibility.[238] The decisions by parties to include *amicus* in their submissions and to open the hearing of the disputes to the public will further this goal.

The reluctance of the WTO dispute settlement bodies to accept and draw on *amicus* briefs is mirrored in other areas of international dispute settlement. There is no provision of *amicus* briefs in the Statute of the ICJ,[239] and the Court has been resistant to attempts to introduce the procedure.[240] In the environmental dispute over the Danube Dam,[241] the ICJ considered NGO material which was adopted by Hungary as part of its own submissions.[242]

E Problems and challenges

My examination of the WTO adjudication of the *US – Shrimp* dispute and other cases reveals several problems and challenges for the interaction between the WTO and other regimes. In the *US – Shrimp* case, it was GATT Article XX that gave rise to regime interaction and various 'balancing' tasks. Other WTO provisions give rise to similar demands and the procedural issues described in this chapter are relevant to a wide range of trade-related fisheries disputes, including those arising under the TBT Agreement, the SPS Agreement and the new fisheries subsidies disciplines. This section gives some tentative conclusions that form part of the general normative and institutional recommendations for regime interaction presented in the next Part.

A focus on regime interaction appears most immediately to presuppose that the WTO adjudicating bodies are the appropriate forums to resolve disputes relating to certain trade-related aspects of fisheries. This is a controversial position, particularly with respect to the

---

[238] See e.g. *Brazil – Retreaded Tyres*, which annexed the two sets of unsolicited *amicus curiae* briefs it received from environmental NGOs: see Panel Report, para 1.8; Exhibits BRA-98 and BRA-99.
[239] Cf Statute of the ICJ, Art. 34(2), which allows the Court to receive information from public international organisations.
[240] Christine Chinkin and Ruth Mackenzie, 'Intergovernmental Organizations as "Friends of the Court"' in Laurence Boisson de Chazournes, Cesare Romano and Ruth Mackenzie (eds.) *International Organizations and International Dispute Settlement: Trends and Prospects* (2002) 135, 140.
[241] *Gabčíkovo-Nagymaros* case (1997) ICJ Rep 7.
[242] Anna-Karin Lindblom, *Non-Governmental Organisations and International Law* (2005) 304.

legitimacy of the adjudicative branch of the WTO ruling on issues of social and political importance. For example, Dunoff has advocated restraint by WTO panels in 'trade and ...' disputes through the adoption of 'passive virtues' commonly used by domestic constitutional courts such as 'ripeness' and 'political question' doctrines.[243] This restraint is aimed at addressing the lack of expertise within the WTO dispute settlement system and the associated problems of competence and legitimacy.

My analysis has led to a different conclusion. Given the wide ambit of trade matters, which I described by reference to the fisheries subsidies example in Chapter 3, I consider that the WTO will have an increasing role in resolving disputes between countries that concern issues such as fisheries sustainability. Instead of calling for WTO adjudicating bodies to restrain from resolving these disputes, I call for appropriate methods of regime interaction that will improve both the competency and expertise of the adjudicating bodies and, relatedly, the legitimacy of their interventions. A similar conclusion is offered by Perez, in the context of his findings on the WTO's 'ecological insensitivities'. Writing specifically about WTO dispute settlement, Perez proposes two institutional changes to 'sensitise' the WTO adjudicating bodies to environmental issues. First, he considers that the WTO should share its 'cognitive' and 'decision-making' roles through extending the participation of IGOs such as UNEP, environmental NGOs and the disputing WTO members themselves.[244] Secondly, he proposes changes to the evidentiary process.[245]

These suggestions may be criticised as increasing the mandate and ambit of the WTO adjudicative branch at the expense of more appropriate political contestation of the correct balance of trade and social issues within the WTO or, indeed, within domestic constituencies. In response, I consider that the adjudicative branch can develop useful modes for ruling upon 'trade and ...' issues which remain accountable and legitimate. This requires, in particular, a more nuanced mode for WTO adjudicating bodies to take account of external regimes, which I describe in this section. Moreover, it requires a legal framework for

---

[243] Jeffrey Dunoff, 'The Death of the Trade Regime' (1999) 10 *EJIL* 733, 757-60.
[244] Oren Perez, *Ecological Sensitivity and Global Legal Pluralism: Rethinking the Trade and Environment Conflict* (2004) 95-107.
[245] Perez considers the burden of proof should shift to the complainant party in claims relating to the *chapeau* of GATT Art. XX: ibid., 107-8.

appropriate regime interaction at all stages of law-making and implementation, which I describe in Part III.

1 Selection of the adjudicators

A clear influence on the use by panellists and AB members of non-WTO sources is their experience and knowledge of other regimes, as I described above. The effectiveness of regime interaction will depend in part on these individuals.

Panel composition is an area of reform that is subject to current negotiations during the Doha round.[246] Various proposals have been made to amend the DSU. The EC, for example, has proposed a permanent body of panel members.[247] The counter-argument to this is that such a reduced pool of panellists decreases the likelihood of obtaining panels with sufficient expertise in a case.[248] Instead, the required qualifications of panellists could extend to at least one other area of international law, and the roster could publicly specify these areas.

More radical proposals on panel composition have been made in the context of the fisheries subsidies negotiations that I examined in Chapter 3. In particular, the proposal to set up an inter-agency group made up of trade and fisheries experts to resolve fisheries subsidies disputes merits attention.[249] Such a panel, which could be made of panellists from the WTO, the FAO and other agencies, and even NGOs, would have the institutional competence to apply and interpret a range of fisheries and other non-WTO sources.

Such proposals for permanent regime collaboration in the adjudication of trade disputes have met with a cool reception from Rules Group negotiators and are unlikely to be accepted by the WTO membership. But even if institutional reform is unlikely, the current framework can still be improved. One such improvement relates to the appointment of panellists. In practice, the Director-General commonly appoints panel members.[250] The interest of current Director-General Lamy in integrating WTO law with the international legal order[251] might indicate a predisposition to selecting panellists with wider competences, although he has not made specific comments in this regard. Moreover,

---

[246] See e.g. WTO Doc. TN/DS/20.   [247] See e.g. WTO Doc. TN/DS/W/38.
[248] See e.g. Frieder Roessler, 'Comment on a WTO Permanent Panel Body – The Cobra Effects of the WTO Panel Selection Procedures' (2003) 6 *JIEL* 230; cf Davey, above n. 118.
[249] See esp. pp. 126–7.   [250] See above n. 118 and accompanying text.
[251] See Pascal Lamy, 'The Place of the WTO and its Law in the International Legal Order' (2006) 17 *EJIL* 969.

even if the qualifications of panellists still remain restricted to trade law, their collaborative powers have been given a broad interpretation. In this regard, the AB's ruling in *US – Shrimp* that the right of panels to 'seek information' extends to the ability to consider unsolicited *amicus* briefs must be supported.[252] According to this interpretation, the introduction of relevant non-WTO sources from non-parties does not depend on the initiative of the adjudicators themselves. Such a finding is important because the panellists may not have the background to approach the most appropriate individuals and bodies who can contribute information about non-WTO sources. A similar outward perspective should be encouraged in the selection of Secretariat staff and in other institutional practices.[253]

2 Framing by the parties

The parties or third parties to a dispute have the most influence in the presentation of non-WTO sources, as described above. Panels and the AB have a duty to give their submissions due consideration. Such a duty is not owed to other individuals or bodies who consult with the panel or submit *amicus* briefs.[254] In addition to the parties' submissions, a Panel may request further information from disputing parties to assist them in the fact-finding process. Like the power to seek information from IGOs and other sources, the panel's discretionary authority to seek information from parties is limited.[255]

Under the current rules, the only individuals or bodies that can participate as parties or third parties are WTO members. As such, the only bodies that can expect their submissions on non-WTO sources to be taken into account in dispute settlement are the WTO members themselves. Expanding these standing rules would increase the likelihood that a panel or AB would be mandated to consider non-WTO sources. Such reform could establish a 'guardian of the ocean' to act for the interest of the global commons in international disputes.[256]

---

[252] See above n. 212 and accompanying text.
[253] See above n. 132 and accompanying text; see also Chapter 4, n. 244 and accompanying text.
[254] AB Report, *EC – Sardines*, para. 166.
[255] DSU Art. 13; see also Panel Report, *Canada – Aircraft*, para. 9.53; AB Report, *Argentina – Textiles and Apparel*, para. 84.
[256] For the idea of an international organisation representing the interests of fisheries, see Christopher Stone, 'Should We Establish A Guardian For Future Generations?' in Christopher Stone, *Should Trees Have Standing? And Other Essays on Law, Morals, and the Environment* (1996) 65, 70.

There is scope to enhance regime interaction even without expanding standing rules in three main areas. First, the ability of WTO members to frame their legal claims according to WTO and non-WTO law will be affected by the level of national policy coordination within the agencies involved in trade litigation. The trade delegations that are often involved in preparing written and oral submissions need to have a good understanding of, and relationship with, other delegations and ministries, such as fisheries agencies. For example, in *Brazil – Retreaded Tyres*, Brazil relied on a range of domestic and international environmental standards to support its claim that its trade restriction on retreaded tyres from the EC was necessary for environmental aims.[257]

Secondly, the ability of WTO members to understand and successfully present non-WTO law will also depend on their capacity. The available resources of WTO members differ widely. This issue has been addressed in part by the establishment of the Advisory Centre on WTO Law which provides legal advice to developing countries.[258] Presumably such legal advice includes issues of relevant non-WTO law, but further initiatives could aim to extend this assistance. Other institutions such as ICTSD already play a significant role in capacity-building in this regard.[259] Moreover, developing countries may increase their capacity if they act jointly.[260] This occurred in *US – Shrimp*, when the developing-country claimants collaborated to some degree.[261]

Thirdly, the emphasis on parties' submissions as the most effective conduits of non-WTO sources reinforces the strategies of some individuals or bodies to submit their *amicus* briefs directly to WTO members, who can then annex them to their own submissions.[262] This has been the practice not only of the NGOs in the *US – Shrimp* case, who were acting upon a restrictive interpretation of DSU Article 13, but also in subsequent cases. In *Brazil – Retreaded Tyres*, Brazil informed the Panel that it would include two *amicus* briefs, which had been sent to the Panel, in exhibits to its own submissions.[263] One outcome of this strategy, however, may be to risk the co-option of NGO positions by states and reduce their independence in advocating on the interests of

---

[257] See e.g. Brazil's reliance on US and Belgian state environmental protection agencies; see further AB Report, *Brazil – Retreaded Tyres*, paras. 203-204.
[258] See www.acwl.ch.   [259] See www.ictsd.org.
[260] On legal issues arising from joint representation, see Panel Report, *EC – Tariff Preferences*, paras. 7.3-7.13.
[261] See, e.g., on selection of experts Panel Report, *US – Shrimp*, para. 5.1.
[262] See also Butler, above n. 230.   [263] Panel Report, *Brazil – Retreaded Tyres*, para. 1.8.

fisheries where such interests are outside the national interests of states.²⁶⁴

Finally, it is important to note that if a respondent party invokes non-WTO law as a defence, such law must be binding on all the disputing parties in order to be applied by a panel or the AB. This has implications for disputes with multiple complainants that are merged by the Panel into a single proceeding.²⁶⁵ The *Biotech* case demonstrates how the merging of a dispute can decrease the likelihood that non-WTO law will be applicable law. In *EC – Biotech*, the CBD was binding on three of the disputing parties but not on the United States. The decision by the Panel to merge the complaints of Argentina, Canada and the United States into a single complaint meant that the CBD could not be regarded as applicable law between the respondent and the complainants.²⁶⁶ As such, the decision to widen the dispute to many parties can paradoxically lead to a reduction in the scope for regime interaction.

3 Parallel membership of treaties and organisations

The scope and methods of regime interaction have indicated that an emphasis on parallel membership of treaties reduces the ability of panels and the AB to take account of non-WTO sources. I pointed to the issue of parallel membership when I described the Doha negotiations under paragraph 31(i) of the Doha Declaration, which are expressly limited to WTO members that are parties to the relevant MEAs.²⁶⁷ In WTO disputes, the issue of parallel membership has arisen in the context of treaty interpretation. The *EC – Biotech* report is the starkest example of how an emphasis on parallel membership reduces the number of non-WTO sources that will be deemed to be 'relevant rules of international law' for the interpretation of WTO treaty terms.

(a) VCLT Article 31(3)(c)

In *EC – Biotech*, the Panel ruled that VCLT Article 31(3)(c) could only apply to rules that were applicable to the WTO membership as a whole.²⁶⁸ Although this does not present a problem for general principles of law,

---

²⁶⁴ Jeffrey Dunoff, 'Border Patrol at the World Trade Organization' (1998) 9 *Yearbook of International Environmental Law* 20, 22–3; see also Robyn Eckersley, 'A Green Public Sphere in the WTO?: The Amicus Curiae Interventions in the Transatlantic Biotech Dispute' (2007) 13 *European Journal of International Relations* 329, 350.
²⁶⁵ DSU Art. 9.   ²⁶⁶ See further Young, above n. 149, 913.
²⁶⁷ See Chapter 2, n. 317 and accompanying text.
²⁶⁸ Panel Report, *EC – Biotech* para. 7.68.

which are by their nature binding on all WTO members, this presents grave difficulties for treaties, which rarely, if ever, bind the entire WTO membership.[269] For those concerned about the 'systemic integration' of international law, the requirement that relevant rules of international law be 'applicable in relations between WTO Members' in order to qualify under Article 31(3)(c) will result in the 'isolation' of multilateral agreements as 'islands' and be contrary to the intent of treaty-makers.[270]

It was on this basis that the ILC Study Group, commenting on the interim report in *EC – Biotech*, was so critical of the case.[271] The Study Group preferred an interpretation of VCLT Article 31(3)(c) that emphasised the treaty membership of the disputing parties, rather than the parties to the treaty under interpretation.[272] The risks of divergent interpretations would be mitigated, according to the Study Group, in two ways. First, the treaty-interpreter could differentiate between 'synallagmatic' treaties that created merely reciprocal obligations between treaty partners and treaties that were more 'interdependent' or 'collective', which created obligations owed *erga omnes partes*. For the former type, divergence in treaty interpretation for sets of disputing parties would be unproblematic. For the latter type, however, the coherence of the treaty would need to be protected by restricting the use of other treaties in interpreting its terms.[273] Secondly, the Study Group considered that a treaty-interpreter should take into account the extent to which another relevant treaty could be said to have been 'implicitly' accepted or tolerated by other parties, notwithstanding non-identical membership.[274] The final conclusions of the ILC Study Group reflected this second qualification.[275] ILC Conclusion (21) suggests that the probative value of a treaty increases according to the degree to which it has been affirmed by states:

Article 31(3)(c) also requires the interpreter to consider other treaty-based rules so as to arrive at a consistent meaning. Such other rules are of *particular relevance*

---

[269] See above n. 100.   [270] ILC Analytical Study, para. 471.   [271] Ibid., para. 450.
[272] Ibid., para. 472.
[273] Ibid. On WTO obligations as reciprocal or integral, see Pauwelyn, above n. 37, 52–6. See also Joost Pauwelyn, 'A Typology of Multilateral Treaty Obligations: Are WTO Obligations Bilateral or Collective in Nature?' (2003) 14 *EJIL* 907 and Chios Carmody, 'WTO Obligations as Collective' (2006) 17 *EJIL* 419.
[274] ILC Analytical Study, para. 472. The Study Group considered the AB Report in *US – Shrimp* to be demonstrative of this approach.
[275] ILC Study Group Conclusions.

where parties to the treaty under interpretation are also parties to the other treaty, where the treaty rule has passed into or expresses customary international law or where they provide evidence of the common understanding of the parties as to the object and purpose of the treaty under interpretation or as to the meaning of a particular term.[276]

On one reading at least, ILC Conclusion (21) appears to endorse a spectrum of 'international consensus' that departs from basing VCLT Article 31(3)(c) on binary questions of consent and non-consent. If conceived to mean that total unanimity of the WTO membership is not required for an extrinsic treaty to be agreed as relevant interpretative context, ILC Conclusion (21) accords, to some degree, with certain institutional provisions in the WTO covered agreements. For example, three-quarters of the total WTO membership may agree to adopt binding interpretations of the WTO agreements.[277] In addition, the WTO agreements envisage that international standards may be binding on WTO members even if they are not agreed by consensus.[278] A spectrum of consensus is also implicit in some cases outside the WTO. For example, the European Court of Justice looked to an international treaty that was not yet binding on the EC in construing the EC Treaty in an early waste treatment case.[279] In the OSPAR arbitration arising from the Mox Plant dispute,[280] a dissenting arbitrator drew on the Aarhus Convention, to which the disputing parties had become signatories, in interpreting the relevant obligations of the parties under the OSPAR Convention.[281]

ILC Conclusion (21) may be read instead as emphasising the need for 'implicit' agreement of treaty terms. This emphasis on implicit agreement as a necessary ingredient to the use of non-WTO sources under Article 31(3)(c) is also supported by other parts of the general rule on interpretation. VCLT Article 31(3)(b) recognises the use by treaty interpreters of the subsequent practice of treaty parties as an interpretative

---

[276] Ibid., 15, para. (21).
[277] Marrakesh Agreement Art. IX(2). For a brief discussion of the limited WTO practice relating to this provision, see Tarcisio Gazzini, 'Can Authoritative Interpretation under Article IX:2 of the Agreement Establishing the WTO Modify the Rights and Obligations of Members?' (2008) 57 *ICLQ* 169.
[278] See e.g. *EC – Hormones*; see further Chapter 3, n. 171.
[279] *Commission* v. *Belgium*, Case C-2/90 (9 July 1992), at para. 35.
[280] Noted in Chapter 1, n. 42.
[281] *Ireland* v. *United Kingdom (OSPAR)* (2003) 42 ILM 1118. Griffith QC drew on rules of interpretation independent of the VCLT: see ibid., 1163.

tool.²⁸² Some authors have suggested that such subsequent practice may be probative even if it is only evidenced by individual parties.²⁸³ For the Appellate Body, the 'implied' agreement of all of the WTO members is necessary to establish subsequent practice, even if the practice has not been engaged in by all parties.²⁸⁴

The *EC - Biotech* Panel did not entertain notions of consensus–spectrums in its reading of VCLT Article 31(3)(c),²⁸⁵ as set out above. However, perhaps in recognition of the tension between the apparent doctrinal correctness of its interpretation of 'the parties' and its restrictive effects, and also because it was still to reconcile the AB's decision in *US - Shrimp*, the Panel went on to consider an alternative aspect of the VCLT rule of interpretation, namely Article 31(1).

(b) VCLT Article 31(1)

The Panel's application of VCLT Article 31(1) removed the requirement of parallel membership of non-WTO and WTO treaties in establishing relevant interpretative context. The Panel considered that VCLT Article 31(1) allowed for the use of rules of international law that were not binding on the parties where those rules provided evidence of the 'ordinary meaning' of the treaty terms.²⁸⁶ This negates the need to establish the consent of treaty parties when taking into account other treaties that, notwithstanding dissimilar treaty membership from the treaty being interpreted, are indicative of 'ordinary meaning'. Of relevance is not whether WTO members have ratified the relevant rule, but whether it is 'informative' because it represents ordinary meaning within the international community, or, at least, to the treaty-interpreter.

The Panel's approach to VCLT Article 31(1) is attractive given that 'ordinary meaning' is not a matter of consent but rather of intersubjectivity. Meaning in language is not dependent on the consent of participants, but rather develops according to social practices within a community. Moreover, given the implied reliance on the concept of

---

[282] VCLT Art. 31(3)(b) provides that subsequent practice may be taken into account in interpreting a treaty if the practice has established 'the agreement of the parties regarding its interpretation'.
[283] Ian Brownlie, *Principles of Public International Law* (2003) 605.
[284] AB Report, *EC - Chicken Cuts*, para. 273.
[285] The *Biotech* Panel considered the construction of Art. 31(3)(b) in *EC - Chicken Cuts* to be supportive of its interpretation of Art. 31(3)(c): see *Biotech*, para. 7.68.
[286] Panel Report, *EC-Biotech*, para. 7.92, noted above n. 92 and accompanying text.

an international community, the Panel's approach may seem appealing to those who call for the systemic integration of international norms.

However, the Panel's use of VCLT Article 31(1) is open to criticism. The ILC Study Group found it to be contrived.[287] Moreover, its claim of doctrinal support for its approach is doubtful; the Panel's reliance on US – Shrimp sits uneasily with the AB's much broader and largely unspecified use of interpretative tools in that case.[288] More importantly, the terms of VCLT Article 31(1) itself do not warrant the recourse to relevant rules of international law to determine 'ordinary meaning'. 'Ordinary meaning' is a seductively simple phrase, suggesting a natural meaning and masking the fact that a different understanding of the meaning of terms is likely to be the root of the conflict.[289] This limitation is recognised by the VCLT. Article 31(1) acknowledges that a treaty's terms cannot be ascertained in the abstract, by requiring that the ordinary meaning of terms be interpreted 'in their context'.[290]

The key question for interpretation is therefore to ascertain which contextual boundaries are imposed by the system itself. In the VCLT, the allowable context is narrowly defined by VCLT Article 31(2) as the body of textual material generated during the conclusion of the treaty.[291] According to the rest of the VCLT's rule of interpretation, the only other relevant extrinsic materials are those developed subsequently by the parties evidencing their common intentions (including with respect to a 'special meaning' to be given to a term),[292] and supplementary means where interpretation under Article 31 leaves the meaning 'ambiguous or obscure' or leads to a result 'which is

---

[287] ILC Analytical Study, para. 450.   [288] See above n. 62 and accompanying text.

[289] Hersh Lauterpacht, *The Development of International Law by the International Court* (1958) 52–60, reproducing in substantial terms his 'A Note on the Doctrine of 'Plain Meaning' (1950), which he submitted to the Institute of International Law: *Annuear*, 42 (1950) (i), 377–90. Indeed, the contextual and contestable nature of meaning has been a preoccupation of many disciplines of academic thought. Most notably, the idea behind deconstruction, as found for example in the works of Derrida, is that words or terms always and necessarily defer to other different terms in a conceivably endless process.

[290] The Institute of International Law's 1956 Resolution, adopted after Lauterpacht's note referred to ibid., came to be influential for the ILC in 1966 in its work on the VCLT.

[291] Although it may seem odd to turn to supplementary means to interpret the phrase 'ordinary meaning', I note that this reading is confirmed by reference to the ILC Commentary on the VCLT. The ILC did not appear to anticipate that 'ordinary meaning' would necessitate reference to any extrinsic texts beyond those texts that established the 'context' of the treaty in Article 31(2): see *Yearbook of the International Law Commission* (1966) Vol II, 221 (para. 12).

[292] VCLT Art. 31(3), (4).

manifestly absurd or unreasonable'.[293] Thus other extrinsic materials, such as informative international law materials and even, perhaps, dictionaries, are not considered to be relevant to establishing 'ordinary meaning'. These contextual boundaries were ignored by the *EC – Biotech* Panel, which instead sought guidance from 'informative' texts to bring to an end its search for 'ordinary meaning'.

In my view, the *EC-Biotech* Panel would have been more convincing if it had relied on the purposive element of VCLT Article 31(1) rather than on 'ordinary meaning'. Treaties that are not binding on all WTO members may still be relevant, for example, to inform a treaty-interpreter about the object and purpose of a treaty.[294] The object and purpose of a treaty regulating the apple trade will more easily be found to exclude the orange trade if a substantial treaty already exists for the trade in oranges, even if their membership is not identical. On this reading, the fact that the CBD parties, most of whom are WTO members, were negotiating a Protocol on Biosafety at the time of the SPS Agreement could be rebuttable evidence that the object and purpose of the SPS Agreement was to exclude rules and disciplines on SPS measures aimed at biotech products.[295] The principles of 'sustainable development' and 'mutual supportiveness' that have been endorsed by WTO members would also be relevant to this kind of purposive inquiry.[296]

Apart from the questionable doctrinal foundations of the *EC – Biotech* Panel's application of VCLT Article 31(1), its use of non-WTO sources to inform the 'ordinary meaning' of a WTO treaty term led to decontextualised and arbitrary reasoning. For example, the Panel found that a WHO definition of 'disease' was more relevant than the World Organization for Animal Health to defining risks of animal or plant diseases.[297] In addition, a common dictionary definition of 'pests' was given primacy over a definition from a standard for plant safety produced by the IPPC.[298] These and other examples suggest the difficulties

---

[293] Ibid., Art. 32.
[294] In advocating the use of Art. 31(1) as part of a process of systemic integration, McLachlan points both to its purposive aspects and the 'ordinary meaning': McLachlan, above n. 37, para. 17.
[295] This evidence would be rebutted by the 'savings clause' in the Biosafety Protocol, which states that WTO rights are not to be affected: see further Sabrina Safrin, 'Treaties in Collision? The Biosafety Protocol and the World Trade Organization Agreements' (2002) 96 *AJIL* 606.
[296] See further below n. 303 and accompanying text.
[297] Panel Report, *EC – Biotech*, paras. 7.277-7.278.   [298] Ibid., para. 7.241.

and dangers of finding an 'ordinary meaning' within a diverse international context.[299]

As explained above, the *EC – Biotech* Panel's interpretation of VCLT Article 31(3)(c) confirmed that 'consent of the WTO Membership' was an entrance condition for the use by a dispute settlement body of international law as interpretative tools. Its use of VCLT Article 31(1), on the other hand, removed the need to establish such consent for non-WTO sources to be used in interpreting WTO terms. In my view, the Panel substituted the entrance condition of 'consent' for 'relevance', so that extrinsic materials could be taken into account if they were informative of the 'ordinary meaning' of WTO treaty terms. This concept of relevance has great promise for improved decision-making, as it does in many areas of domestic law.[300] Yet the Panel's interpretation of Annex A of the SPS Agreement was problematic in its seemingly arbitrary selection of non-WTO sources. The Panel needed to be more open as to how it ascertained the 'ordinary meaning' within the international system, and in particular how it decided whose social practices constructed it. The counter to the dilemma of parallel membership is thus the dilemma of possible over-inclusiveness of non-WTO sources in adjudication. The following section examines how adjudicators can take account of non-WTO sources in a disciplined and appropriate way.

4 Legitimacy and the need for guidance in the use of exogenous sources

Discussions on treaty interpretation have focused on the question of whether adjudicators can consider non-WTO sources while remaining within the judicial function,[301] as encapsulated by the respect for state consent implicit in the requirement for parallel membership in treaty interpretation. My analysis of WTO cases, particularly the recent *EC – Biotech* Panel, reveals the need to move beyond the question of whether non-WTO sources will be taken into account in regime interaction. Instead, it is necessary to consider *how* such sources should be taken

---

[299] I have described elsewhere several problematic examples of the reasoning of the Panel in interpreting Annex A of the SPS Agreement by recourse to 'informative' materials: see Young, above n. 149, 922–5.

[300] Special regard may be had to principles familiar to many administrative lawyers, such as the judicial reviewability of failures by decision-makers to take relevant considerations into account.

[301] See e.g. Duncan French, 'Treaty Interpretation and the Incorporation of Extraneous Legal Rules' (2006) 55 *ICLQ* 281.

into account in order to avoid the unrestrained and decontextualised reasoning that I criticised above.

Panels and the AB are obviously already assessing non-WTO sources using their own means. For example, in *US – Shrimp*, the scientists assisting the Panel drew on non-WTO sources in making their findings. It is unclear how the AB ranked these sources when it decided to refer to CITES, rather than the IUCN or the CMS, in its assessment of the threatened nature of sea turtles.[302] Similarly, in *EC – Biotech*, the Panel's use of non-WTO sources was dictated by whether they deemed such sources to be 'informative'.

Implicit in this approach is the notion of an 'international system' or 'international community' providing the context for an assessment of a rule's relevance. Questions must be asked about any biases that result from this conception of the international community. For example, in *EC – Biotech* this 'international community' was a scientific and technical community that had been active in developing rules and guidelines. Yet absence of rule-making by parts of this scientific community may be attributed to an absence of scientific inquiry (due to diverted research funding, epistemic weaknesses, etc) rather than an absence of collective concern. Silence in the international system may be 'informative' for many reasons. As such, the Panel's pronouncement on which non-WTO sources seem the most 'informative' is too limited, particularly given that all the disputing parties argued at various points that there was 'consensus' for the interpretation that they were advancing.

Ideas for the appropriate use of non-WTO sources may be provoked by an examination of some of the WTO's institutional provisions for consultation and coexistence with international organisations. For example, the concept of 'mutual supportiveness' has been incorporated in a number of WTO instruments such as the Decision on Trade and Environment and the Doha Declaration.[303] Accordingly, members have agreed to negotiate and formulate policy with this concept in mind. If the *EC – Biotech* Panel's consultations had been driven by this concept rather than its restrictive notions of treaty interpretation, it arguably would have been able to take into account a much wider scope of non-WTO sources, including the Biosafety Protocol and the CBD,[304] and better understand

---

[302] AB Report, *US – Shrimp*, para 132.
[303] WTO Decision on Trade and Environment (1994); WTO Doha Declaration (2001).
[304] For criticism that the Panel in *Biotech* failed to incorporate the concept of mutual supportiveness, see Nathalie Bernasconi-Osterwalder, 'Interpreting WTO Law and the Relevance of Multilateral Environmental Agreements in *EC-Biotech*' Background Note

the relationship between these non-WTO sources themselves in order to determine their appropriate use *relative to each other*.

The status of international organisations in the SPS and TBT Agreement may also provide some guidance to adjudicators. Relevant standard-setting bodies are identified in the SPS Agreement as the Codex, the IPPC and the OIE, although further international bodies can be identified through the SPS Committee provided they are open for membership to all WTO members.[305] The TBT Agreement goes further and endorses standards developed by international bodies that are open to the relevant bodies of all WTO members.[306] There is no need for consensus in the development of the standards for them to be relevant.[307] However, the standard-setting bodies are encouraged to operate with open, impartial and transparent procedures.[308] Moreover, international bodies may apply for observer status to the relevant committees.[309] Accessibility for the WTO membership, rather than parallel membership, is therefore the main theme of the WTO's institutional co-existence with other international organisations under this framework. Such accessibility might therefore be one factor in determining that the work of an IGO ought to be taken into account and given probative value by a WTO panel.

Other factors that might assist in providing adjudicators with guidance on the use of non-WTO sources might relate to the breadth of an organisation's support, and the balance of its membership between developing and developed countries.[310] Procedures for transparency and cooperation between secretariats, and openness to non-state actors such as NGOs, might be further factors that reinforce the 'relevance' of the norms developed under the auspices of IGOs. Additional considerations such as the concepts of subsidiarity and flexibility will also be helpful and demand further attention.[311]

---

to presentation at BIICL WTO Conference, May 2007, available on the website of the Center for International Environmental Law (www.ciel.org).

[305] SPS Agreement Annex A:3.   [306] TBT Agreement Annex 1:4.
[307] See AB Reports in *EC – Hormones* and *EC – Sardines*, noted in Chapter 3, n. 171 and accompanying text.
[308] See TBT Committee Decision noted in Chapter 3, n. 212.
[309] There has been a long-standing request by the CBD and the Biosafety Protocol for observer status to the SPS Committee, which has been delayed on political grounds: see further Joanne Scott, *The WTO Agreement on Sanitary and Phytosanitary Measures: A Commentary* (2007) 63.
[310] For a similar idea in relation to treaty interpretation, see Asif Qureshi, *Interpreting WTO Agreements: Problems and Perspectives* (2006) 114–59, 120.
[311] Joanne Scott, 'International Trade and Environmental Governance: Relating Rules (and Standards) in the EU and the WTO' (2004) 15 *EJIL* 307, 346.

These ideas call into question the current judicial tools of interpretation. The VCLT itself could be interpreted in an evolutionary fashion to incorporate a spectrum of 'international consensus' as opposed to binary questions of consent and non-consent. Recognising the increased involvement of international organisations and non-state actors in law-making adds a further shade to this spectrum and allows institutional questions of openness and accessibility to be raised at the interpretative stage. As such, the *process* by which a treaty comes into being may be assessed to determine its probative value in interpreting other treaty obligations.

The current restrictive practices for panel consultation also need to change if panels are to follow this guidance in using non-WTO sources. For example, panels may need to consult with other international institutions in a proactive way that does not rely on the disputing parties.[312] IGOs themselves may choose to submit unsolicited *amicus* briefs rather than wait to be consulted.[313] A greater willingness to include IGOs and NGOs in the dispute settlement process will enhance the ability of WTO adjudicators to take account of non-WTO sources and deliberate in the wider context of the international system.

Again, regime collaboration through increased consultation calls for guidance of its own. As I described above, the 'sources of non-WTO sources' (i.e. those individuals or bodies that make submissions and provide information) are already informally ranked by adjudicating bodies. Unlike parties, IGOs and NGOs do not have the right to have their submissions taken into account by panels or the AB. I do not disclaim the need to assess the relative weight of different participants in this way. Instead, I call for more transparency in this process.

In this regard, it is clear that improved regime collaboration requires more open assessment of the *amicus curiae*. Up until now, panels and the AB have been silent on the status and background of the parties submitting *amicus* briefs[314] and have merely provided blanket statements that they do not 'find it necessary to take into account' their briefs. A better system would apply public criteria to assess the usefulness and

---

[312] For an assessment of an enhanced judicial role in participation, information-sharing and principled decision-making, see Joanne Scott and Susan Sturm, 'Courts as Catalysts: Rethinking the Judicial Role in New Governance' (2007) 13 *Columbia Journal of European Law* 565.

[313] Steve Charnovitz, 'WTO Cosmopolitics' (2002) 34 *NYU Journal of International Law and Politics* 299, 353.

[314] See, e.g., AB Report *EC – Sardines*, above n. 235 and accompanying text.

relevance of such submissions. The need for such guidance has been expressed within the WTO itself[315] and was perhaps the instigating factor for the special appellate working procedures adopted in *EC – Asbestos*, which required applicants for leave to file *amicus* briefs to specify the nature of their interest in the appeal.[316] Internal commentators have stressed the need for substantive criteria to enable adjudicating bodies to assess the source of the *amicus* brief, such as (i) the public interest of the brief (especially for groups that would have no direct financial interest in the outcome of the case); (ii) the representation and accountability of the NGOs; and (iii) demonstrated expertise in the area.[317] I agree that regime collaboration is more likely to be achieved where the sources of *amicus* briefs are transparent about their funding and interests, broadly accountable and have demonstrated expertise. This is dealt with in greater depth in Part III of the book.

In a similar way, panels and the AB must be open in their interrogation of IGOs. I considered above such factors as internal transparency and accessibility of IGOs that would be useful in establishing the relevance of norms produced under their auspices. Such questions will often arise during the consultation stage, calling for adjudicating bodies to establish from the IGOs their interest and accountability to their respective members. The trend towards public hearings of disputes[318] will assist in this process.

These ideas would have benefited the adjudicators in *US – Shrimp* and *EC – Biotech*. The next Part combines these findings with my analysis of negotiations and implementation and concludes on implications for fisheries governance and the fragmented international legal order as a whole.

F Conclusions

This chapter has investigated the regime interaction that occurred in the resolution of a fisheries dispute in *US – Shrimp* and in other cases. The relevant forum was the WTO dispute settlement mechanism, for reasons provided in earlier chapters: the WTO regime contains superior enforceability to UNCLOS and FAO regimes as well as to international environmental law. Although UNCLOS itself provides for binding decisions by bodies such as ITLOS, the jurisdiction of such bodies is

---

[315] Consultative Board, above n. 1, para. 260. [316] See above n. 221.
[317] Marceau and Stilwell, above n. 220, 180. [318] See above n. 116.

curtailed, especially when states parties have agreed on specific and non-enforceable obligations in regional fisheries regimes.[319] In addition, the lack of compulsory jurisdiction in the ICJ or other tribunals over environmental wrongs has led to calls for the establishment of a World Environment Court,[320] which are as yet unmet.

I have described how WTO adjudicating bodies draw on sources from other regimes. Such sources are relevant as applicable law, tools for treaty interpretation or as relevant facts. The direct conduits for such sources have been the panellists, AB members and the WTO Secretariat, the disputing parties, scientific experts, IGO secretariats and NGOs and individuals submitting *amicus* briefs.[321] There is, however, continuing uncertainty in dealing with IGO secretariats. In addition, the particular reluctance of WTO dispute settlement bodies to accept and utilise *amicus* briefs is manifest elsewhere in international dispute settlement, including the ICJ.

The chapter suggested that regime interaction can be made more effective through more discerning selection of the adjudicators, better coordination in disputing parties' submissions and a more nuanced approach to treaty interpretation. In particular, the notion that 'any relevant rules of international law applicable in the relations between the parties' includes as interpretative context only rules agreed by all WTO members, as suggested by the *EC – Biotech* Panel's reading of VCLT Article 31(3)(c), must be rejected. However, regime interaction divorced from parallel membership gives rise to its own problems and challenges, and I have discussed the need for guidance in the use of sources from other regimes. There needs to be enhanced transparency in the consultations between the WTO and other IGOs. Similarly, the adjudicating bodies need to assess the credibility of any entity seeking to act as an *amicus curiae*, a move that may well increase the flow of important knowledge and expertise from other regimes. These ideas indicate a need for adjudicating bodies to scrutinise the contributions from other regimes as a necessary part of regime interaction. A legal framework for appropriate regime interaction is explored in greater depth in the next Part.

---

[319] *Southern Bluefin Tuna Case (Australia and New Zealand* v. *Japan) (Award)* (2000) (Annex VII Tribunal), discussed further in Chapter 2.

[320] See e.g. Daniel Esty's proposal for a Global Environment Organisation: *Greening the GATT: Trade, Environment, and the Future* (1994); 'Stepping Up to the Global Environmental Challenge' (1996–97) 8 *Fordham Environmental Law Journal* 103.

[321] Indirect conduits are limitless and include newspapers and the internet: see Mavroidis, above n. 213.

PART III

Towards Regime Interaction

6 From fragmentation to regime interaction

Interagency cooperation and coordination (under development) [1]

The three case studies canvassed in this book provide clear examples of regime interaction in fisheries governance. In Part II, I investigated regime interaction that occurred during the negotiation, implementation and adjudication of norms. Chapter 3 analysed the making of new rules on fisheries subsidies, Chapter 4 assessed the implementation of existing commitments within the CITES and FAO regimes, and Chapter 5 examined the adjudication of trade disputes surrounding fisheries at the WTO. In this Part, I draw on my findings to make significant claims about regime interaction in international law. The phenomenon of fragmented fisheries governance provides important insights into the interests and motivation for regime interaction. Based on the factors that promote and obstruct regime interaction, I make normative arguments for how appropriate regime interaction *should* occur, and I offer tentative ideas for a legal framework for regime interaction. This legal framework has relevance beyond fisheries governance, and I conclude on its implications for legal doctrine and theoretical debates in international law.

In the current chapter, I document the move away from forum shopping and towards regime interaction. In the different contexts of my case studies, certain factors either promoted or obstructed regime interaction. Inter-regime learning and coordination at both the national and international level were significant factors that allowed different

---

[1] An inactive link on the website of the UN Department of Ocean Affairs and the Law of the Sea: www.un.org/Depts/los/index.htm (last accessed 4 February 2008, subsequently removed).

regimes to address problems of fisheries over-exploitation. On the other hand, an emphasis on delimited competences and mandates, inaccessible decision-making and parallel state membership significantly obstructed regime interaction. The chapter compares these factors, pointing to significant processes, actors and attitudes.

An awareness of these factors necessitates further inquiries that enable me in the next chapter, Chapter 7, to offer a legal framework in support of regime interaction. This framework builds on existing doctrines of institutional law, including the principle of implied powers of international organisations. At a theoretical level, the framework requires new conceptions for international law's legitimacy, which move from state consent to a participatory model of international law. Chapter 8 concludes with examples of a legal framework operating in the context of fragmented fisheries governance. It notes the significant departure from the ILC Study Group on Fragmentation, and offers some closing remarks about the general implications of a legal framework for regime interaction in fragmented fisheries governance and beyond.

A From forum shopping to interaction

The case studies reveal important struggles over the forum and form in which fisheries problems are addressed. Often, states and other interested parties sought exclusivity within one regime. This occurred at all stages of the spectrum of law-making, implementation and adjudication. In efforts to discipline fisheries subsidies through new rules, for example, attempts were made to draw clear boundaries between 'trade' issues and other issues.[2] Some states suggested that the WTO was an inappropriate forum given that the FAO regime already sought to manage and conserve fisheries resources. For these states, the existence of the FAO regime signified *exclusivity* in addressing a global resource problem, notwithstanding that the problem has multiple economic, environmental and social dimensions.

The implementation of existing global rules to conserve endangered marine species also involved regime struggles.[3] Some states sought exclusivity for the FAO regime and denied the ability of CITES to restrict trade in marine species. Early drafts of the Memorandum of Understanding between the CITES Secretariat and the FAO revealed

[2] See further Chapter 3. [3] See further Chapter 4.

attempts to entrench hierarchy between regimes. Questions of forum were also evident when states sought to settle their fisheries disputes (as well as trade and environmental issues surrounding biotechnology) at the WTO.[4]

On their face, forum queries seem to be motivated by an underlying need to point to legitimacy in international law-making. In treaties, negotiated textual outcomes give rise to obligations for state participants.[5] New participants are expected to sign on to treaties according to this existing structure.[6] Moreover, some treaties are negotiated in very general 'framework' terms and are updated by regular meetings of states parties.[7] Conceived in this manner, it is a short step to suggest that these international treaties 'cover the field' of policy-making and jurisdiction as set out in their textual mandates; trade issues are to be dealt with under existing trade treaties; environmental problems are addressed by environmental treaties, and so on. The most extreme form of this model warns against such 'self-contained regimes' traversing into each other's fields of competence. Concerns about bias and institutional competence flow from these assumed boundaries.

The idea of pre-defined and exclusive competences is particularly strong in trade circles.[8] Questions about whether the WTO is the 'right forum' to incorporate environment and other issues, such as human rights, labour, competition law and investment law, have been asked since its establishment.[9] Similar questions were asked during negotiations for an International Trade Organization, which were

---

[4] See further Chapter 5.
[5] See Gerald Fitzmaurice, 'Some Problems Regarding the Formal Sources of International Law' in *Symbolae Verzijl* (1958) 153, 157 (noting the difficulty of referring to treaties as a source of law because they only bind the parties to the treaty) cited in Duncan Hollis, 'Why State Consent Still Matters – Non-State Actors, Treaties, and the Changing Sources of International Law' (2005) 23 *Berkeley Journal of International Law* 1, 6; see also *pacta tertiis* noted in Chapter 1, n. 98.
[6] New participants may also have to agree to additional requirements: see, e.g., the WTO accession process.
[7] This is notable in the environmental area: see Robin Churchill and Geir Ulfstein, 'Autonomous Institutional Arrangements in Multilateral Environmental Agreements: A Little-Noticed Phenomenon in International Law' (2000) 94 *AJIL* 623.
[8] On discussion of the WTO as a self-contained regime, see ILC Analytical Study 87–91; see also Jagdish Bhagwati, 'Afterword: The Question of Linkage' (2002) 96 *AJIL* 126, 129 ('We really must pause before we drag into WTO jurisprudence and practice others' differences in domestic policies and institutions, claiming in effect that any such differences abroad are to be the stuff of which "unfair trade" charges of denial of negotiated market access are made.')
[9] As discussed in the 'trade and…' literature, see Chapter 1, n. 63 and accompanying text.

abandoned in the 1940s in favour of the GATT.[10] Many trade negotiators argue that the WTO is a self-contained regime because they wish to ensure that WTO members keep to the commitments to which they have agreed in their schedules of concessions. A unilateral trade measure that has protectionist effects, for example, which a respondent WTO member may claim to be justified by reference to a non-trade objective contained in GATT Article XX, is met with deep suspicion.[11] The simplest way to limit divergences from the 'incomplete contract' of the WTO is to construct strict boundaries surrounding 'trade' and 'non-trade' issues.

The need to guard the boundary of the WTO is based on the very reason that many seek to open it: the fact of the binding nature of trade commitments vis-à-vis other international norms. With its compulsory dispute settlement system, the WTO contains a stronger compliance system than other regimes and has become a 'policy magnet'.[12] In addition, given that its norms are self-enforcing, the WTO has much stronger influence on domestic legal orders than regimes such as CITES.[13] There are other strategic advantages of linking non-trade issues with the WTO, such as the ability of states to increase their relative bargaining power in negotiations by drawing on a wider variety of issues.[14]

The fisheries subsidies negotiations confirmed the strategies involved in law-making amongst varying effectiveness of existing regimes. Chapter 3 demonstrated that at least part of the motivation of addressing fisheries subsidies at the WTO is to enable trade rules to be enforced

---

[10] The UN Economic and Social Council (ECOSOC) was established in 1946 and adopted a resolution for the convening of a United Nations Conference on Trade and Employment. However, plans for an International Trade Organization failed due to stalling by the US Congress and only the GATT was adopted: see John Jackson, William Davey and Alan Sykes, *Legal Problems of International Economic Relations* (2002) 57–8; 211–18.

[11] Bhagwati, above n. 8.

[12] See e.g. Sylvia Ostry, 'World Trade Organization: Institutional Design for Better Governance' in Roger B Porter, Pierre Sauve, Arvind Subramanian and Americo Beviglia Zampetti (eds.) *Efficiency, Equity, and Legitimacy: The Multilateral Trading System at the Millennium* (2001) 361.

[13] See e.g. the discussion of how the WTO diminishes the usual ways in which states consult national values in their international actions: Claire Kelly, 'The Value Vacuum: Self-enforcing Regimes and the Dilution of the Normative Feedback Loop' (2001) 22 *Michigan Journal of International Law* 673.

[14] David Leebron, 'Linkages' (2002) 96 *AJIL* 5, 12. Leebron warned however, with a presentiment borne out by the current stalling of the Doha single undertaking, that linking too many issues could inhibit agreement between states: at 25.

to bring about wider economic, environmental and even social reform. By contrast, FAO instruments are non-binding and there are significant contingencies to UNCLOS's compulsory dispute settlement mechanisms. Opposition to new rules on fisheries subsidies therefore comes from those who wish to restrict fisheries conservation and management issues to other (apparently relatively ineffective) regimes, and from those who are wary of diluting the WTO's essence of economically driven trade policy.

Similarly, the implementation of existing CITES rules in the fisheries context encountered severe opposition from many who claimed that endangered marine species ought to be managed within the FAO regime. In Chapter 4, it was clear that arguments for regime exclusivity undermined attempts to draft a Memorandum of Understanding between the FAO and the CITES Secretariat. In dispute settlement too, the cases demonstrate prior forum shopping before WTO members filed their complaints at the WTO. Strategic choices over hard and soft regimes are clearly at play in decisions to take cases to the WTO. Such decisions are clearly affected by the absence of a long-mooted World Environment Court with equal powers and enforceability.[15]

In legal terms, these concerns can be addressed by turning to textual mandates. Indeed, Chapter 3 concluded that the WTO's role in fisheries subsidies was supported by its wide textual mandate and by the potential for new disciplines under the SCM Agreement to promote fairness, domestic restraint and institutional effectiveness. Chapter 4's investigation into the appropriateness of CITES as a forum to protect endangered marine species confirmed that the listing of such species was recognised in its constitutive documents. The fisheries disputes discussed in Chapter 5, including the parallel dispute settlement proceedings filed at both the WTO and ITLOS, also indicate that mandates are generally non-exclusive.

Within the case studies discussed, a different question emerged about governance. The imperative question was *not* the question of forum (which remains indeterminate on certain issues) but rather the question of how existing regimes *interact*. Given the open-ended nature of mandates and competences, the non-exclusivity of regimes and the absence of priority of one regime over another, there is a need to consider whether regimes *interact* in ways that are legitimate. If the WTO has an ability to govern fisheries subsidies, how does it interact with

---

[15] See further Chapter 1, n. 64 and accompanying text.

existing regimes? If CITES restricts trade in endangered marine species, how does it interact with other regimes which allow, or even facilitate, such trade?

Interaction between regimes during law-making must still answer queries about legitimacy. The rationale of legitimacy may continue to lie in a conception of state sovereignty that seeks to ascertain whether states have consented to the norms imposed on them by regime interaction. There are alternative conceptions, however, which are developed in this chapter and the following one. For example, the legitimacy of WTO interaction with other regimes to discipline fisheries subsidies may reside in the adaptive ability of the WTO to make new rules that address the inevitable political wrangling in an open and accountable way.

The move from forum shopping to regime interaction has implications for the scholarly literature on linkages between 'trade' and 'non-trade' issues. Instead of categorising policy candidates for inclusion into the WTO as a priori belonging to an existing and fixed regime of 'environment', 'investment', 'competition' or 'human rights', it may be more useful to consider how global governance can be reformed to address urgent areas of human concern.[16] Reform must acknowledge the current asymmetry in international governance, and that there are no relevant environmental or fisheries treaties that possess the 'teeth' of the WTO's compulsory and very active dispute settlement system.[17] As such, the question moves from asking 'what is the right forum?', to asking how international governance structures might meet the policy problem at hand. This occurs through regime interaction during negotiations, as described in Chapter 3, or during the implementation of established rules, as discussed in Chapter 4, or during the settlement of international disputes, the topic of Chapter 5.

The capacity of international law to respond to global problems like fisheries in a legitimate way may depend on a wider group of forum issues than strict mandate inquiries. It could involve recognition of the

---

[16] Sara Dillon, 'A Farewell to "Linkage": International Trade Law and Global Sustainability Indicators' (2002) 55 *Rutgers Law Review* 87, 92-3 ('The real debate is not about linkage. It is about the shape of a growing interest in an imagined global governance.') See also Andrew Lang, 'Reflecting on "Linkage": Cognitive and Institutional Change in the International Trading System (2007) 70 *Modern Law Review* 523.

[17] See e.g. Joel Trachtman, 'Institutional Linkage: Transcending "Trade and..."' (2002) 96 *AJIL* 77, 91 ('Given that the WTO exists, and no World Environmental Organization yet exists, actions such as adding functional responsibility to the WTO may make sense that would not make sense were circumstances different.')

diversity of policy problems, actors and policy solutions and the need for inclusivity and accountability within relevant institutions. Although it is important to inquire about the mandate of organisations in the development of new rules, it is equally important to determine the capacity for organisations to learn and adapt to new forms of governance.[18] This is, in part, a quest for an appropriate legal framework for regime interaction, as developed in the next chapter. But first it is important to canvass the factors that either promoted or obstructed regime interaction in the context of the case studies.

B The promotion of regime interaction

Regime interaction is promoted at a national level by domestic policy coordination and, at an international level, by mutual learning and information-sharing between regimes.

1 National policy coordination

In various ways, the case studies demonstrate the need for national policy coordination in international trade and environmental governance. In the WTO Doha negotiations on trade and environmental synergies, some WTO members have directed attention to domestic procedures to enhance coherence between trade and environmental policy.[19] An awareness of the need for national policy coordination has been expressed by WTO members outside of the context of the Doha negotiations, such as in ongoing committee work.[20] Similarly, states in the forum of the FAO recognise the importance of domestic policy coordination. During the development of the MOU between CITES and the FAO, many states acknowledged the need for their domestic agencies to cooperate on CITES listings of marine species and perceived the lack of inter-agency coordination as an obstruction to effective regime interaction. This acknowledgment has reached the level of obligation in the new Port State Measures Agreement, which obliges parties to 'take measures to exchange information among

---

[18] See Steve Charnovitz, 'Triangulating the World Trade Organization' (2002) 96 *AJIL* 28, 53. For a review of the literature on adaptive governance and its application to the WTO, see Rosie Cooney and Andrew Lang, 'Taking Uncertainty Seriously: Adaptive Governance and International Trade' (2007) 18 *EJIL* 523.
[19] See Chapter 4, n. 241 and accompanying text.
[20] See e.g. WTO Doc. G/TBT/M/43 (TBT Committee).

relevant national agencies and to coordinate the activities of such agencies in the implementation of this Agreement'.[21]

The staff members of many IGOs have reiterated this view. The FAO considered that improved national policy coordination could overcome the problems it faced in finalising the MOU with its CITES counterparts. In the framing of new rules for fisheries subsidies, UNEP sought to encourage national policy coordination by inviting domestic trade, fisheries and environmental delegations to its workshops. In consultations for improving general ocean governance, the UN's Under-Secretary- General for Legal Affairs recognised the need to consider national coordinating processes:

> Cooperation and coordination are essential also at the national level, where sometimes a multiplicity of agencies responsible for oceans-related matters generate problems rather than solutions, thus exacerbating the fragmentation of ocean governance. Coordinated and strategic national government actions and initiatives in different international forums are crucial for the development of the required interlinkages and synergies, which could in turn produce meaningful results and avoid pitfalls.[22]

The recognition of the need for coordination is reflected more generally in international law, with commentators quipping that '[c]ordination begins at home'.[23]

Although this emphasis on inter-agency collaboration is important, there are two things to note about procedures for national policy coordination. First, they are not solely concentrated on domestic agencies, but rely also on the contribution of a number of other stakeholders. Secondly, national policy coordination is but one tool for the promotion of regime interaction and should not be overstated.

As to the first point, many countries have emphasised the importance of involving external stakeholders when coordinating national policies. In the Doha negotiations, most of the WTO members' submissions on their domestic implementation of MEA trade obligations revealed that they do not limit their internal processes to inter-agency collaboration. Instead, domestic officials relied on members of the public and other stakeholders to provide views on negotiations and implementation.[24] In the United States, for example, a public advisory

---

[21] Port State Measures Agreement 2009, Art. 5(c).   [22] UN Doc A/55/274, Annex I, para. 15.
[23] Henry Schermers and Niels Blokker, *International Institutional Law* (2003), 1107 (para. 1739).
[24] See, e.g., WTO Doc. TN/TE/W/72/Rev.1, para. 9.

committee comprising business, academic and non-profit organisations supports the United States government work relevant to the CTE, while less formal and ad hoc consultative processes also exist.[25] Extensive consultation with the public is also said to accompany policy coordination on CITES implementation; the US Fish and Wildlife Service seeks public comment in advance of a CITES COP.[26]

Similarly, Australia, Switzerland and the EU have reported to the WTO on their own inclusive procedures in policy coordination. The EU emphasised, in particular, the European Commission's White Paper on European Governance.[27] The White Paper endorses principles of openness and participation to promote policy coherence.[28] Emerging EU practices of conducting sustainability impact assessments in the trade sector also seek to incorporate the views of stakeholders.[29]

It is difficult to judge from the submissions to the CTE how WTO members' domestic stakeholder consultations operate in practice. Notwithstanding these official endorsements of openness, the participation of the citizens and organisations that are affected by trade policies may be rather limited. For example, although Australia's submission was supportive of stakeholder input, the summary of its domestic implementation of CITES did not refer to any public consultation or participation. When the relevant Australian department assesses applications for permits, it engages in inter-agency consultation and consultation with scientific experts but does not engage the public.[30] This suggests inconsistencies between Australia's endorsements of inclusive domestic policy coordination and actual practice.

For the purposes of my study it is sufficient to note that regardless of the effectiveness of public participation in policy coordination between trade and environmental governance, WTO members are increasingly aware that stakeholder consultation is part of attempts to ensure mutual supportiveness between their policies. This awareness is relevant to the legal framework of regime interaction considered in the following chapter.

---

[25] WTO Doc. TN/T/W/40W40, para. 10.   [26] Ibid., paras. 15–18.
[27] COM(2001) 428 final.   [28] WTO Doc. TN/TE/W/53, para. 6.
[29] See European Commission *Communication on Impact Assessment* (2002) (COM(2002) 276), which includes stakeholder consultation in its extended impact assessment: see 19; see further on trade SIA: European Commission *Handbook for Trade Sustainability Impact Assessment* (2006) 23-7. See also Dillon's calls for sustainability indicators in the field of trade law: Dillon, above n. 16, 127.
[30] WTO Doc. TN/TE/W/45, para. 18.

Aside from the necessity for external stakeholder input, the second point of note with respect to national policy coordination is that it is but one of a range of methods for achieving regime interaction in fragmented fisheries governance. International law develops due to collective problems *between* states. Therefore, coherence within individual states is not sufficient to address global problems such as the sustainability of fisheries. Moreover, the resources of states vary in international law. The capacity of a least-developed-country member of the WTO to coordinate on domestic policies will be very different from the capacity of a developed-country member. For example, some developing countries claim that the growth of regimes erodes their 'policy space'.[31] Moreover, enhanced policy coordination cannot remedy the situations when states deliberately argue different positions in different forums.[32]

WTO members tend to overstate the importance of national policy coordination in achieving mutual supportiveness between trade and environment governance. In response, other members have asserted that national policy coordination is but one aspect of mutual supportiveness. The EU, for example, has stressed that institutional coordination and cooperation between international bodies are equally important.[33]

The EU considers that the basis for international coordination in 'member-driven organisations' such as the WTO is the recognition by WTO members of its importance and a resultant 'willingness to foster it'.[34] The EU considers that members can achieve this coordination

---

[31] 'Policy space' is a shorthand term to describe how international agreements and global markets reduce the scope for national regulators to make policies related to trade, investment and industrial development: see UNCTAD Doc TD/410 'Sao Paulo Consensus', para. 8. The issue of policy space has become a controversial issue between developed and developing countries: see Martin Khor, 'Debate on Policy Space Dominates UNCTAD Review' TWN Info Service on WTO and Trade Issues (May06/10) (13 May 2006).

[32] See Doaa Abdel Motaal, 'The Trade and Environment Policy Formulation Process' in Adil Najam, Mark Halle and Ricardo Meléndez-Ortiz (eds.) *Trade and Environment: A Resource Book* (2007) 17, 23 (giving the example of African countries that demanded strict trade measures in MEAs such as the Basel Convention, but have been opposed to the application of MEAs by the WTO; 'Some countries may simply try to obtain the most they can out of different fora – even if this means certain inconsistencies – since different regimes play different roles.')

[33] WTO Doc. TN/TE/W/53, paras. 30–34, see also WTO Doc. TN/TE W/58, para. 16 (Switzerland).

[34] Ibid., para. 31.

through institutionalising information-sharing and observership between MEA secretariats and the WTO[35] and ensuring effective information exchange in dispute settlement procedures.[36] While my case-study analysis is compatible with this position, it also features a significant difference. The notion that institutional collaboration can only occur if members of the relevant organisations agree to it underlies the EU's submission. As I have discussed, this idea is foundational to the alleged requirement that relevant rules of international law can assist in the interpretation of a treaty only if all the members of that treaty have also adopted the relevant rule.[37] It is also the basis for the requirement that negotiations under paragraph 31(i) of the Doha agenda do not affect WTO members that are not parties to the relevant MEAs.[38]

Contrary to these presumptions, my case studies revealed several circumstances where international institutions acted outside of a state consent framework. Instead, cooperative arrangements have arisen due to the practical necessity for learning and understanding within the system of international law. These activities throw into doubt the implicit assumption of the EU's submissions on institutional collaboration and have significant implications for the legal framework for appropriate regime interaction considered in the next chapter.

2 Learning and information-sharing

Institutional learning and information-exchange are necessary to counter fragmented fisheries governance. Formal and ad hoc procedures between secretariats enhance collaboration and mutual supportiveness between regimes. For example, the FAO and the CITES Secretariat were cooperating on the listing criteria for marine species well before the MOU was debated by states. The Director-General of the WTO expressed continuing optimism in the ability of his organisation to work with other IGOs in the framing of the fisheries subsidies rules.[39]

Importantly, learning processes were not limited to interplay between relevant secretariats. The case studies demonstrated the important role of non-state actors. For example, Chapter 3 documented the important influence of civil society groups on the fisheries subsidies negotiations. In particular, it noted the tangible influence of informal

---

[35] WTO Doc. TN/TE/W/39.   [36] WTO Doc. TN/TE/W/53 para. 34.
[37] VCLT Art. 31(3)(c), as discussed in Chapter 5.   [38] As discussed in Chapter 2, n. 317.
[39] Consultation by author with Director-General Lamy during online chat discussion hosted by www.wto.org, 16 November 2007.

workshops convened by UNEP and NGOs, which promoted learning opportunities for trade delegates, environmental delegates and other stakeholders. This influence was documented through public statements, private interviews and the Rules Group submissions themselves. The finding on the importance of civil society has theoretical significance. It updates existing literature that has traced the effect of NGO advocacy on the *formation* of regimes,[40] and demonstrates the important role of civil society in regime *interaction*.

Processes of learning and information-exchange have arisen because of the need for participants within the international legal system to understand its constituent parts. Such learning was sometimes entrenched through the use of standards, Memoranda of Understanding between IGOs and even peer review. Yet the case studies demonstrated that these processes were often stymied by states parties to regimes. For example, in its investigation of the learning processes during the negotiations of new subsidy disciplines, Chapter 3 described how some members of the WTO Rules Group rejected the suggested outreach to IGOs such as the FAO as well as NGOs such as WWF. This control over regime interaction is reminiscent of the current negotiations over specific trade obligations in MEAs, which are strictly limited in both conduct and effect to the states parties of relevant regimes.[41]

Yet although states parties attempted to control or limit inter-regime learning, the examples studied demonstrated that their consent was not always contingent. By contrast, the fisheries case studies suggest that institutional collaboration does *not* depend upon the consent of the parties to a regime. For example, even without formalised information-sharing within the Rules Group, Chapter 3 demonstrated how the activities of the FAO, UNEP and NGOs influenced member submissions. Due to successful informal learning arrangements, the learning about fisheries subsidies by negotiators was *not* dependent on the consent by all Rules Group participants.

Similarly, my discussion of the interaction between the FAO and the CITES Secretariat in Chapter 4 demonstrated that expertise was shared between regimes even outside the parameters agreed by participating states. This interaction occurred largely outside of the development of

---

[40] See e.g. Oran Young, 'The Politics of International Regime Formation: Managing Natural Resources and the Environment' (1989) 43 *International Organization* 349, 353–4 (using CITES as an example).

[41] See further Chapter 2, n. 315 and accompanying text.

the MOU, as was clear by the FAO's involvement both in revisions to the CITES listing criteria and in the proposed amendments to list particular marine species in the CITES Appendices. Moreover, while states had a major role in the evolution of the MOU, the IGOs themselves were the final signatories, which could give rise to a greater future role. Although enhanced by state involvement, the learning about the listing of marine species did not depend upon states.

Such sharing of information was a main theme of chapter 5, which considered the methods used by the Panel and the AB in *US – Shrimp* and other cases to establish facts relevant to the status of sea turtles and to assist in treaty interpretation. Notwithstanding a tendency for panels and the AB to restrict their use of external sources to that agreed by the disputing parties, I concluded that the DSU empowers the WTO adjudicating bodies to act independently of the disputing parties. Moreover, I found that consultation by adjudicating bodies extends beyond findings bounded by the rules on treaty interpretation. Instead, a WTO panel has the freedom to consult in order to achieve an objective assessment of the matter before it, which in fisheries cases will often entail consultation with IGOs and NGOs.

These findings concur with early proposals for international lawyers to manage the constant creation of new treaties through close inter-regime consultation. Jenks appealed to both states and IGOs to take an active role in avoiding future conflicts in international law through '[t]he practice of close consultation prior to and during the drafting of instruments between those responsible for the drafting of instruments dealing with related subjects which are negotiated under the auspices of different bodies'.[42] The findings also resonate with contemporary literature from political science and international relations that gives increasing attention to the building of networks of 'epistemic communities' within international governance. Haas's observation of the sharing of scientific and technical expertise within international environmental regimes was one of the first contributions to this area.[43] In shaping international norms, these epistemic networks are considered to have a governance role.[44] Because this is

---

[42] Wilfred Jenks, 'Conflict of Law-Making Treaties' (1953) 30 *BYIL* 401, 452.
[43] Peter Haas, 'Introduction: Epistemic Communities and International Policy Coordination' (1992) 46 *International Organization* 1.
[44] See further Peter Haas 'Addressing the Global Governance Deficit' (2004) 4 *Global Environmental Politics* 1 (suggesting reforms to a model of global governance that is based on diffuse networks of diverse actors performing multiple and overlapping functions).

governance by other than state actors, some commentators have labelled it 'post-sovereign'.[45]

The notion that learning and information-exchange between regimes shapes international norms, and that such governance may occur without the consent of member states, has significant implications. First, it gives rise to questions about whether IGOs that collaborate with other IGOs do so according to their mandates and, relatedly, in an accountable and legitimate way. Moreover, if institutional learning is not limited to the mere provision of information, but also influences the development of international law, international law practitioners need guidance to be able to scrutinise and contest the information that they receive. In particular, this gives rise to possible new conceptions about the concurrent duties of stakeholders within regimes. These issues influence the legal framework for appropriate regime interaction.

3 *Allocation of resources*

In line with the emphasis on institutional processes developed in this chapter, I note that inter-regime coordination is enhanced if resources are directed to this issue, although more research is required to measure this.[46] For example, it appeared that the MOU's express allocation of resources,[47] instigated by the CITES Secretariat, allowed the FAO and CITES to plan the funding of their collaboration. The WTO Secretariat has also recognised this issue. For example, a consultative body to the WTO Director-General has called for greater resources and capacity for 'horizontal coordination'.[48] And perhaps rather ominously, the new UN body coordinating oceans activities continues to remind Members of their duties to provide funding and support.[49] Although it is early to judge the efforts of the UN Consultative Process[50] and the associated

---

[45] Bradley Karkkainen, 'Post-Sovereign Environmental Governance' (2004) 4 *Global Environmental Politics* 72, 75 (defining post-sovereign governance as non-exclusive, non-hierarchical and post-territorial).
[46] For the impacts on efficiency of increasing numbers of actors in treaty negotiations, see José Alvarez, *International Organizations as Law-Makers* (2005) 367. An exemplar of problems of accommodating wide numbers of participants is the 2010 Copenhagen Conference on the United Nations Framework Convention on Climate Change.
[47] MOU, Art. 9.
[48] See Consultative Board, *The Future of the WTO: Addressing Institutional Challenges in the New Millennium* (2004) 40.
[49] See e.g. UN Doc A/63/174 (2008) para. 132.
[50] The UN Consultative Process on Oceans and the Law of the Sea has been operating for seven years: see further Chapter 2, n. 150 and accompanying text.

Oceans and Coastal Areas Network (UN-Oceans),[51] these initiatives do not appear to have made a significant contribution to collaboration in fisheries governance, at least in the context of my case studies.

This recognition of the need for resources for regime interaction is in line with broader literature which recognises the connection between active treaty management and compliance. In international environmental law, strong secretariats and dedicated resources have allowed IGOs to apply pressure to non-complying members.[52] As such, specialised funding from bodies such as the Global Environment Facility (GEF) may enhance joint activities between secretariats.[53]

Absent other factors, however, resources and capacity may have only a minor effect on effective regime interaction. A view that the success of inter-regime coordination depends on the level of resources devoted to the particular IGOs is simplistic. Moreover, the allocation of resources and capacity-building is not linear. My description of the successful collaboration in the fisheries subsidies negotiations revealed a major governing role of UNEP (which would have drawn heavily on its own budget), which did not depend on enhanced resources of the WTO Secretariat.

Finally, it is important to remember that it is not only states that allocate resources to regime interaction. Civil society groups contribute important technical and organisational skills for regime interaction. For example, WWF plays a prominent role in the fisheries subsidies negotiations and together with other NGOs, such as Greenpeace, provides important monitoring functions in fisheries.[54] ICTSD, an independent non-profit and non-governmental organisation, aims to promote dialogue between a broad range of stakeholders.[55] Its publication *Bridges*,

---

[51] UN-Oceans was established in 2004: see further Chapter 2, n. 158 and accompanying text.

[52] Antonia Handler Chayes, Abram Chayes and Ronald B Mitchell, 'Active Compliance Management in Environmental Treaties' in Winfried Lang (ed.) *Sustainable Development and International Law* (1995) 75, 88.

[53] The GEF was established in 1991 to assist developing countries to fund environmental protection projects: see www.gefweb.org. The GEF is the financial mechanism for the CBD, among others. See further Work Plan for Implementation of Joint Activities between the CITES Secretariat and the Convention on Biological Diversity: CITES, 'Synergy between CITES and CBD, see www.cites.org/eng/dec/valid13/13-02_05.shtml.

[54] Note also the crucial role of NGOs such as TRAFFIC in providing scientific data to CITES: Holly Edwards, 'When Predators Become Prey: The Need for International Shark Conservation' (2007) 12 *Ocean and Coastal Law Journal* 305, 353.

[55] One of the three aims of ICTSD's fisheries project is to '[s]upport disadvantaged stakeholders, including those making and influencing policies, to engage more effectively in the ongoing WTO negotiations on fisheries subsidies'. The project is

an information resource that is influential for collaboration between trade and environmental groups on issues such as fisheries subsidies, receives funding from NGOs as well as from states.[56] The resources of NGOs, which have already been recognized as enhancing treaty compliance,[57] affect regime interaction and also impact upon issues of accountability, which are explored below.

C Impediments to regime interaction

My case studies demonstrated several impediments to regime interaction, which included attempts to deny the role of one regime vis-à-vis another through the questioning of competence, a lack of transparency and openness within regimes, and the futile quest for parallel membership of regimes.

1 Exclusivity of forum

The clearest impediment to regime interaction is the attempt to limit or reduce the role of one regime vis-à-vis another. Such attempts are familiar to international lawyers and represented a political struggle over fisheries governance. For example, Chapter 3 described early attempts by some WTO members to dispute the mandate of the WTO to regulate fisheries subsidies by arguing *inter alia* that the FAO and RFMOs were more appropriate international bodies. Chapter 4 demonstrated that one of the main factors that stalled collaboration between CITES and the FAO was the attempt by certain states within the FAO to limit the role of CITES in listing marine species by drafting special provisions in the MOU. Questions of forum also arose in Chapter 5, albeit in a different way. Calls for the WTO adjudicating bodies to exercise restraint in 'trade and...' issues emanated from concerns about the separation of powers and the need for deference.[58]

The idea that there should be an a priori determination of competence for an international organisation to address a particular policy issue concords with some perspectives of international institutional law. In *Use of Nuclear Weapons*, the ICJ, by eleven votes to three, rejected

---

funded by Dutch and New Zealand support: see www.trade-environment.org/page/ictsd/projects/fish_desc.htm.
[56] See further http://ictsd.org/news/bridgesweekly/.
[57] Abram Chayes and Antonia Handler Chayes, *The New Sovereignty: Compliance with International Regulatory Agreements* (1995) 21.
[58] Jeffrey Dunoff, 'The Death of the Trade Regime' (1999) 10 *EJIL* 733, 757–60; see further Chapter 5, n. 243 and accompanying text.

a request for an advisory ruling on the legality of nuclear weapons from the WHO. The court found that the WHO's request was not a question which related to the scope of the WHO's activities.[59] The majority considered that international organisations, unlike states, do not possess general competence but are governed by the 'principle of speciality'; accordingly organisations 'are invested by the States which create them with powers, the limits of which are a function of the common interests whose promotion those States entrust to them'.[60] The majority considered that peace and security issues associated with nuclear weapons fell within the competences of the United Nations and outside its specialised agencies.[61]

The notion that competences are separated between international organisations seems to be warranted to avoid the practical problems of increased implementation costs associated with obligations arising from multiple regimes or regime interaction. Japan advanced this notion in its criticisms of the role of both CITES and the WTO in fisheries issues. Commentators such as Leebron agree that it is preferable to improve an ineffective regime governing a particular problem rather than to link sanctions with other regimes.[62]

My analysis rejects the compartmentalisation of regimes in this way. I have found that classifying and locating various policy problems within bounded legal structures and institutions can be counterproductive. Such an approach can also fail in resolving forum queries; although I conclude that the WTO is appropriately mandated to deal with fisheries subsidies, I acknowledge that the FAO may also have a useful role in encouraging plans of actions to reduce capacity. Other regimes, such as subsidy programmes within the OECD, may also have useful contributions to the objective of capacity-reduction. Similarly, the fact that the WTO adjudicated the dispute in US – Shrimp does not mean that other regimes such as CITES are not relevant to import restrictions on fisheries. There is a situation of continuing diversity and imbalance in international law and emerging problems require different institutional and normative responses.

These findings are similar to the view of international law expressed by Judge Weeramantry in his dissent in the WHO advisory opinion.

---

[59] (1996)(I) ICJ Rep 66. The Court noted the difference in language between 'activities' and 'competence' but considered this did not have implications for the present case: see 74 (para. 18).
[60] Ibid., 78 (para. 25).   [61] Ibid., 80 (para. 26).   [62] Leebron, above n. 14, 27.

Judge Weeramantry would have allowed the request for an advisory opinion and felt that the WHO's question was within its legitimate and constitutional sphere of interest. In response to the majority's application of the principle of speciality, his Honour recalled that institutions, particularly within the United Nations family, are connected and interdependent. Accordingly, they contain several areas of overlapping competences, from which they may contribute different areas of specialisation.[63] In relation to the UN, '[t]o expect otherwise would be contrary to the essence and rationale of a complex organization which straddles all facets of human activity'.[64] Judge Weeramantry has continued to address institutional interdependencies extrajudicially.[65]

I do not conclude that forum queries are meaningless. Nor do I offer conclusions about which international forums are the most appropriate to deal with trade-related aspects of fisheries. I have balanced my analysis of the WTO's role in negotiating and adjudicating on fisheries issues with an analysis of CITES and a continuing awareness of the role of the FAO. In particular, my suggestions of the way the WTO regime should incorporate other perspectives should not be read as an argument that the WTO regime is validly assuming a global governance role.[66] Instead, I describe attempts to enhance the governance of fisheries through multiple regimes and point to ways to manage the resulting overlaps.

My views are different from responses to mandate struggles that seek to establish the intent of states parties. For example, in response to the claim by some WTO members that the WTO had no mandate to negotiate on fisheries subsidies, regard could be had to the Doha Declaration and accompanying legal processes. Against attempts by some states to limit the role of CITES, the VCLT prevents the *inter se* modification of CITES except in special circumstances.[67] The implication here is that

---

[63] *Use of Nuclear Weapons in Armed Conflict* (1996) ICJ Rep 66, 150–1.   [64] Ibid., 151.
[65] See Marie-Claire Cordonier Segger and C G Weeramantry (eds.) *Sustainable Justice: Reconciling Economic, Social and Environmental Law* (2005), which I reviewed in (2007) 56 ICLQ 209.
[66] See, e.g., the significant differences in views about human rights and their achievement by the WTO: Ernst-Ulrich Petersmann, 'Time for a United Nations "Global Compact" for Integrating Human Rights in the Law of Worldwide Organizations: Lessons from European Integration' (2002) 13 *EJIL* 621; Robert Howse, 'Human Rights in the WTO: Whose Rights, What Humanity? Comment on Petersmann' (2002) 13 *EJIL* 651; Philip Alston, 'Resisting the Merger and Acquisition of Human Rights by Trade Law: A Reply to Petersmann' (2002) 13 *EJIL* 815; Ernst-Ulrich Petersmann, 'Taking Human Dignity, Poverty and Empowerment of Individuals More Seriously: Rejoinder to Alston' (2002) 13 *EJIL* 845.
[67] See Chapter 4, n. 221 and accompanying text.

parties to a particular regime must agree before the substance of the regime can be modified. This is conceptually similar to the ILC Fragmentation Study Group's recommendations to assist in situations of conflicting norms between different regimes. As I discussed in Chapter 1, the Study Group made recommendations based on a number of legal rules to determine which particular regime will have primacy in a dispute, such as the principles of *lex posterior* and *lex specialis*. Like international treaty and institutional laws on mandate and modification, these principles operate in the context of finding the true intent of the parties.

A major finding of my analysis of fragmented fisheries governance, however, is that regime interaction does not depend on the agreement of all participating states, whether express or implied. Instead, the mutual supportiveness of regimes may be required through fact-finding or treaty interpretation in the context of the system of international law. Attempts to limit the role and jurisdiction of particular regimes must be evaluated in light of this view.

Accordingly, there will be examples where the operational space of a particular regime such as CITES will appropriately diminish vis-à-vis another regime even without the intent of its parties. For example, the increasing attention given to ecosystem management in fisheries conservation may lead to a corresponding decrease in the value of CITES' species-centred approach. The preference for ecosystem management is discernible, for the moment at least, in the views of some scientists, NGOs and states, but mainly in the forums of the FAO and the CBD. The majority of CITES parties do not yet advocate ecosystem management within CITES. Yet if FAO members, appropriate scientific peers and other stakeholders discuss and implement an ecosystem approach in an open and accountable way, as developed below, there may, in theory, be appropriate grounds to reduce CITES' relative involvement in the protection of marine species.

2 Lack of transparency and openness

In all three case studies, a lack of transparency and openness within a particular regime inhibited the collaboration by actors from other regimes. Chapter 3 described how observers were unable to attend meetings of the WTO Rules Group, although some information-exchange and informal participation was achieved due to the online accessibility of the WTO members' submissions and the ability to contribute to external forums. Chapter 5 explained that WTO dispute settlement proceedings were closed to non-parties,[68] and that the disputing parties reviewed panel reports and suggested changes before they were published but that other stakeholders were not able to.[69] The ability of non-parties to submit *amicus* briefs and engage in consultation opened these proceedings to some degree, although practices of the adjudicating bodies suggested that such openness was more apparent than real.[70] The most vivid example of the negative impact of a lack of transparency and openness in regime interaction was Chapter 4's description of the stalled collaboration between the FAO and CITES. The use of closed 'working groups' in the FAO Sub-Committee on Fish Trade and the exclusion of the CITES Secretariat representative from working group meetings contributed to the delay in the MOU between the two IGOs.

In contrast to these examples of IGO insularity, the constituent members of regimes have consented to greater openness and transparency in a number of areas. The FAO recognises the contribution of IGOs and NGOs in a number of its instruments and policies[71] and as a general principle of fisheries management.[72] The parties of CITES also open the COPs to external participants unless one-third of them object.[73] There is recognition of the need to develop criteria for observer status by WTO members, although negotiations are ongoing.[74] As such, states are making important contributions to the legal framework of regime interaction.

3 Need for parallel membership

A fixation on the need for parallel membership of regimes inhibits regime interaction and effective scrutiny of relevant norms. This was

---

[68] Unless the parties agreed to open hearings to the public: see Chapter 5, n. 116.
[69] Cf the leak of the interim review *Biotech* report by one of the parties, which was subsequently published online by one of the NGOs that had submitted a joint *amicus* brief: see further Margaret A. Young, 'The WTO's Use of Relevant Rules of International Law: An Analysis of the Biotech Case' (2007) 56 *ICLQ* 907, 910–11.
[70] See Chapter 5, n. 264 and accompanying text.
[71] See FAO 'Cooperation with International Governmental Organizations'; FAO, 'Policy concerning NGO relations'; FAO, 'Guiding Lines Regarding Relationship Agreements Between FAO and IGOs'; FAO, 'Cooperation with NGOs'; FAO, 'Granting of Observer Status'. See also FAO, *Guidelines for the Eco-labelling of Fish and Fish Products* (2005) esp. para. 54.
[72] FAO Code of Conduct, Art. 7.1.6; see further Chapter 2, n. 121 and accompanying text.
[73] CITES Art. XI(7); see further Chapter 2, p. 59.
[74] WTO Doha Declaration (2001) para. 31(ii); cf criticisms of negotiations at Chapter 2, n. 338.

apparent in the approach of WTO adjudicating bodies to draw on certain non-WTO sources as interpretative tools only if they were binding on all WTO members.[75] The need for parallel membership also shaped the mandate of the Doha negotiations on the relationship between WTO rules and MEAs, which expressly excluded non-MEA members,[76] and was invoked as a reason to reject the involvement of the FAO in the implementation of the WTO fisheries subsidies disciplines.[77] This quandary of parallel membership is well known in the context of conflicting international norms, when tribunals have to determine which treaties take priority in the event that disputing parties are members of one but not both treaties.[78] Yet this book has demonstrated that the quandary arises even if regime interaction does not directly affect states' obligations, such as in the use of information and epistemic resources from other regimes by the WTO Rules Group, or the use of FAO data in the listing of marine species on the CITES Appendices.

Contrary to this perceived requirement of parallel membership for regime interaction, I have demonstrated that membership by the same constituent states of two sets of regimes is not necessary for those regimes to influence the norms of each other in fisheries governance. However, it is important to acknowledge the underlying theory that gives rise to this fictitious need for parallel membership by states of interacting regimes. This theory is associated with an emphasis on the sovereignty of states in international law.[79] According to this theory, if a state is not a member of a regime, yet a second regime of which it *is* a member is influenced by that regime, any normative effects of this regime interaction invalidly alter the rights and duties of this state.

Although the impetus for parallel membership as a basis for regime interaction has theoretical resonance, in practice it has minor relevance for states. Chapter 3 demonstrated that the states that were most sceptical of the impact of environmental and fisheries norms in the WTO Rules Group negotiations on fisheries subsidies were members of *all* relevant regimes. Similarly, Japan's rejection of CITES' role in restricting trade in endangered marine species was not based on its

---

[75] Panel Report, *EC – Biotech* (interpreting VCLT Art. 31(3)(c)), Chapter 5, n. 268 and accompanying text; see also AB Report, *EC – Chicken Cuts* (interpreting VCLT Art. 31(3)(b)), Chapter 5, n. 284.
[76] See Chapter 2, n. 317 and accompanying text.   [77] See p. 121.
[78] See ILC Analytical Study, paras. 257ff.
[79] As introduced in Chapter 1 by reference to Kelsen's description of the 'primacy of national laws': see Chapter 1, n. 97 and accompanying text.

non-membership of CITES. It seems that in many cases where trade and environmental regimes overlap, states that are affected by such overlap ensure that they sign up to the relevant regimes.

This evidence suggests that states may often question the mandates of IGOs not because of their lack of consent to norm-setting processes, but because of their perceived lack of *control* over those processes. In these cases, the issue is not to protect against backdoor normativity but rather to ensure accountability of the relevant processes. If a country such as Japan has the right to be heard in the decision-making that affects it, along with the others that are affected, this is a better prescription for legitimate decision-making in international law than formalistic restrictions of IGO mandates.

Of course, this will not be the case for all regime overlap, as demonstrated by some of the examples of WTO adjudication. In the *EC – Biotech* dispute, the EC claimed that the Biosafety Protocol was relevant to the Panel's adjudication of the complaint by the United States, Argentina and Canada about access for GM products to the EU's market. Unlike the EU, none of the complainants were parties to the Biosafety Protocol. Thus, unlike Japan in CITES, they had not formally agreed to the relevant regime. However, these states did influence the development and operation of the Biosafety Protocol at least to some degree. The United States, Argentina and Canada all participated in the negotiations in the lead-up to the adoption of the Biosafety Protocol[80] and continue to participate as national focal points in the Biosafety Clearing House. Even in this example, the relevant states retain some control in the overlapping regimes.

A further point about membership of interacting regimes is that a prior requirement for parallel membership effectively gives non-parties a veto over the evolution of international law. In the global fisheries context, there is an increasing awareness about impending ecological disasters caused by overfishing and overcapacity. It is right that such awareness feeds in to the interpretation of treaties, the fact-finding by adjudicating bodies and the scientific, technical and general epistemic engagement between IGOs, notwithstanding that a minority of states have opted out of relevant regimes.

This point is strengthened if regard is had to those states that are currently non-members of the particular regimes I studied. These states effectively enjoy a notional 'veto' over the normative effect of regime

---

[80] See e.g. UN Doc UNEP/CBD/ExCOP/1/2 (15 February 1999).

interaction. For example, Angola, which is a member of the WTO and the FAO but not CITES, has a notional veto against the influence of CITES norms on the operation of the WTO and FAO regimes. Uzbekistan, which belongs to CITES and the FAO but not the WTO, has a notional veto against the use of trade norms in elaborating the legal activities and influence of CITES and the FAO. Singapore, which is a member of the WTO and CITES but not the FAO, has a notional veto over the normative influence of the FAO on fisheries issues as they are enforced in the WTO and CITES regimes. The point is not that these countries are actively obstructing the interaction between the regimes – they are mainly on the fringes of regime-interplay. Nor do I claim that the views of these states are not important. The point is rather that the model of international law that requires two regimes to have parallel membership before they can influence each other's norms, if applied rigorously, allows a minority of states – even a minority of one – to veto the development of international law.

A counter-argument to this view denies that the rights and duties of states are unduly affected by regime interaction. For example, the ILC Study Group on Fragmentation makes the point that treaty interpretation does not alter rights or obligations 'that would exist in some lawyers' heaven where they could be ascertained "automatically" and independently of interpretation'.[81] Instead, treaty interpretation simply allows the meaning to be constructed according to a treaty's normative environment.[82] This normative environment will include recourse to exogenous sources if states have consented to a process of normative development upon adoption of the relevant treaty. State consent to a process of normative development has been considered as an emerging feature of international law, as suggested by 'common-interest' type commitments.[83] The ILC preferred, however, to confine consent to normative development to specific legal contexts that are indicated by 'evolving' treaty terms[84] and relational concepts such as necessity in GATT Article XX.[85] The ILC favoured this approach because it avoided

---

[81] ILC Analytical Study, para. 447.   [82] Ibid.
[83] See e.g. Ellen Hey, *Teaching International Law: State-Consent as Consent to a Process of Normative Development and Ensuing Problems* (2003) 11–25 (booklet citing examples from a range of commitments including with respect to human rights and international criminal law).
[84] ILC Analytical Study, para. 478; see in particular *Gabčíkovo-Nagymaros* case (1997) ICJ Rep 7, 76–80; see also 113–15 (Separate Opinion of Judge Weeramantry).
[85] ILC Analytical Study, para. 478.

the difficult question of inter-temporal law,[86] namely whether treaties should be interpreted according to the law in force at the time of adoption, or according to the present-day conditions of its application. The VCLT does not contain provisions about inter-temporal law.[87]

If, on the other hand, it is conceded that treaty interpretation and institutional collaboration do indeed alter the rights of states because states have not consented to a process of normative development with respect to either specific or general rights, it is important to ascertain whether there is a legitimate basis for such alteration. This is an issue of appropriate regime interaction, which is the subject of the next chapter.

D Conclusions

This chapter has summarised the need, in the context of fisheries governance, to turn from forum shopping and towards regime interaction. It analysed the factors that promote regime interaction, which centre on inter-regime learning and coordination at both the national and international level. Conversely, the attempts to provide for a priori regime exclusivity and secrecy impeded regime interaction, especially when combined with an emphasis on parallel state membership between the relevant regimes.

The analysis does not end, however, with the claim that regime interaction should occur, free and unobstructed from such considerations. Underlying efforts to obstruct or limit regime interaction are some real and important concerns surrounding the legitimacy of international law. States need to understand and maintain their international treaty obligations, and clearly wish to ensure that their treaty partners do the same. If norms and practices from other regimes affect these obligations, confusion and disorder may result and the *legitimacy* of the system may be threatened.

It is not enough, therefore, to promote regime interaction without addressing legitimacy concerns. The next chapter sets out to do this by proposing a legal framework for regime interaction.

---

[86] Ibid.
[87] See Rosalyn Higgins, 'Time and the Law: International Perspectives on an Old Problem' (1997) 46 *ICLQ* 501, 519 (noting that the absence from the VCLT of a general rule on the inter-temporal question reinforces the need to consider the intentions of the parties and the object and purpose of the relevant treaty).

7 A legal framework for regime interaction

> As more and more of the responsibility of the national public realms is exteriorised and communalised in what we may call international intergovernment, the more urgent becomes the problem of its theoretical justification.[1]

This chapter develops a legal framework for regime interaction. It first points to the various ways in which regime interaction proceeds, whether based on the intent of the same states parties to the relevant regimes, or something else. Where regime interaction is *not* based on the consent of member states, questions of legitimacy become more important, and the chapter investigates the link between sovereignty of states and the behaviour of international organisations. Such an approach falls short of proposing a form of constitutionalism, but instead engages in principles of international institutional law, including implied powers, to ascertain the capacities and limits of modern international organisations in addressing global problems in a fragmented legal order.

A The multiple bases of regime interaction

The book's investigation into numerous examples of regime interaction demonstrates that there is no single basis on which such interaction occurs. Instead, there are multiple bases for regime interaction. The strictest conception requires all member states of each regime to have adopted mutual norms before they can have influence ('parallel membership'). One example is the complaint by a non-UN member against

---

[1] Philip Allott, *The Health of Nations: Society and Law beyond the State* (2002) 379.

the FAO involvement in WTO fisheries subsidies disciplines. A less extreme version of the requirement for consent from members of both regimes envisages that states can agree to allow normative and institutional interplay, notwithstanding that not all members have adopted the relevant regimes ('mutual agreement'), as proposed in the draft subsidies rules, and as already occurs in the TBT and SPS Agreements' endorsement of international standards from exogenous regimes. Consent to regime interaction can also be entrenched through agreement between intergovernmental organisations ('institutional arrangements'), as occurred with the Memorandum of Understanding between the CITES Secretariat and the FAO. However, such institutional arrangements are often informal, and are based instead on the need for international institutions to learn and collaborate. The following section describes each of these bases of regime interaction in greater detail, noting that as categories they should be considered as points on a spectrum rather than as discrete occurrences.

1 Parallel membership

The normative and institutional interaction between regimes to which the same states are party is a powerful idea which appears to grant full assurances as to legitimacy. Yet the practical examples of such regime interaction are rather limited.

One example comes from the Doha round of WTO negotiations, where WTO members have sought to impose specific limits on the normative interaction between the WTO and other legal regimes. As described in Chapter 2, negotiations on the relationship between WTO rules and specific trade obligations set out in MEAs are expressly limited to WTO members that are already parties to the relevant MEAs.[2] WTO members have thus agreed to negotiate on the boundaries between trade and other regimes but only for situations where WTO members have already consented to the other regimes. In consequence, there is an express refusal to consider regime interaction that occurs in situations where there is not uniform membership between the WTO and other regimes.

In the fisheries subsidies negotiations, Chapter 3 demonstrated that there were few attempts to require parallel membership between relevant regimes such as the WTO and FAO in the new disciplines. However, in rejecting the proposed rules, one WTO member did draw on the idea

---

[2] See Chapter 2, n. 317 and accompanying text.

of parallel membership in order to complain against the involvement of the FAO regime. Taiwan, which is unable to formally join the FAO due to its lack of state recognition, objected to the involvement of that organisation in the proposed disciplines.

A suggested need for inter-regime unanimity for interaction between the WTO and other regimes also arises in the context of WTO dispute settlement, as described in Chapter 5. In applying the rules of treaty interpretation contained in VCLT Article 31(3)(c), some adjudicating bodies have taken account of rules of international law exogenous to the WTO covered agreements only if all WTO members have consented to them.[3]

Chapter 6 argued that a fixation on the need for parallel membership obstructed regime interaction and led to grave problems in the ability of international law to meet new and emerging challenges. At its most extreme, a requirement that regime interaction be contingent on parallel membership gives states that are not parties to regimes a power of veto over the evolution of international law. These arguments do not need to be restated. Instead, I merely note here that parallel membership should be regarded as sufficient but not necessary to regime interaction.

2 Mutual agreement

An alternative legal basis for regime interaction still requires the consent of states, but allows regime interaction to occur without uniform membership between relevant regimes.

There are many examples of regime interaction through mutual agreement. Such regime interaction often stems from separate and stand-alone agreements. Chapter 2 offered a range of examples, including the WTO Declaration on Coherence in Global Economic Policy Making, which provides a general mandate for joint work between IGOs,[4] and the FAO member resolution for IGO relationships.[5]

Alternatively, regime interaction is entrenched in the treaties themselves. CITES opens its regular conferences to qualified representatives from IGOs and non-governmental organisations.[6] The proposed fisheries subsidies rules provides innovative ways in which future interaction between the WTO, FAO and other regimes would assist in disciplining trade-distorting and ecologically harmful subsidies.

---

[3] See further Chapter 5's analysis of the Panel Report, *EC – Biotech*.
[4] See Chapter 2, n. 313. [5] See Chapter 4, n. 124. [6] CITES Art. XI(7).

UNCLOS expressly envisages that states parties will enter into separate legal regimes, although it assumes that any deviation by those parties will not derogate from its basic principles.[7]

Regime interaction through mutual agreement appears at first glance to satisfy legitimacy concerns, given the implicit consent of states parties to resulting normative influence. Yet examples from the SPS and TBT Agreements described in Chapter 3 demonstrate that concerns continue to arise. Those agreements draw on standards from exogenous regimes in the following way: in general terms, if members' trade measures conform to international standards, they may be exempted from onerous WTO commitments or presumed necessary for domestic objectives such as the protection of human health. In WTO disputes, parties have argued against this use of standards, especially when the standards were not adopted by consensus within the relevant exogenous regime.[8] This dilemma has led to new ideas about the role of dispute settlement bodies in scrutinising standards from exogenous regimes,[9] which are informative for the discussion of legitimacy and accountability outlined below.

3 Institutional arrangements

A third basis for regime interaction comes from formal or informal arrangements within and between the international institutions themselves. Formally, such arrangements might be subject to careful control and consent by states parties, as was exemplified in Chapter 4 by the MOU between the CITES Secretariat and the FAO. However, such arrangements may also be informal. Arrangements may emerge because in carrying out their functions, IGO secretariats and adjudicating bodies find it necessary to learn, share and adopt information from external sources. Such institution-led collaboration occurred when the FAO was involved in updating the listing requirements in CITES and when the AB in *US – Shrimp* drew on understandings of international environmental law in interpreting WTO rules. The collaboration often has normative influence, as occurred in the *Biotech* dispute when the Panel consulted with international organisations and drew on reference material in order to interpret terms in the SPS Agreement. Sometimes, however, the normative influence is less direct. Examples of IGO

---

[7] UNCLOS Art. 311(2).   [8] See e.g. AB Report, *EC – Sardines*, para. 225.
[9] Joanne Scott, 'International Trade and Environmental Governance: Relating Rules (and Standards) in the EU and the WTO' (2004) 15 *EJIL* 307.

collaboration that only indirectly affected legal norms are UNEP's workshops for WTO members and other states and NGOs as part of the Doha negotiations on fisheries subsidies.

Although divorced from the consent or demands of states parties, regime interaction through informal institutional arrangements may have normative influence. Is it possible that such influence is legitimate?

B Legitimacy of regime interaction

If, through its influence on international norms, regime interaction alters the rights of states, it is important to have regard to its legitimacy. Yet locating the source of legitimacy will differ depending on the bases from which regime interaction has emerged, namely parallel membership, mutual agreement or institutional arrangements.

1 Consent and sovereignty

Where regime interaction occurs through parallel membership or mutual agreement, any consequent impact on the rights of affected states appears to be legitimate. As explained above, conceptions of sovereignty emphasise the supreme authority of states, and hold that states should not be subject to specific international laws, or the effects of regime interaction, except by consent. In its most extreme form, most commonly encountered in the context of the WTO, conceptions of sovereignty require regimes to be 'self-contained'.

Regime interaction through institutional arrangements may also conform to traditional conceptions of sovereignty, especially when such arrangements are made according to express powers of IGOs. Yet the appropriateness of IGO collaboration *without* the direct consent of IGO members rests on other factors. Such factors need to be carefully considered and articulated. As long emphasised by Kingsbury, the theoretical and practical deficiencies of sovereignty do not themselves found radical proposals for change; alternative means must be located and developed first.[10] According to my research into regime interaction, these factors relate to the legal capacities of the particular IGOs, the substance of the intersecting norms and the fulfilment of several procedural safeguards for openness and accountability, as I describe below.

---

[10] Benedict Kingsbury, 'Sovereignty and Inequality' (1998) 9 *EJIL* 599.

2 Express and implied powers

The capacity of IGOs to make institutional arrangements with other regimes is sometimes expressed in the relevant constitutive instruments. For example, there is express provision for the WTO to cooperate with the IMF and the World Bank in its founding instrument,[11] and an express power of the FAO to enter an MOU with other IGOs.[12]

By contrast, member states sometimes choose to retain the power to make consultative arrangements with other IGOs and NGOs. WTO members, for example, retain the responsibility for making 'appropriate arrangements' for effective cooperation and consultation with IGOs and NGOs.[13] On this basis, the WTO General Council has entered into agreements with the UN and five other IGOs.[14] By corollary, it could be suggested that such provisions deny the ability of IGOs to collaborate independently with external sources.

The legal basis for informal, institution-led collaboration resides, instead, in implicit powers. IGOs such as the WTO may have the implied power to take account of external sources because it is necessary for them to do so as part of their functions. This argument builds on the *UN Reparations* case, in which the ICJ pointed to several necessary UN functions in assessing whether the UN had legal personality and the capacity to bring its claim to the court.[15]

In assessing the capacity of the UN to bring an international claim, the Court found that its legal personality and rights and duties were different from those possessed by a state.[16] Instead, the rights and duties of an entity like the UN 'must depend upon its purposes and functions as specified or implied in its constituent documents and developed in practice'.[17] In the context of the UN's claims for reparations, the Court found that the UN's functions 'are of such a character that they could not be effectively discharged if they involved the concurrent action, on the international plane, of fifty-eight or more Foreign Offices'.[18] On this basis, the Court concluded that the members of the UN had endowed it with the capacity to bring international claims 'when necessitated by the discharge of its functions'.[19]

---

[11] Marrakesh Agreement Art. III(5).
[12] FAO Constitution Art. XIII(1); cf the absence of such power for the CITES Secretariat: see further Chapter 4, n. 129 and accompanying text.
[13] Marrakesh Agreement Art. V; see further Chapter 2, n. 322 and accompanying text.
[14] See Chapter 2, n. 323.   [15] (1949) ICJ Rep 174.   [16] Ibid., 179.   [17] Ibid., 180.
[18] Ibid.   [19] Ibid.

The centrality of the test for the UN's capacities was the question of 'necessity'. The Court decided that the UN had the capacity to claim reparations for the injury of one of its agents, both for the damage caused to the UN and to the victim itself.[20] This was found to be necessary to enable the UN to 'entrust its agents with important missions to be performed in disturbed parts of the world'.[21] By contrast, it was not considered necessary by the majority in the advisory opinion on *Use of Nuclear Weapons in Armed Conflict* for the WHO to have the competence to address the legality of the use of nuclear weapons, 'even in view of their health and environmental effects'.[22] For the Court, the principle of implied powers was subject to the principle of speciality of international organisations. In dissent, Judge Weeramantry considered that the principle of speciality should not be used to disavow the multiplicity and complexity of organisations within the UN family.[23]

Although sharply limited by the majority approach in *Use of Nuclear Weapons in Armed Conflict*, the principle of implied powers is highly relevant to the phenomenon of fragmentation in international law. In circumstances where regimes are dependent on other regimes, such as the WTO's proposed reliance on standard-setting and benchmarking functions in the fishery subsidy disciplines, IGOs will have the discretion to learn and apply norms and facts from external sources because such collaboration is necessary for their functions.

In fisheries governance, it is functionally necessary for IGOs to take account of multiplicious interests and institutions. These interests are variously economic, consumerist, ecological and ecocentric. Where states do not represent these interests, it may be functionally necessary for IGOs to collaborate with non-state actors that do, including IGOs and NGOs. This argument finds support in studies that demonstrate how a lack of awareness of important interests that fall outside traditional structures of representation leads to flawed decision-making. For example, the lack of standing of natural objects such as trees or rivers means that ecological concerns are often not 'matters for judicial cognizance', leading to cases skewed against environmental

---

[20] Ibid., 180–1; 184.
[21] Ibid., 193; cf the dissent by Judge Hackworth that the exercise of the power to make private claims had not been shown to be necessary for the exercise of the UN's powers: ibid., 198.
[22] (1996) ICJ Rep 66, 79 (para. 25).   [23] Ibid., 150.

issues.[24] Ethnographies of domestic and supranational legislative processes have also exposed a need for enhanced representation of non-dominant interests.[25]

An assessment of what is 'functionally necessary' for international organisations to carry out their functions in fisheries governance does not depend solely on issues of representation. Such an assessment also flows from an awareness of the complexity of problems associated with achieving sustainability and the need for open deliberation about these issues. Central to this idea is that legitimacy may flow from a particular process rather than membership or another reflection of the concept of state sovereignty.[26]

My case studies demonstrate that the problems associated with fisheries sustainability are not limited to the problem of obtaining cooperation between representative states. There is a range of additional complexities, not least of which are the scientific and technical limitations of predicting stock behaviour in the context of the 'eco-system'. Although the drafters of UNCLOS were conscious 'that the problems of ocean space are closely interrelated and need to be considered as a whole',[27] my case studies show that the problems of ocean space link with problems of land – in short, with the full ambit of human activity. In these terms, the management and conservation of fisheries fulfil conditions of a 'wicked problem'[28] where there can be no definitive formulation or solution.[29] Co-existence of multiple variables makes it impossible to determine whether fisheries sustainability depends on trade policies that target supply and demand, environmental regulation, rights-based approaches or even policies that address a completely different issue like climate change.

There is no 'right' solution of fisheries sustainability that will be determined when the correct decision-makers are in charge. Instead,

---

[24] See Christopher Stone, 'Should Trees Have Standing? Toward Legal Rights for Natural Objects' in Christopher Stone, *Should Trees Have Standing? And Other Essays on Law, Morals, and the Environment* (1996) 1, 10.
[25] See e.g. Francis Snyder, 'Thinking about "Interests": Legislative Process in the European Community' in June Starr and Jane Collier (eds.) *History and Power in the Study of Law: New Directions in Legal Anthropology* (1989) 168.
[26] On the connection between legitimacy and 'right process', see Thomas Franck, *The Power of Legitimacy Among Nations* (1990) 41–195.
[27] UNCLOS, Preamble.
[28] The term comes from Horst Rittel: see, e.g., Horst Rittel and Melvin Webber, 'Dilemmas in a General Theory of Planning' (1973) 4 *Policy Sciences* 155.
[29] As compared, for example, to a problem of chess or mathematics: see ibid., 160.

there is need for open cross-forum experimentation seeking to address current and future hazards to sustainable fisheries. My case studies have documented how trade policy has been harnessed as a possible tool to ensure the effectiveness of fisheries management strategies. I described how the emerging emphasis on trade-related mechanisms for fisheries sustainability has led to the need for IGOs like the WTO and CITES to work with other IGOs and NGOs with expertise and interest in fisheries. To address the sustainability of fisheries, there is a need for learning and peer review within a wide range of international organisations, not just the ones with traditional mandates for fisheries management. Moreover, there is a need for this learning to take place at the international level, and not just within domestic agencies and domestic stakeholders as advocated by proponents of 'national policy coordination'.

The awareness of the need for diverse and pluralistic solutions accords with literature that downplays the need for representation in governance and explores instead the practical need for diverse perspectives.[30] Even if NGOs do not have representative functions, for example, some commentators call for their participation in WTO decision-making to ensure the inclusion of ideas that are 'overlooked or undervalued by governments'.[31] Enhanced participation can also aid in the practical implementation of the resulting rules, as is demonstrated by studies of international environmental commitments.[32] In the field of systems theory, analysts consider that regularised forms of openness are required in order to counteract the 'closure', or the fixing of processes, caused by the creation of institutions.[33]

Apart from these practical imperatives for openness and transparency, the moral content of environmental issues has relevance for demonstrating why it is necessary for IGOs involved in fisheries

---

[30] See in relation to the setting of product standards, Harm Shepel, *The Constitution of Private Governance: Product Standards in the Regulation of Integrating Markets* (2005) 412-13; and in relation to EU governance, Deirdre Curtin, 'Private Interest Representation or Civil Society Deliberation? A Contemporary Dilemma for European Union Governance' (2003) 12 *Social and Legal Studies* 55.

[31] Steve Charnovitz, 'WTO Cosmopolitics' (2002) 34 *NYU Journal of International Law and Politics* 299, 343.

[32] David Victor, Kal Raustiala and Eugene Skolnikoff, 'Introduction and Overview' in David Victor, Kal Raustiala and Eugene Skolnikoff (eds.) *The Implementation and Effectiveness of International Environmental Commitments: Theory and Practice* (1998) 1, 21-4 (noting the role of open participation in implementation).

[33] Philip Selznick, 'Self-Regulation and the Theory of Institutions' in Gunther Teubner, Lindsay Farmer and Declan Murphy (eds.) *Environmental Law and Ecological Responsibility: The Concept and Practice of Ecological Self-Organization* (1994) 395, 398-9.

governance to collaborate with external sources.[34] Some writers base this moral content on an important dimension of human welfare.[35] Others ascribe an 'ecocentric' value that is independent of humans,[36] or merge the utilitarian and moral aspects of environmental issues to claim that 'ecological communication', rather than representation, is required to ensure the policy space for environmental needs.[37]

I have demonstrated that the legal framework for regime interaction must allow IGOs to take into account external sources in the negotiation, implementation and dispute resolution of fisheries matters where they have express or implied capacities to do so. These capacities arise from the necessity of either enhanced representation or deliberation. As such, the capacities of IGOs to collaborate are often dissociated from the consent of the members.

3 The risk of managerialism

The independence of IGOs to collaborate with external regimes without reference to the consent of their members may lead to abuse. At its extreme, such independence may signal a 'technicalisation' of international affairs, where managerial procedures and decision-making by unseen experts begins to control global politics.[38] Given the biases of experts within regimes, allowing them greater latitude in regime interplay has significant risks.[39]

---

[34] On the moral content of environmental issues, see Christopher Stone, *The Gnat is Older than Man: Global Environment and Human Agenda* (1993). Cf claims for animal welfare based on utilitarianism (Peter Singer, *Animal Liberation: A New Ethics for our Treatment of Animals* (1976)), and the 'capabilities approach' (Martha Nussbaum, *Sex and Social Justice* (1999)).

[35] See e.g. the capabilities approach, which recognises the principle that humans should be 'able to live with concern for and in relation to animals, plants, and the world of nature': Martha Nussbaum, *Women and Human Development: The Capabilities Approach* (2000); for complementary formulations see Amartya Sen, *Commodities and Capabilities* (1985).

[36] Robyn Eckersley *Environmentalism and Political Theory: Toward an Ecocentric Approach* (1992).

[37] John Dryzek, 'Political and Ecological Communication' in John Dryzek and David Schlosberg (eds.) *Debating the Earth* (1999) 584 (extending Habermas's ideas of communicative action).

[38] Martti Koskenniemi, 'International Law: Constitutionalism, Managerialism and the Ethos of Legal Education' (2007) *European Journal of Legal Studies* (online); see also Martti Koskenniemi, 'The Fate of Public International Law: Between Technique and Politics' (2007) 70 *Modern Law Review* 1; see also Stephen Toope, 'Emerging Patterns of Governance and International Law' in Michael Byers (ed.) *The Role of Law in International Politics: Essays in International Relations and International Law* (2000) 91, 106.

[39] Koskenniemi, 'The Fate of Public International Law', above n. 38, 6–9.

Adjudicating bodies, for example, are currently restrained by rules of treaty interpretation and consultation that are based on the need to interpret treaty texts according to intentions of the parties. Lifting these rules by allowing IGOs to ascertain 'ordinary meaning', for example, can lead to arbitrary decision-making. There was a clear lack of effective scrutiny of the sources of non-WTO law by the *EC – Biotech* Panel when it sought to find the ordinary meaning of terms of the SPS Agreement pursuant to VCLT Article 31(1). The need for accountable regime interaction was also apparent in the development of the MOU between the FAO and CITES, where there were discrepancies in the access accorded to NGOs. Within the FAO Sub-Committee on Fish Trade, NGOs were blocked when discussion was moved from the Plenary and into private working groups.

A different problem existed in the CITES Standing Committee, which was open to NGOs. There, the NGO that seemed to influence the draft MOU was one that sought to represent the rights of states, while the NGO that was representing the rights of animals had no apparent influence in the draft text.[40] Some might contend that access to the finite public realm of the CITES Standing Committee was therefore depleted for groups promoting interests other than sovereignty-based concerns. Proponents of a 'representation' theory and a 'deliberation' theory of NGO engagement might, to different degrees, contend that the sovereign interests of states have adequate voice in this forum, and that other interests, such as animal rights NGOs, should be permitted a greater degree of participation. Others might respond that inquiring into the values of particular NGOs would undermine the goal of openness. For ideals of discursive democracy, it is the substance of the ideas, rather than the status of the speaker, that matters.[41] In the case studies I considered, however, there was a clear need to assess the background and credentials of NGOs.

These examples point to a need for IGOs to scrutinise and review the 'sources' of external regimes (the norms themselves, including the degree to which they are supported internationally) and the 'sources of the external sources', including the NGOs that wish to gain access to relevant regimes, in order for regime interaction to be accountable. This

---

[40] See Chapter 4, n. 233 and accompanying text.
[41] Robyn Eckersley, 'A Green Public Sphere in the WTO?: The Amicus Curiae Interventions in the Transatlantic Biotech Dispute' (2007) 13 *European Journal of International Relations* 329, 337.

need exists for all types of regime interaction, although it is more significant when regime interaction stems from informal institutional arrangements, rather than parallel membership or the mutual agreement of states. As such, the framework for regime interaction attempts to address the dangers of managerialism by emphasising accountability. The following section explains these ideas in more detail.

C Accountable regime interaction

The accountability of regime interaction depends upon a number of mechanisms through which IGOs, in extending accessibility to other regimes, also scrutinise the relevant norms and the actors that convey them. Such procedures involve constant and continuous inter-regime scrutiny, without any hierarchy between groupings. This removes the impediments to regime interaction described in Chapter 7, but also ensures that such regime interaction is accountable. Procedural safeguards to ensure openness, transparency and participation are significant amongst these mechanisms.

Such scrutiny and review occurs in some areas of fisheries governance. For example, proposals to list marine species are subject to a vote by the CITES parties. This process opens the collaboration between the FAO and CITES to political scrutiny, at least to some degree. Moreover, where there has been domestic stakeholder consultation in the development of countries' positions in the COP, of the kind outlined by the United States, there will be an enhanced ability to include and contest a number of scientific, cultural and ecological values. In addition, the MOU between the FAO and CITES Secretariat contains several provisions to ensure openness and transparency in regime interaction. The provision in the MOU for information-sharing, observership and periodic reporting are positive examples of the potential for mutual learning to be revised and critiqued. However, the fact that shared observership is not routine, but is rather restricted to meetings that are deemed to discuss issues of 'common interest' weakens this process. It is unclear from the MOU which participants determine the 'common interest' and how they should do so.

Some procedures for regime interaction between the WTO and standard-setting bodies have been agreed upon by WTO members, as described in Chapter 3. For example, international standard-setting bodies are not guaranteed observer status to the TBT Committee, but their standards may still be endorsed within the WTO provided that such bodies are open to all WTO members.[42] There is no need for consensus in the development of these standards for them to be relevant,[43] but the standard-setting bodies are encouraged to operate with open, impartial and transparent procedures.[44] Thus, for the norms of a standard-setting body to be incorporated into WTO law (and raised in a WTO dispute), the body must be accessible for the WTO membership and must itself have observable procedures for good governance.

This example points to a gatekeeper role for IGOs to use norms that are exogenous to their own regime. I have demonstrated by reference to the multiple bases of regime interaction that there is no requirement for parallel membership or mutual agreement of states before norms can exert influence on another regime. However, IGOs must have regard to whether there is a high degree of international consensus to those norms. This may include inquiries into whether the norm has been agreed by a range of developing countries as well as developed countries. Moreover, whether the norm was *itself* developed in an open and accessible way is relevant in a decision to accord it influence in another regime.

My analysis of the adjudication of fisheries trade disputes also demonstrates potential procedures that enhance accountability of regime interaction. Chapter 5 referred to the existing powers of WTO adjudicating bodies to assess evidence and legal submissions from both IGOs and NGOs, and pointed to a need to better understand the modes of scrutiny of such information and norms. Ideas for reform were also contained in Chapter 3, which described a recommendation for the establishment of a body of WTO, FAO and NGO experts to resolve fisheries subsidies disputes.

Some of these procedures involve peer-level review where experts learn from one another in order to extend their competencies. There is scope in a legal framework for appropriate regime interaction to improve techniques for peer review. In the context of expert evidence presented to legal tribunals like WTO panels, for example, it may be useful to develop adversarial techniques that subject experts to

---

[42] TBT Agreement Annex 1:4.
[43] AB Report, *EC – Sardines*, para. 225; see further Chapter 3, n. 171 and accompanying text.
[44] WTO Doc. G/TBT/1/Rev.8 (Decision of the TBT Committee on Principles for the Development of International Standards, Guides and Recommendations); see further Chapter 3 n. 212 and accompanying text. Similar ideas are emerging in the FAO fisheries management regime: see FAO, *Guidelines for the Eco-labelling of Fish and Fish Products* (2005) para. 3.

cross-examination by their peers rather than rely on effective scrutiny by panellists themselves.[45] Even more promising are procedures that call for wider participation of stakeholders. The combined UNEP/WWF workshops on fisheries subsidies were an example where both experts and interested stakeholders deliberated together.[46] This concept of stakeholder review is supported by literature in diverse fields.[47] Theories of democratic experimentalism, for example, are based on the idea that affected stakeholders are in the best position to review information.[48] Related literature has drawn on scholarship on US civil republicanism to advocate the integration of stakeholders in the WTO.[49] In this vein, Shell proposes that groups that are 'broadly representative of diverse citizen interests'[50] should be included in WTO decision-making.[51]

The framework of accountable regime interaction is also influenced by checks on state authority developed domestically. Principles such as the doctrine of procedural justice in administrative decision-making and the inclusion of stakeholders in legislative rule-making[52] provide useful sources for study, in addition to the EU open models of governance.[53] These legal developments have emphasised 'process norms' such as transparency and reason-giving to enhance legitimacy in decision-making. They have also highlighted the role of ongoing review and open deliberation to address complex policy issues. As I discussed above, the dynamic and complex nature of fisheries clearly calls for

---

[45] See e.g., reference by Heerey J of the Federal Court of Australia to a 'hot tub' of experts in 'Recent Australian Developments' (2004) 23 *Civil Justice Quarterly* 386.
[46] See Chapter 3.
[47] See e.g., literature on 'new governance' noted in Chapter 1 n. 72 and accompanying text; see Michael Dorf and Charles Sabel, 'A Constitution of Democratic Experimentalism' (1998) 98 *Columbia Law Review* 267; see further below pp. 290–2.
[48] Dorf and Sabel, ibid.
[49] G Richard Shell, 'Trade Legalism and International Relations Theory: An Analysis of the World Trade Organization' (1995) 44 *Duke Law Journal* 829; see also G Richard Shell, 'The Trade Stakeholders Model and Participation by Nonstate Parties in the World Trade Organization' (1996) 17 *University of Pennsylvania Journal of International Economic Law* 359.
[50] Shell, 'Trade Legalism', above n. 49, 910.
[51] Shell draws on examples from a range of other forums to describe a 'Trade Stakeholders Model', although specific forms of stakeholder participation in the WTO are outside the scope of his work: ibid., 924.
[52] See 'reg neg' literature, e.g. Anonymous, 'Rethinking Regulation: Negotiation as an Alternative to Traditional Rulemaking' (1981) 94 *Harvard Law Review* 1871.
[53] See further Gráinne de Búrca and Joanne Scott, 'Introduction: New Governance, Law and Constitutionalism' in Gráinne de Búrca and Joanne Scott (eds.) *Law and New Governance in the EU and the US* (2006) 1.

such institutional activity, not only in addressing 'wicked problems' but also in alleviating the so-called 'tragedy of the commons'.[54]

A similar theme is taken up in the work of scholars of 'global administrative law', who seek to locate accountability structures in non-traditional sites of globalised law-making, including by market actors and international committees of IGOs.[55] Such accountability has real and potential effects on global politics as well as law.[56] Although the relationship between international institutions is not a focus of global administrative law, I consider many of the emerging principles of accountability to be relevant to regime interaction.

As is evident, the legal framework proposed here falls short of a kind of constitutionalism, which has been forcefully argued as the ideal form for situating public power within international society.[57] Although goals of a self-constituting international society are attractive in the context of globalising social and political structures, the trade and environmental quandaries presented in my case studies manifested a strong resistance to hierarchical norms. In the interaction between regimes, arguments for exclusivity within regimes or trumping norms gave way to a need to continuously engage and contest social and political solutions and governance structures. As such, the legal framework builds on existing structures and processes to ensure ongoing and open contestation between stakeholders rather than entrenched ideals.[58]

My studies in fisheries governance show that accountability through stakeholder participation in regime interaction requires safeguards of its own.[59] I have demonstrated how both IGOs and NGOs can influence regime interaction through participation in negotiation, implementation and dispute settlement. Most clearly, these stakeholders must be open to scrutiny. Yet the forms of accountability of these stakeholders will be different from traditional forms that focus on legitimate representation, especially when their participation extends from recognition

---

[54] See further Ostrom, above n. 26.
[55] See further Benedict Kingsbury, Nico Krisch and Richard Stewart, 'The Emergence of Global Administrative Law' (2005) 68 *Law and Contemporary Problems* 15.
[56] Joshua Cohen and Charles Sabel, 'Global Democracy?' (2005) 37 *NYU Journal of International Law and Politics* 763.
[57] Allott, above n. 1, 342–79.
[58] Cf Allott, who sees the notion of governance as 'disintegrate[ing] the integrity of society': ibid., 161–2.
[59] This is especially necessary to respond to the type of critique that may be implied by Allott's work: ibid.

of their deliberative contributions rather than their representative functions.

For IGOs that participate in regime interaction as stakeholders, for example, assessments of their accountability will focus on two main aspects: their functions and the way they operate. As to the first aspect, the norms which an IGO is responsible for advancing will be scrutinised. This includes the questions about the level of membership of states, and the degree to which such membership is spread across developing and developed countries, that I identified above. The OECD, for example, which represents only thirty developed countries, may have less weight in appropriate regime interaction than a body such as the WTO. Moreover, a norm will demonstrate a higher degree of consensus if it has been developed according to principles of openness and transparency. As to the second aspect, the openness, accessibility and transparency of procedures *within* the IGO will be relevant in determining the appropriateness of its participation. For example, if the FAO is to act as a peer reviewer for the conduct of states in the proposed WTO subsidies disciplines, the WTO should continually assess the FAO's procedures. Such assessments of IGO accountability could fit in with existing procedures. An example is the FAO's 'Guiding Lines' regarding relationship agreements between the FAO and other IGOs, which set out criteria according to which an IGO is to be recognised as having intergovernmental character.

This form of scrutiny of IGOs will be different from the scrutiny needed to determine whether NGOs can appropriately participate in regime interaction. NGOs differ from both states and IGOs in their functions, constituencies and intended beneficiaries. As such, checks on whether they are credible and accountable need to acknowledge these differences.[60] Forms of accountability will not necessarily depend on the degree of consensus of the norms the NGOs represent. More important will be the issue I identified as the second aspect of IGO accountability: namely, an inquiry into the open, accessible, transparent and participatory way in which the NGO operates.

The challenges of assessing the credibility of NGOs are apparent in other regimes within the international system. The UN, for example, requires NGOs to reveal the sources of voluntary contributions and

---

[60] See Sasha Courville, 'Understanding NGO-based Social and Environmental Regulatory Systems: Why We Need New Models of Accountability' in Michael W Dowdle (ed.) *Public Accountability: Designs, Dilemmas and Experiences* (2006) 271.

explain other sources of funding.⁶¹ WIPO, on the other hand, accredits observers without inquiring into their sources of funding.⁶² CITES requires observers to its COP to be 'qualified in protection, conservation or management of wild fauna and flora' and to have legal personality and an 'international character, remit and programme of activities'.⁶³

The work of groups such as the International Social and Environmental Accreditation Alliance (ISEAL)⁶⁴ is relevant to this issue. ISEAL has developed a voluntary code of good practice for NGOs that set environmental and social standards. The Maritime Stewardship Council, an NGO responsible for setting certain standards for the labelling of fish products, has agreed to this code. Moreover, requiring NGOs to declare their sources of funding should be a first step for all institutional collaboration. The guidelines issued by the Appellate Body Secretariat in the *Asbestos* appeal, which required *amicus* briefs to state the nature of their interest, is a good example of how such ideas can be implemented, notwithstanding that the guidelines engendered criticism amongst WTO members.

Outside of my case studies, there are several other examples of IGOs that are shaping the way they assess the credibility of the actors that seek to influence regime interaction. A striking example is the decision by the WHO to restrict the involvement of tobacco lobbies in the development of the Framework Convention on Tobacco Control. The Framework Convention was developed in the context of existing policies of other international organisations, including the FAO, which had an interest in promoting legal and profitable agricultural activities such as tobacco. The Framework Convention was bound to affect a number of actors with significant economic stakes in tobacco, including the tobacco industry. In the course of developing the Convention, the WHO determined that these tobacco companies were engaging in 'surreptitious activities' to undermine WHO tobacco control programmes.⁶⁵ The WHO also documented how tobacco companies had

---

⁶¹ NGO status is based on UN Charter Art. 71 and on ECOSOC Resolution 1996/31 adopted in 1996; see further esa.un.org/coordination/ngo/new/index.asp.
⁶² See www.wipo.int/members/en/admission/observers.html.
⁶³ CITES Art. XI(7) elaborated in Resolution Conf. 13.8; see further Chapter 2, n. 194 and accompanying text.
⁶⁴ The International Social and Environmental Accreditation Alliance (ISEAL Alliance): see www.isealalliance.org.
⁶⁵ See Report of the Committee of Experts on Tobacco Industry Documents, Tobacco Company Strategies to Undermine Tobacco Control Activities at the World Health Organization (July 2000).

attempted to shape the relationship between the WHO and the FAO, relying on previous alliances.[66] Based on this information, the WHO actively decided upon its own 'rules of engagement' with external sources. It resolved that the practices of the tobacco industry were not open, transparent or ethical.[67] On this basis, access by certain stakeholders to regime interaction was denied.

Drawing on these examples, my ideas to improve accountability as part of a legal framework for regime interaction are exploratory and in some instances speculative. Yet they represent an effort to understand the progressive development of international law in the context of fragmentation and an attempt to improve the way fisheries governance adapts to complexity and pluralism. They lead to new conceptions about capacities and duties of IGOs and have implications for international law's legitimacy.

D Duties to take others into account

The idea that the relevance of regimes is dictated by both their substance (if the regimes are necessary for epistemic links and to avoid decontextualised normative development), and their procedural accountability gives rise to further questions about IGO activities. I have claimed that there are specific capacities of IGOs as part of a legal framework for appropriate regime interaction. A related question is whether this extends to a duty to take account of sources from other regimes.

Some states consider that IGOs should possess express duties to take account of other sources. For example, in the WTO/MEA negotiations, the EU has made a number of proposals relating to the overlap between trade and environmental policy. In a submission made pursuant to the negotiating mandate of paragraph 31(i) of the Doha Declaration, the EU proposed that WTO members agree to an express duty for WTO committees and panels to take account of certain IGOs. The EU proposed the adoption of the following text:

Where a WTO Committee examines issues with an environmental content, relating to a particular MEA, that Committee shall call for and defer to, in the relevant points, the expertise of the MEA in question.

---

[66] Ibid., 39–42.   [67] Ibid., 244.

Where a WTO panel examines issues with an environmental content, relating to a particular MEA, the panel shall call for and defer to, in the relevant points, the expertise of the MEA in question.[68]

Other WTO members have expressed disagreement with these provisions. The suggestion for deference by WTO panels has been opposed by some developing countries as well as Australia, Argentina and the United States.[69] Moreover, ideas to link MEAs and WTO agreements through procedures for regular information-exchange between MEA secretariats and the relevant WTO committees, and the criteria for the granting of observer status (paragraph 31(ii) of the Doha Declaration) have been considered as a dangerous extension of the mandates granted to such institutions.[70]

It seems unlikely, at least at this stage, that WTO members will agree to express duties for the WTO to take account of other non-WTO sources in either negotiations or in dispute settlement.

Alternatively, there may be implied duties for IGOs to take account of external sources. Tietje has advanced an argument for an implied duty of institutional cooperation on the basis of good faith obligations of IGOs.[71] His argument analogises the AB's reasoning in *US – Shrimp* regarding the duty of states to cooperate in situations of overlapping jurisdiction and transposes this duty onto international organisations.[72]

Such a duty appears at first glance to be positive for enhanced regime interaction. An implied duty to collaborate could have led to the inclusion of UNEP, the FAO and other IGOs in the Rules Group negotiations on developing rules on environmentally harmful fish subsidies. Such a duty may also have prevented the FAO Sub-Committee on Fish Trade from excluding the CITES representative from their closed working-group discussions on the MOU. It could also have ensured that the Panel in *EC – Biotech* took account of relevant rules of international law such as the Biosafety Protocol regardless of the lack of parallel membership.

---

[68] WTO Doc. TN/TE/W/68 para. 3.   [69] (2007) 11:38 *Bridges Digest* (7 November 2007).
[70] Some countries such as China and Thailand have reportedly preferred any information exchange between the WTO and MEA Secretariats to be 'needs driven' rather than institutionalised: see (2007) 11:3 *Bridges Digest* (31 January 2007).
[71] Christian Tietje, 'Global Governance and Inter-Agency Co-operation in International Economic Law' (2002) 26 *JWT* 501.
[72] Ibid., 513.

There are two immediate problems with this construction of an implied duty to collaborate. First, in applying the analogy from *US – Shrimp*, the conception fails to engage with crucial differences between states and IGOs, and particularly the differences in the constituents and functions of IGOs. It sits uneasily with the *Reparations* case, which involved implied capacities, rather than implied duties, of IGOs. Secondly, and relatedly, such an unqualified duty does not provide the needed restrictions and guidance for collaboration that led to my conception of necessary forms of accountability outlined above.

Indeed, when certain organs of IGOs *do* have a duty to take account of other regimes, this duty is fettered. The book's analysis of trade disputes involving fisheries demonstrates that such a duty may exist for an adjudicating body if norms and information from another regime are presented by states. The duty of a WTO panel to consider non-WTO sources varies according to the method by which such facts are presented.[73] Panellists have a duty to take the submissions of the parties into account but have discretion to take account of facts presented by experts, and any misrepresentation of expert opinion will not normally violate the DSU.[74] Notwithstanding these differences, mandatory institutional collaboration by WTO adjudicating bodies may emerge for certain fisheries disputes if the draft Chair's text on fishery subsidy disciplines is accepted.[75]

Accordingly, there is an embryonic and qualified duty of IGOs to take external sources into account in certain circumstances. This may evolve into a comprehensive duty for IGOs to take account of other sources in issues of regime interaction, and will be developed by the need for knowledge and accountability within the international system. The main findings of my case studies reinforce the need for IGOs to accommodate views that are not always put forward by the state participants of one or both of the relevant overlapping regimes. However, my findings go further to demonstrate a need for such accommodation to be subject to checks. This is not just because of the practical problems in accommodating all affected interests of international law (which may be attenuated to a small degree by information technology), but also because of the bias that might already be present in the international

---

[73] See e.g. AB Report, *US – Shrimp*, para. 101; see further Chapter 5, n. 139 and accompanying text.
[74] AB Report, *EC – Hormones* paras. 138, 144; *EC – Asbestos*, para. 152; see further Chapter 5, n. 174 and accompanying text.
[75] Art. VIII.4, Chair's text, reproduced at Appendix A below; see further Chapter 3.

system.[76] Such issues give rise to significant further implications, which I explore in the next chapter.

E Conclusions

This chapter has argued that regime interaction can be broadly categorised as stemming from parallel state membership of regimes, mutual agreement by states or autonomous institutional arrangements. Collaboration by IGOs is often necessary to deal with the legal, political and social requirements of multiple international regimes, and I have claimed that their legal powers should be read on this basis. As a result, sovereignty's presumption of freedom for states is rebutted.[77] This has required a reconception of legitimacy in international law, which rests, at various moments of regime interaction, on the capacity for stakeholder participation and deliberation rather than on the consent of states.

The powers of international organisations to confront the challenges of fragmentation are not without their limits. There is a risk of managerialism inherent in any enhancement of the role of experts and institutional bodies. In response, I have argued that regime interaction should be constrained by procedural safeguards to ensure openness, transparency, participation and ongoing scrutiny and review. The next chapter provides some tentative remarks on the operation of such regime interaction in practice, and concludes with some further implications for international law.

---

[76] See further my discussions of the *EC – Biotech* Panel's conception of international community in Chapter 5.
[77] See further Crawford's categorisation of the *Lotus* presumption in James Crawford, *The Creation of States in International Law* (2nd edn, 2006) 42.

8 Implications for international law

> Complex problems have ramifications in many specialized directions...[1]

The book has outlined detailed examples of the inter-regime activities resulting from law's response to the global fisheries crisis. In response to the detailed legal and political analysis contained in Part II, it has offered, in this Part, suggestions for deliberation and oversight according to a legal framework of regime interaction. This has several major implications for international law, and this chapter reflects on the operation of the legal framework in practice. It then discusses further implications, both for the conclusions of the ILC Study Group on Fragmentation and for fisheries governance and beyond. The chapter pre-emptively replies to possible critiques that the legal framework endorses managerialism or makes otherwise invalid assumptions, and concludes by calling for greater attention to the future participants of regime interaction in the context of the fragmented legal order.

A Appropriate regime interaction in practice

One of the major implications of the legal framework for regime interaction is that it operates across the full spectrum of law-making, implementation and adjudication. Regime interaction in negotiating new laws promotes the possibility of entrenched procedures. Ongoing regime interaction in the implementation of rules will be relevant to the resolution of disputes. Conversely, in drawing on rules and

---

[1] *Use of Nuclear Weapons in Armed Conflict* (1996) ICJ Rep 66, 151 (Dissenting Opinion of Judge Weeramantry).

standards from exogenous regimes, dispute settlement bodies will consider how such standards were themselves developed, and the history of participation in negotiations will again be relevant. Reflecting further on some of the examples encountered in the book leads to some observations about appropriate regime interaction in practice.

1 Law-making

For law-making, the implication of a legal framework of regime interaction is that new laws must be made in full recognition of the existing content of regimes and of the existing and often politicised commitments to those regimes. This bears on the role of states as conduits for information (and their need to coordinate national policy), but also places significant onus on learning and collaboration within and between IGOs and external stakeholders.

On the one hand, such collaboration may appear natural in the context of learning and problem-solving. On the other hand, the fisheries subsidies negotiations demonstrate the strict control imposed by states. WTO members have maintained the inaccessibility of the Rules Group to external participants, due to their wish that the WTO remain a 'Member-driven organisation'. Some WTO members have distrusted data from other IGOs and have been concerned about WTO adjudicating bodies ruling on WTO member compliance with exogenous instruments such as UNCLOS.

As is discussed in the previous chapter, an attempt to frame these concerns according to principles of international law might refer to the view that states should not be subject to norms from regimes to which they have not consented. This view is manifested in the suggested need for parallel membership between the members of the relevant regimes as a precursor to normative interplay.[2] However, the idea of parallel membership as a precursor to regime interaction is misplaced, especially in the context of the framing of new rules on fisheries subsidies. Although there were some complaints about the FAO involvement in fishery subsidies disciplines by non-FAO members, in general the WTO members wary of regime interaction are members of the exogenous regimes. This reinforces findings that IGOs have different sets of economic interests and political pressures – including, for the WTO,

---

[2] See e.g. VCLT Art. 31(3)(c) as interpreted by *EC – Biotech*, which I consider at greater length in Chapter 5. See also the disclaimer by third-party states in the WTO Doha Declaration (2001) para. 31(i) noted in Chapter 2, n. 317 and accompanying text.

'ecological insensitivities'[3] – and suggests that it is these differences, rather than non-identical regime membership, that are often at issue. Moreover, in real terms, there is no uniformity of membership between the WTO and *any* of the data-producing IGOs, including the FAO, OECD, and World Bank.

Instead of directing the quest for legitimacy in regime interaction in law-making to state consent, the legal framework offers alternatives which ensure accountability and openness to stakeholders. Such alternatives go further than WTO reform proposed by commentators. Guzman, for example, has suggested restructuring the WTO to take account of 'non-trade' interests of states.[4] Guzman recommends that the WTO adopt 'departments' with different competencies, such as environment or competition, where expertise would evolve. Negotiations would take place within 'departmental rounds' so that manageable discussions would be possible. [5] Although undoubtedly useful, I consider that these reforms are underinclusive because of their narrow conception of interests – in Guzman's suggested departments, for example, all relevant interests are assumed to be conveyed by states. By contrast, examples of inclusive decision-making for state and non-state parties are informative. Dorf and Sabel, for example, argue that encouraging mutual learning among the decentralised bodies of the United States public administration, and involving citizens in decisions that affect them, enhances the accountability and efficiency of decision-making.[6] Such insights have been applied to the polyarchy of the EU.[7] Cohen, Dorf, Sabel and other commentators call attention to

---

[3] Oren Perez, *Ecological Sensitivity and Global Legal Pluralism: Rethinking the Trade and Environment Conflict* (2004); see also further suggestions about the role of strong and weak states as powers underlying different treaty bodies: see Eyal Benvenisti, 'The Interplay between Actors as a Determinant of the Evolution of Administrative Law in International Institutions' (2006) 68 *Law and Contemporary Problems* 319, 326 (noting that 'the comparative study of administrative law in international institutions must be undertaken with great sensitivity to the[se] factors').

[4] Andrew Guzman, 'Global Governance and the WTO' (2004) 45 *Harvard International Law Journal* 303.

[5] See e.g. ibid., 335.

[6] Micheal Dorf and Charles Sabel, 'A Constitution of Democratic Experimentalism' (1998) 98 *Columbia Law Review* 267.

[7] Joshua Cohen and Charles Sabel, 'Sovereignty and Solidarity: EU and US' in Jonathan Zeitlin and David Trubek (eds.) *Governing Work and Welfare in a New Economy: European and American Experiments* (2003) 345. See further in relation to international harmonisation, Harm Shepel, *The Constitution of Private Governance: Product Standards in the Regulation of Integrating Markets* (2005) 412.

'governing councils' which collect and distribute experiences gathered within a variety of networks to promote mutual learning.

In the framing of new rules on fisheries subsidies, UNEP and WWF acted as 'governing councils' that shared information and promoted learning to further democratic experimentalism. Their deliberations were forwarded to the Rules Group. WTO members relied on the support of these deliberations and a range of non-state actors in crafting their proposals.

In the course of the negotiations, WTO members themselves acted as agents of peer review, expressing concerns about FAO data and the bias of the OECD. The role will continue in the implementation of the rules. The Chair's text places a heavy reliance on WTO members themselves to bring fisheries issues to the attention of the SCM Committee, through their own notifications, through requesting information from other members, or through filing information from 'pertinent outside sources' as to any apparent IUU fishing.[8]

It is uncertain whether the regime interaction in the implementation of the rules will meet these participatory and experimentalist ideals. The FAO will act as peer reviewer of WTO members' fisheries management systems, but there is no peer review of the FAO instruments themselves, and no role for the considerable monitoring and scrutiny undertaken by NGOs. The Chair's text departed from novel proposals for authority-sharing arrangements between experts from the FAO, UNEP, RFMOs and NGOs and instead proposed a framework for consultation in fisheries issues that was purposefully ambiguous about the role of IGOs and silent as to NGOs. These structures are a rather more limited form of 'post-sovereign environmental governance' observed in other forums.[9] Moreover, without institutional structures for ongoing regime interaction, the 'crisis' advocacy network that has developed around the issue of fisheries subsidies may move on to address other sustainability goals after the completion of the negotiations.[10] There

---

[8] For an analysis of the SPS Committee acting as agent of peer review, see Joanne Scott, *The WTO Agreement on Sanitary and Phytosanitary Measures: A Commentary* (2007) 50ff.
[9] See, e.g., Bradley Karkkainen, 'Post-Sovereign Environmental Governance' (2004) 4 *Global Environmental Politics* 72, 81.
[10] See, on the growth of networks to address 'crisis' situations, and an admonishment not to rely on such networks in everyday governance, Helen Hershkoff and Benedict Kingsbury, 'Crisis, Community, and Courts in Network Governance: A Response to Liebman and Sabel's Approach to Reform of Public Education' (2003) 28 *NYU Review of*

remains significant scope for the proposed rules on fisheries subsidies to follow an appropriate legal framework for regime interaction in both their formulation and implementation.

2 Implementation

Two prominent examples considered in the book highlight possibilities for appropriate regime interaction in the implementation of existing commitments. The listing of endangered marine species on the CITES Appendices generated a major debate over the interaction between the CITES and FAO regime, and was even presented by some as a problem of normative conflict. Instead of relying on the ILC's tool-box, however, institutional arrangements between the relevant secretariats allowed for interplay and collaboration. According to the legal framework of regime interaction, the Memorandum of Understanding between the CITES Secretariat and the FAO was an agreement which was functionally necessary for them to conclude. Such a conclusion is significant for the CITES Secretariat at least, which, unlike the FAO, lacks express powers to conclude agreements with other IGOs.

The provisions of the MOU allowed for information-sharing and observership, capacity-building and consultation and review of listing proposals. However, uncertainty remains, especially in the involvement of external stakeholders, the allocation of resources and the scope for joint enforcement. There is further scope for the MOU to allow for the participation of interested parties, and to expressly scrutinise the background and credibility of such groups.

The MOU has been mooted as a model for collaboration in other areas of fisheries governance, including in the relationship between the WTO and FAO. So far, however, the fisheries subsidies negotiations have proceeded without such formal arrangements. Yet the implementation of the proposed rules promises much by way of ongoing regime interaction, and provides the second major example of the possibility of appropriate regime interaction considered in the book.

Law & Social Change 319. This danger is exacerbated when one considers issue areas, such as the high seas, which lack institutional support: on the absence of a global institution on management of marine areas beyond national jurisdiction, see Rosemary Rayfuse and Robin Warner, 'Securing a Sustainable Future for the Oceans Beyond National Jurisdiction: The Legal Basis for an Integrated Cross-Sectoral Regime for High Seas Governance for the 21st Century' (2008) 23(3) *The International Journal of Marine and Coastal Law* 399; Robin Warner, *Protecting the Oceans Beyond National Jurisdiction: Strengthening the International Law Framework* (2009) 233-44.

The proposed use of fisheries standards in the draft fisheries subsidies rules draws on precedents from the SPS Agreement and the TBT Agreement. As identified in Chapter 3, the SPS and TBT Agreements use standards from exogenous regimes to determine whether WTO members can be exempt from disciplines because they are 'necessary' for a legitimate domestic objective, including the protection of human health. In the proposed fisheries subsidies rules, the proposed 'red-box' of prohibited subsidies includes subsidies used for illegal, unreported or unregulated (IUU) fishing.[11] The provision directly incorporates the definition of IUU fishing from the FAO's international plan of action. In other parts of the rules, exemptions to the proposed 'amber-box' of actionable subsidies can be based on adherence to fisheries instruments including the Fish Stocks Agreement, the FAO Code of Conduct, the FAO Compliance Agreement, technical guidelines, and plans of action.[12]

Entrenched regime interaction in the fisheries subsidies rules has led to complaints by several Rules Group participants, who argue that fisheries standards should not have a direct role in the disciplines because of uncertainty about standard-setting fisheries organisations. Such concerns may be addressed by the legal framework for regime interaction. In particular, the legal framework emphasises procedures that ensure that authority of standards is checked and contested.[13] For example, WTO members have proposed to subject the OECD Exports Credit Arrangement to notification and surveillance requirements. FAO guidelines on eco-labelling and the work of the TBT Committee both address the need for transparency within organisations involved in setting technical standards. In addition, the WTO Secretariat has provided workshops to promote transparency in the use of SPS measures. These techniques suggest an emerging role for regime interaction by WTO bodies in the implementation of rules.

For example, the use of the FAO standard for IUU fishing, proposed by the Chair's red-box text, may be appropriate if the standard remains transparent and subject to ongoing scrutiny. The FAO standard's current emphasis on participation, coordination and non-discrimination is

---

[11] Chair's text, Art. I.1(h), footnote 81. See further pp. 116–17.
[12] Ibid., Art. V. See further pp. 119–20.
[13] This idea draws on Scott's analysis of the EU approach to standard-setting: see Joanne Scott, 'International Trade and Environmental Governance: Relating Rules (and Standards) in the EU and the WTO' (2004) 15 *EJIL* 307, 307 ('authority and contestability go hand in hand'); see further Chapter 3, n. 215 and accompanying text.

of particular interest.[14] The absence of a requirement for consensus in the adoption of the FAO IUU standard does not eliminate its role; rather, such an absence might lead to a more in-depth scrutiny of its internal procedures by an adjudicating body. This also addresses the risk that the seemingly impossible aim of reaching consensus for their adoption leads to a vacuum of fisheries and conservation standards.

The reference in the Chair's text to the Fish Stocks Agreement, the FAO Code of Conduct and the Compliance Agreement is also immediately attractive not only because these instruments were developed in an accessible and transparent way but also because they seek to promote substantive openness and transparency in fisheries management. The Code of Conduct, for example, was developed in a spirit of collaboration and is subject to ongoing review,[15] and emphasises the need for ongoing collaboration by stakeholders in the Code's implementation.[16] Less convincing, perhaps, are the references by the Chair's text to technical guidelines and plans of actions for the implementation of these instruments. The development of these standards is a rather more closed affair.

Contestation of such standards is envisaged in several ways by the Chair's text. First, there is a distinct role for the SCM Committee. As set out in Chapter 3, the SCM Committee maintains surveillance over the peer review by the FAO and other fisheries bodies of WTO members' fisheries management regimes.[17] There is some scope to assess the strength of these fisheries management regimes by transparency requirements. WTO members are required to notify the SCM Committee of any measure that they consider to be exempt from disciplines due to their conformity with provisions relating to the green-box or special and differential treatment.[18] Other members have the right to be kept informed about these measures.[19] The terms of access-rights for fisheries are to be made public and notified to the SCM Committee.[20]

---

[14] IPOA-IUU Fishing, para. 9.
[15] Details on the development of the FAO Code of Conduct are set out in Annex 1 of the Code. On the review of the Code, see FAO Code of Conduct, Art. 4.3.
[16] See ibid., Art. 4.1; see also Art. 11.3.2.
[17] Chair's text, Art. V.1. The Chair's text is reproduced at Appendix A.   [18] Ibid., Art. VI.
[19] Ibid., Art. VI.5. (requesting information); Art. V.2. (enquiry points). I note possible implications for the future allocation of the burden of proof in disputes: See AB Report, *EC – Sardines*, paras. 269–283; see further Robert Howse, 'The Sardines Panel and AB Rulings – Some Preliminary Reactions' (2002) 29(3) *Legal Issues of Economic Integration* 247, 254.
[20] Chair's text, Art. VI.2.

In recognition of the costs of compliance with the disciplines, the Chair's text urges members to provide technical assistance to developing countries, either bilaterally or through appropriate international organisations.[21]

These provisions are useful starting points for the regime interaction between the WTO and other IGOs in the fisheries subsidies disciplines. There is further scope, however, for the SCM Committee to contest the authority of the fisheries organisations that are instituting the relevant fisheries 'best practices'. For example, there should be a requirement that relevant instruments are themselves publicly accessible. The Chair's text is currently silent on requiring instruments such as technical guidelines to be transparent or inclusive for all relevant WTO members. Moreover, the Chair's text is currently open to international instruments aimed at recognising fisheries best practices without regard to the inclusivity of those instruments. Certain bodies closed to many WTO members, such as RFMOs, create standards relating to fisheries management programmes. One issue of relevance to the SCM Committee is that such bodies are not open to all WTO members. The probative weight of standards issued by the bodies may be lessened accordingly, unless there is evidence of other ways in which WTO members can participate and contribute. Similarly, the SCM Committee should be aware of any problems in internal procedures or accountability of the relevant standard-setting body.

These ideas demonstrate regime interaction in the implementation of ongoing legal obligations. The depth and success of this regime interaction will also impact on the resolution of disputes. The fisheries subsidies example demonstrates how the level of entrenchment of standards impacts on the discretion of adjudicating bodies. Paradoxically, leaving only bare reference to external regimes (or none at all) may give WTO panels full discretion to interact with regimes in the settlement of disputes. The next section considers examples of the framework for such interaction.

3 Dispute settlement

The adjudication of disputes brings further scope for the scrutiny and contestation of exogenous regimes according to a legal framework. Proposals for new fisheries subsidies rules included innovative suggestions of authority-sharing arrangements between WTO dispute

---

[21] Ibid., Art. III.4.

settlement bodies and external participants.[22] Such an arrangement would have provided full scope for the interrogation of standards of fisheries management, which might then be relevant in condemning or absolving the provision of fisheries subsidies. Even the currently mooted role of the PGE provides discretion to scrutinise fisheries standards to some degree.

It is perhaps useful to illustrate the legal framework for appropriate regime interaction in the context of an existing dispute. The following example is taken from the facts of the *EC – Biotech* case. In that case, the *EC – Biotech* Panel was asked to consider the relevance of treaties and principles of environmental law, including the Biosafety Protocol, in interpreting the EU's obligations under the SPS Agreement. The Panel found that it could not refer to the Biosafety Protocol because the WTO membership as a whole had not ratified it. The approach outlined in this chapter would instead note the discretion of an adjudicating body to take account of other areas of law and other regimes. It would point to the consultative powers that already exist for WTO dispute settlement bodies and claim that such consultative powers should be liberally used to inquire into the position of all stakeholders, as put forward by IGOs and NGOs. However, this discretion would be fettered; in assessing the relevance of these contributions, a panel would have due regard to the substance of the norms and the process by which they have come into being as well as the credibility of the sources.

Contrary to the *EC – Biotech* Panel's refusal to consider the Biosafety Protocol, a legal framework for appropriate regime interaction would give the panel discretion to determine its relevance by recourse to a number of substantive and procedural aspects. Accordingly, a panel would inquire into the development of the Biosafety Protocol and specifically assess the level of collaboration between relevant stakeholders in the trade and environmental regimes. It would determine whether the Biosafety Protocol was open to all WTO members and had been endorsed by a majority of them. It would also assess how the negotiations for the Biosafety Protocol 'took account' of the WTO regime.[23] A panel would also consider whether the Protocol negotiations were transparent, accessible to a range of actors (including the parties to the dispute, the members of the WTO and participants from other

---

[22] WWF proposed a joint body comprising the WTO and including the FAO, UNEP, RFMOs and NGOs: see further pp. 126–7.

[23] See, e.g., the provision for 'mutual supportiveness' in the Biosafety Protocol.

IGOs and NGOs) and provided for mutual learning between the trade and environment regimes: the presence of these factors would be decisive for the panel's interpretation of relevant WTO obligations with reference to the Biosafety Protocol.

The framework also has implications for the way panels exercise their consultative powers. In *EC – Biotech*, the Panel consulted with a wide range of IGOs, although such consultation was conducted with the consent of the disputing parties. In the framework suggested here, this consultation would aim to be accountable by reference to the process norms described in this section. For example, if some WTO members were denied access to another IGO, a panel could legitimately discount the views of that IGO.

This example demonstrates the implications of my approach for the way WTO panels exercise their consultative powers. It also points to the need for an evolutionary interpretation of the customary norms of interpretation and the VCLT, as I discussed in Chapter 5. The main implication is that treaties will be interpreted with reference to one another; consent is relevant to this interpretation but not determinative. Moreover, the openness with which a treaty comes into being will be relevant in determining the degree of consensus which accompanies that treaty. This 'openness' will be assessed according to the participatory model of international law described above. Openness towards external stakeholders will be relevant 'treaty practice'[24] to determine the relevance of the treaty's norms to the international system.

The *EC – Biotech* Panel accepted the *amicus* briefs that were filed in the case but found that it was not necessary to take them into account, without providing further details. According to the legal framework for appropriate regime interaction that I propose, the panel would require relevant NGOs to state their legal interest in the case and their sources of funding. If an NGO refused to disclose its funding sources, the panel could legitimately refuse to admit its *amicus* brief. A panel that accepted an *amicus* brief but decided not to take it into account would also publish reasons for its decision as part of the panel record.

This section has provided some examples of regime interaction according to a legal framework, and demonstrated the link between law-making, implementation and adjudication. In doing so, it has emphasised the ongoing need to access and contest sources from overlapping regimes. My conclusions call for regime interaction that

---

[24] This is an intended analogy from the concept of 'state practice'.

continually scrutinises, for example, the operation of FAO norms in the WTO disciplines. These conclusions are in major contrast to the work of the ILC Study Group on Fragmentation. More generally, the emphasis on the contestation of norms may seem at variance with, or even counter-productive for, a legal system. Authority is usually given as a guarantee of order and law's role is to provide stability and predictability. The following section considers these themes in the context of the further implications of regime interaction for international law, for fisheries governance and beyond.

B Further implications

1 The ILC fragmentation study

As I discussed in Chapter 1, the ILC Study Group developed a series of ideas to address the problem of overlapping jurisdictions and conflicting norms in international law. The mandate of the Study Group excluded an analysis of institutions.[25] By contrast, my study in fisheries governance demonstrates that the activities of institutions and the ways in which they take account of each other's activities change the prescriptions for fragmentation.

The ILC Study Group advocated the use of international law to 'coordinate and organize the cooperation of (autonomous) rule-complexes and institutions',[26] but its substantive prescriptions were very different from mine. It pointed to the continued relevance of norms such as *lex specialis*, *lex posterior* and *lex superior*, while acknowledging that law must leave room for political processes to resolve pluralistic problems. In my view, of greater *legal* relevance to fragmentation is the use of norms of accountability that strive to include the many voices affected by global problems such as fisheries depletion and therefore enhance the possibility for democratic political engagement over regime struggles.[27]

The case studies of fisheries subsidies, trade restrictions on endangered marine species and import restrictions on fish products presented here demonstrate that process-norms such as the accountability models identified above are more relevant to the solving of global problems than the substance-norms of hierarchy that are often discussed in the context of fragmentation. Significantly, although my case studies deal

---

[25] ILC Analytical Study, para. 13. [26] Ibid., para. 487.
[27] The Study Group acknowledged the increased attention to 'global law' but excluded it from the scope of its work: ibid., para. 490.

with issues of global public interest such as the conservation of fisheries stocks, the policy problems are not framed in the language of *peremptory norms*, either in state submissions, institutional commentary or, with limited exceptions,[28] in scholarly observation. For example, there is no peremptory norm that prohibits the payment of subsidies to the fishery sector. Similarly, there is no trumping rule on the correct management techniques for vulnerable marine species. More generally, this absence of normative hierarchy is reminiscent of the complexity and balance inherent in the notion of 'sustainable development'.[29] My case studies demonstrate the difficulty in raising the language of normative hierarchy in areas involving significant economic and development interests. This reinforces my conclusion that models that account for a plurality of interests and a complexity of solutions are more relevant for regime interaction than models that entrench certain values. The accountability and legitimacy of fisheries trade must be measured according to principles of governance rather than closed indicators that trump all others.

2 Beyond fisheries governance

I have proposed a legal framework to facilitate appropriate regime interaction in fisheries governance. The global and complex problem of fisheries sustainability has called for policy responses from trade law, environmental law and the law of the sea. Rather than ask how the mandates of these regimes might be made to cohere, this study is based on an acceptance of continuing diversity, and focuses instead on the institutional and normative interaction between relevant organisations, other actors and their overarching sets of rules and principles. I have found that the intersections between regimes are being managed in complex and novel ways by the IGOs themselves (e.g. FAO–CITES), by non-state actors (e.g. WWF in fish subsidies) as well as by states. I have referred to the multiple bases for regime interaction, namely interaction that occurs where there is parallel membership of the intersecting treaties, interaction that occurs through mutual agreement of states, and interaction that occurs through independent IGO collaboration. To ensure that such regime interaction is appropriate, I call for attention to

---

[28] Jost Delbrück, '"Laws in the Public Interest" – Some Observations on the Foundations and Identification of *erga omnes* Norms in International Law' in Volkmar Götz, Peter Selmer and Rüdiger Wolfrum (eds.) *Liber Amicorum Günther Jaenicke: Zum. 85. Geburtstag* (1998) 17; see further Chapter 1, n. 58 and accompanying text.

[29] See e.g. Steven Bernstein, *The Compromise of Liberal Environmentalism* (2001).

the methods by which the norms and information of regimes have influence, which must be open, transparent, participatory and subject to ongoing scrutiny and review.

This model of regime interaction has promise beyond fisheries governance. Other complex global problems call for a variety of policy responses. One obvious issue will be the trade-related aspects of climate change. Many commentators consider that international trade law presents the most promising regime for targeting greenhouse gas emissions, and the United States, China and France are currently considering the relevance of the WTO regime in this context.[30] Lessons from regime interaction in fisheries governance will have relevance for disciplines on energy subsidies,[31] as well as possible litigation at the international level.[32] In addition, the interplay between the UNFCCC and diverse regimes such as trade and heritage protection requires further attention.[33]

As well as ensuring that regime interaction is appropriate, this model can also counter the disincentives that international law-makers currently face in addressing global problems that relate to multiple international institutions. Commentary on domestic law-making in the United States has suggested that if there are too many regulators from different policy areas, these regulators will be deterred from addressing complex social problems because of a 'regulatory commons' type tragedy.[34] This regulatory inertia can be addressed, to some degree, by the generating and sharing of information between institutions.[35] At an international level, the regulatory commons dynamic is complicated further by differences in state membership of particular regimes. It is likely, however, that the prescriptions for transparency and open IGO deliberation that I propose as part of the framework for appropriate legal interaction will similarly counter the disincentives to address

---

[30] See Joseph Stiglitz, 'A New Agenda for Global Warming' (2006) 3:7 *The Economists' Voice* available at www.bepress.com/ev/vol3/iss7/art3; Jochem Wiers, 'French Ideas on Climate and Trade Policies' (2008) 1 *Carbon and Climate Law Review* 18.

[31] Sadeq Z Bigdeli, 'Will the "Friends of Climate" Emerge in the WTO? The Prospects of Applying the "Fisheries Subsidies" Model to Energy Subsidies' (2008) 1 *Carbon & Climate Law Review* 78.

[32] William C G Burns, 'Potential Causes of Action for Climate Change Imapcts under the United Nations Fish Stocks Agreement' in William C G Burns and Hari M Osofsky (eds.) *Adjudicating Climate Change: State, National and International Approaches* (2009) 314.

[33] See further Margaret A. Young, 'Climate Change Law and Regime Interaction' (2011) *Carbon & Climate Law Review* (forthcoming).

[34] William Buzbee, 'Recognizing the Regulatory Commons: A Theory of Regulatory Gaps' (2003) 89 *Iowa Law Review* 1.

[35] Ibid., 63.

complex global problems in a fragmented 'regulatory commons', and there is scope for further empirical studies to support this.

The suggestion that the legal framework for appropriate regime interaction will apply beyond fisheries governance gives rise to further implications. One evolves from the recognition that regime interaction often occurs without the direct consent of states: even if it is unlikely that states 'consent' to regime interaction, could the conception of accountability in regime interaction be a general principle of law?[36] Could the principles of openness, transparency, participation and review in regime interaction come to embody a principle of good governance that could be applied along with the traditional sources of international law? Such questions reveal scope for further studies to assess the developing normative status of the legal framework for appropriate regime interaction. This project would include an awareness of the risk of idealising systematic unity.[37]

In the meantime, recognising the broader potential of the legal framework for appropriate regime interaction gives international lawyers a different lens through which to view the making of law. In negotiations between states, attention moves from efforts to bargain national interests towards participatory models that incorporate multiple interests and perspectives. In the implementation of existing norms, attention moves from mandate struggles and forum shopping towards structures for ongoing accountability and peer review between international institutions and other actors. In the settlement of disputes, adjudicating bodies such as WTO panels will be freed from current restrictive norms of interpretation that are based on the notion of self-contained regimes and consent in international law. In addition to ascertaining state consent in the resolution of norm conflicts, these adjudicating bodies will have regard to existing norms from international institutions based on whether such institutions are open to participatory models. Instead of ascertaining a hierarchy of norms, they will seek to discover the breadth and type of support for norms for a large number of actors.

This potential for openness and participation in regime interaction is a basis for renewed optimism in international law. It may indicate that

---

[36] ICJ Statute, Art. 38(1)(c).
[37] See Alexander Somek, 'Kelsen Lives' (2007) 18 *EJIL* 409, who critiques the sociological approaches to regime collisions in the works of Fischer-Lescano and Teubner (cited in Chapter 1, n. 67) and others from the perspective of legal positivism.

international law can meet the need, expressed by many, to accommodate the views of those who are affected by it. This highlights the need to situate the participants of international law, as discussed below. Moreover, there is scope for further work to understand the challenges and opportunities presented by regime interaction in diverse practical and theoretical contexts.[38]

3 Confronting managerialism

The legal framework for regime interaction might be subject to two major criticisms. First, the legal framework might be read as forestalling political contestation. The answer to this critique exposes the work to a second one: that the legal framework unjustifiably destabilises law. These ideas deserve greater explanation and an attempt at pre-emptive reply.

The first criticism may be implied from the work of Koskenniemi and other critical legal theorists. Koskenniemi warns against the promotion of expert authority, especially in the context of inter-regime cooperation.[39] This book's emphasis on collaboration by expert bodies such as the FAO and CITES Secretariat might be treated with scepticism. In particular, where law is envisaged 'beyond the state', and as 'an instrument for particular values, interests, preferences', there is a grave risk of managerialism.[40] Moreover, Koskenniemi is wary of pragmatic proposals for institutional reform which fail to recognise the diversity of human preferences and the relevance of a range of projects, both legal and non-legal.[41]

This book has sought to address these insights in both its methodology and prescriptions. It recognises that institutions, and regimes, are indeterminate and should not be conceived as 'natural' for obtaining specific functional objectives such as fisheries preservation. It engages

---

[38] See Margaret A. Young (ed.) *Regime Interaction in International Law: Facing Fragmentation* (forthcoming). Contributing chapters provide analysis and critique of regime interaction in a range of contexts including human rights, trade, climate change and investment law, as well as insights from diverse fields such as private law and domestic constitutional theory.
[39] 'International Law: Constitutionalism, Managerialism and the Ethos of Legal Education' (2007) 1 *European Journal of Legal Studies* (online); see also Martti Koskenniemi, 'The Fate of Public International Law: Between Technique and Politics' (2007) 70 *Modern Law Review* 1; see also David Kennedy, 'The Mystery of Global Governance' (2008) 34 *Ohio Northern University Law Review* 827.
[40] Koskenniemi, 'International Law' and 'The Fate of Public International Law', n. 39 above.
[41] See Martti Koskenniemi, *From Apology to Utopia: The Structure of International Legal Argument* (2006) 603–15.

in the politics of regime definition[42] and asks which voices are included in proposed rules targeting the economics of fishing, and which structures have obstructed or facilitated their admittance. This has required close empirical study within a number of regimes and accompanying institutions, including the WTO, the FAO and CITES.

In terms of critiquing the role of international law in the fisheries crisis, this project could have focused on other areas. The potential support by international law for a type of fisheries consumption that is agnostic about environmental effects is one example.[43] The implied legal hierarchy in the allocation of tariff preferences on fish products by large markets such as the European Union is another.[44] My choice to dwell on the challenges of fragmentation for fisheries governance within particular regimes should not be seen as a promotion of one set of preferences within one set of institutions, but rather a grounded analysis of the political wrangling that is currently occurring over perverse incentives, endangered species and import restrictions in the fishing sector.

In addition to the mapping of state and non-state participants in the processes, this book takes a normative step in proposing a framework of contestation and inter-regime engagement. My call for a legal framework for regime interaction seeks to ensure that a range of perspectives are included. Whether the processes are conceived as wrangling between regimes, or as political contestations about the definitions of regimes themselves, the book suggests a need for a greater awareness of how knowledge is produced and by whom, and how legal oversight will continue and by whom. Models calling for democratic experimentalism in domestic contexts are particularly instructive.[45] I consider that such models can confront the risk of managerialism and expert-rule by endorsing the participation of those affected by rules. This is consistent with arguments from political scientists about the democratic legitimacy of trade governance depending on a transnational sphere.[46]

---

[42] The phrase is from Koskenniemi, 'The Fate of Public International Law', above n. 39, 27; see further Chapter 1, n. 92 and accompanying text.

[43] See, e.g., through legal restrictions on labelling for sustainability objective. For current policy initiatives on labelling within the EU, see Mar Campins Eritja (ed.) *Sustainability Labelling and Certification* (2004), which I reviewed in (2006) 18 *Journal of Environmental Law* 176.

[44] For an exemplar case, see Margaret A. Young, 'WTO Undercurrents at the Court of Justice' (2005) 30 *European Law Review* 211.

[45] See e.g. pp. 280–1.

[46] Patrizia Nanz, 'Democratic Legitimacy and Constitutionalisation of Transnational Trade Governance: A View from Political Theory' in Christian Joerges and Ernst-Ulrich

The call for greater deliberative procedures in regime interaction leads to a second possible critique. Responses to the political and legal agenda proposed by democratic experimentalists are useful here. Scheuerman, for example, supports the focus on democratic procedures for producing law, but is concerned that democratic experimentalists risk abandoning stabilising democratic institutions such as the 'rule of law'.[47] He concludes that the democratic experimentalists 'aspire to synchronize the temporal logic of political and legal decision making with the temporal dynamics of capitalism'.

An immediate answer might distinguish the liberal democracies of Scheuerman's study from the international legal system, and argue that a departure from domestic constitutional ideals and towards managerial procedures of open and accountable regime interaction is justified. Even so, Scheuerman's reminder about the assumptions of time inherent in notions of deliberation remains valid for all systems. I acknowledge that further work will need to confront these assumptions, and the possible risk that a legal framework for regime interaction prioritises particular sets of 'crises' and particular sets of stakeholders for whom time, resources and advocacy skills are plentiful, as discussed below.

4 Situating the participants

The ideas for appropriate regime interaction have extended the notion of 'participants' in international law, and call for greater attention to the activities both of states (and specifically, of national delegations) and of non-state actors, including IGO secretariats and NGOs. Further development of the problem of articulating all the interests of those 'affected by' modern forms of governance can extend current ideas in philosophy and political theory.[48] From a different field of inquiry, research based on actor network theories and other performance-based approaches could be useful to develop further understanding

---

Petersmann (eds.) *Constitutionalism, Multilevel Trade Governance and Social Regulation* (2006) 60, 80; see also Inger-Johanne Sand, 'Polycontextuality as an Alternative to Constitutionalism' in Christian Joerges, Inger-Johanne Sand and Gunther Teubner (eds.) *Transnational Governance and Constitutionalism* (2004) 41, 61–5.

[47] William Scheuerman, 'Democratic Experimentalism or Capitalist Synchronization? Critical Reflections on Directly-Deliberative Polyarchy' (2004) XVII *Canadian Journal of Law and Jurisprudence* 101, 127.

[48] See Robert E Goodin, 'Enfranchising All Affected Interests, and Its Alternatives' (2007) 35 *Philosophy & Public Affairs* 1; Eric Cavallero, 'Federative Global Democracy' (2009) 40 *Metaphilosophy* 42; see further Chapter 1, n. 121 and accompanying text.

about the ongoing role, activities and influence of the participants in fisheries governance.[49]

Greater attention to the role of states in regime interaction will concentrate on the national delegations for each regime, particularly with respect to their role in promoting national policy coordination. Further work could identify the differing levels of capacity within national delegations, and perhaps point to further strategies to address imbalances within country resources.

For non-state actors, my analysis of fragmented fisheries governance demonstrates the importance of IGOs and NGOs in regime collaboration. This extends previous literature pointing to the driving force of IGOs and NGOs in regime formation,[50] and demonstrates the central role of these bodies in regime interaction. It shows, in particular, that IGOs can bring about dynamic stakeholder participation more effectively than states. These findings underscore the need for further research on institutional design in the context of the fragmentation of international law. There is a need to reassess the structures of these organisations according to the findings set out in this chapter. Such work will focus not only on collaborative procedures (such as MOUs), but on the openness, transparency and participation *at the time such collaborative procedures were developed*. Ongoing regime collaboration between such bodies as the WTO and FAO in fisheries governance is more likely to be successful if the regimes collaborated in the drafting of relevant rules.

Such work would also continue to identify the relative institutional strengths of regimes. My case studies demonstrate the influence of trade law in fisheries governance. Chapter 5, in particular, demonstrated the influence of the WTO on other regimes. Further study will identify other areas where the laws of the WTO influence other regimes, with similar normative effects for international law.[51] My work on fisheries

---

[49] Elements of the work of Bruno Latour, particularly in relation to conceptualizing the agency of non-humans, are one such useful direction: Bruno Latour, *Reassembling the Social: An Introduction to Actor-Network-Theory* (2005).

[50] See Oran Young, 'The Politics of International Regime Formation: Managing Natural Resources and the Environment' (1989) 43 *International Organization* 349; see also Ralph B Levering and Miriam L Levering, *Citizen Action for Global Change: The Neptune Group and the Law of the Sea* (1999); Anna-Karin Lindblom, *Non-Governmental Organisations and International Law* (2005).

[51] See also Claire Kelly, 'Power, Linkage and Accommodation: The WTO as an International Actor and Its Influence on Other Actors and Regimes' (2006) 24 *Berkeley Journal of International Law* 79.

governance also highlights the institutional weaknesses in certain policy areas, such as the lack of an adequate secretariat for UNCLOS.

The ideas of this book go further than merely accommodating the views of NGOs, but actually see regime collaboration as dependent on them. This recognition could lead to practical suggestions for state-based funding of NGOs in appropriate cases. Such suggestions have been made in the domestic context to achieve pragmatic decision-making over ecological resources.[52] More far-reaching proposals have been made about the establishment of a 'guardian of the ocean',[53] a body to represent the interests of ecology in fisheries governance. Such a body would address some of the imbalances identified in the current international institutions, and be given greater strength than newly established efforts within the UN, such as the UN Consultative Process.

Finally, the influence of critical commentators as participants must be acknowledged. As I described in Chapter 1, the 'trade and ...' literature has often portrayed trade law as a damaging regime that has simply needed more balance from 'corrective' regimes of environmental law or human rights. Such criticism has probably been extremely useful in opening up the WTO as an international organisation, which is arguably far more accessible for stakeholders and onlookers than many of the MEA regimes.[54] I have shown that the issue is rather more complex, and that analysts need to look more closely both at other regimes and the techniques by which collaboration is achieved. Those affected by international law are only just beginning to see themselves as part of an international system, and there is much work to be done to ensure effective and ongoing collaboration between its constituent parts.

---

[52] Daniel Farber, *Eco-Pragmatism: Making Sensible Environmental Decisions in an Uncertain World* (1999) 204.

[53] Christopher Stone, *Should Trees Have Standing?: And Other Essays on Law, Morals, and the Environment* (1996) 65, 70. See related proposals for a global institution to oversee the conservation and management of marine areas beyond national jurisdiction: Rayfuse and Warner, above n. 10; Warner, above n. 10, 233–44.

[54] The WTO website is perhaps the most accessible and innovative of all IGO websites, with innovations such as the DG online chat forum and the NGO portal.

Appendices

A: DRAFT CONSOLIDATED TEXT OF THE PROPOSED FISHERIES SUBSIDIES DISCIPLINES

Circulated as proposed Annex VIII of the SCM Agreement by the Chair of the Rules Group, Ambassador Guillermo Valles Games, on 30 November 2007:[*]

Annex VIII

Fisheries Subsidies
Article I
   Prohibition of Certain Fisheries Subsidies

   I.1   Except as provided for in Articles II and III, or in the exceptional case of natural disaster relief[77], the following subsidies within the meaning of paragraph 1 of Article 1, to the extent they are specific within the meaning of paragraph 2 of Article 1, shall be prohibited:
   (a) Subsidies the benefits of which are conferred on the acquisition, construction, repair, renewal, renovation, modernization, or any

---

[*] Reproduced at WTO Doc TN/RL/W/213, 87-93. Footnote references are from original.
[77] Subsidies referred to in this provision shall not be prohibited when limited to the relief of a particular natural disaster, provided that the subsidies are directly related to the effects of that disaster, are limited to the affected geographic area, are time-limited, and in the case of reconstruction subsidies, only restore the affected area, the affected fishery, and/or the affected fleet to its pre-disaster state, up to a sustainable level of fishing capacity as established through a science-based assessment of the post-disaster status of the fishery. Any such subsidies are subject to the provisions of Article VI.

other modification of fishing vessels[78] or service vessels[79], including subsidies to boat building or shipbuilding facilities for these purposes.

    (b) Subsidies the benefits of which are conferred on transfer of fishing or service vessels to third countries, including through the creation of joint enterprises with third country partners.

    (c) Subsidies the benefits of which are conferred on operating costs of fishing or service vessels (including licence fees or similar charges, fuel, ice, bait, personnel, social charges, insurance, gear, and at-sea support); or of landing, handling or in- or near-port processing activities for products of marine wild capture fishing; or subsidies to cover operating losses of such vessels or activities.

    (d) Subsidies in respect of, or in the form of, port infrastructure or other physical port facilities exclusively or predominantly for activities related to marine wild capture fishing (for example, fish landing facilities, fish storage facilities, and in- or near-port fish processing facilities).

    (e) Income support for natural or legal persons engaged in marine wild capture fishing.

    (f) Price support for products of marine wild capture fishing.

    (g) Subsidies arising from the further transfer, by a payer Member government, of access rights that it has acquired from another Member government to fisheries within the jurisdiction of such other Member.[80]

    (h) Subsidies the benefits of which are conferred on any vessel engaged in illegal, unreported or unregulated fishing.[81]

I.2 In addition to the prohibitions listed in paragraph 1, any subsidy referred to in paragraphs 1 and 2 of Article 1 the benefits of which are conferred on any fishing vessel or fishing activity affecting fish stocks that are in an unequivocally overfished condition shall be prohibited.

---

[78] For the purposes of this Agreement, the term "fishing vessels" refers to vessels used for marine wild capture fishing and/or on-board processing of the products thereof.

[79] For the purposes of this Agreement, the term "service vessels" refers to vessels used to tranship the products of marine wild capture fishing from fishing vessels to on-shore facilities; and vessels used for at-sea refuelling, provisioning and other servicing of fishing vessels.

[80] Government-to-government payments for access to marine fisheries shall not be deemed to be subsidies within the meaning of this Agreement.

[81] The terms "illegal fishing", "unreported fishing" and "unregulated fishing" shall have the same meaning as in paragraph 3 of the International Plan of Action to Prevent, Deter and Eliminate Illegal Unreported and Unregulated Fishing of the United Nations Food and Agricultural Organization.

Article II
General Exceptions
Notwithstanding the provisions of Article I, and subject to the provision of Article V:

(a) For the purposes of Article I.1(a), subsidies exclusively for improving fishing or service vessel and crew safety shall not be prohibited, provided that:
   (1) such subsidies do not involve new vessel construction or vessel acquisition;
   (2) such subsidies do not give rise to any increase in marine wild capture fishing capacity of any fishing or service vessel, on the basis of gross tonnage, volume of fish hold, engine power, or on any other basis, and do not have the effect of maintaining in operation any such vessel that otherwise would be withdrawn; and
   (3) the improvements are undertaken to comply with safety standards.
(b) For the purposes of Articles I.1(a) and I.1(c) the following subsidies shall not be prohibited: subsidies exclusively for: (1) the adoption of gear for selective fishing techniques; (2) the adoption of other techniques aimed at reducing the environmental impact of marine wild capture fishing; (3) compliance with fisheries management regimes aimed at sustainable use and conservation (e.g., devices for Vessel Monitoring Systems); provided that the subsidies do not give rise to any increase in the marine wild capture fishing capacity of any fishing or service vessel, on the basis of gross tonnage, volume of fish hold, engine power, or on any other basis, and do not have the effect of maintaining in operation any such vessel that otherwise would be withdrawn.
(c) For the purposes of Article I.1(c), subsidies to cover personnel costs shall not be interpreted as including:
   (1) subsidies exclusively for re-education, retraining or redeployment of fishworkers[82] into occupations unrelated to marine wild capture fishing or directly associated activities; and
   (2) subsidies exclusively for early retirement or permanent cessation of employment of fishworkers as a result of government policies to reduce marine wild capture fishing capacity or effort.
(d) Nothing in Article I shall prevent subsidies for vessel decommissioning or capacity reduction programmes, provided that:
   (1) the vessels subject to such programmes are scrapped or otherwise permanently and effectively prevented from being used for fishing anywhere in the world;

---

[82] For the purpose of this Agreement, the term "fishworker" shall refer to an individual employed in marine wild capture fishing and/or directly associated activities.

(2) the fish harvesting rights associated with such vessels, whether they are permits, licences, fish quotas or any other form of harvesting rights, are permanently revoked and may not be reassigned;

(3) the owners of such vessels, and the holders of such fish harvesting rights, are required to relinquish any claim associated with such vessels and harvesting rights that could qualify such owners and holders for any present or future harvesting rights in such fisheries; and

(4) the fisheries management system in place includes management control measures and enforcement mechanisms designed to prevent overfishing in the targeted fishery. Such fishery-specific measures may include limited entry systems, catch quotas, limits on fishing effort or allocation of exclusive quotas to vessels, individuals and/or groups, such as individual transferable quotas.

(e) Nothing in Article I shall prevent governments from making user-specific allocations to individuals and groups under limited access privileges and other exclusive quota programmes.

Article III
Special and Differential Treatment of Developing Country Members

III.1 The prohibition of Article 3.1(c) and Article I shall not apply to least-developed country ("LDC") Members.

III.2 For developing country Members other than LDC Members:

(a) Subsidies referred to in Article I.1 shall not be prohibited where they relate exclusively to marine wild capture fishing performed on an inshore basis (i.e., within the territorial waters of the Member) with non-mechanized net-retrieval, provided that (1) the activities are carried out on their own behalf by fishworkers, on an individual basis which may include family members, or organized in associations; (2) the catch is consumed principally by the fishworkers and their families and the activities do not go beyond a small profit trade; and (3) there is no major employer-employee relationship in the activities carried out. Fisheries management measures aimed at ensuring sustainability, such as the measures referred to in Article V, should be implemented in respect of the fisheries in question, adapted as necessary to the particular situation, including by making use of indigenous fisheries management institutions and measures.

(b) In addition, subject to the provisions of Article V:

(1) Subsidies referred to in Articles I.1(d), I.1(e) and I.1(f) shall not be prohibited.

(2) Subsidies referred to in Article I.1(a) and I.1(c) shall not be prohibited provided that they are used exclusively for marine

wild capture fishing employing decked vessels not greater than 10 meters or 34 feet in length overall, or undecked vessels of any length.

(3) For fishing and service vessels of such Members other than the vessels referred to in paragraph (b)(2), subsidies referred to in Article I.1(a) shall not be prohibited provided that (i) the vessels are used exclusively for marine wild capture fishing activities of such Members in respect of particular, identified target stocks within their Exclusive Economic Zones ("EEZ"); (ii) those stocks have been subject to prior scientific status assessment conducted in accordance with relevant international standards, aimed at ensuring that the resulting capacity does not exceed a sustainable level; and (iii) that assessment has been subject to peer review in the relevant body of the United Nations Food and Agriculture Organization ("FAO")[83].

III.3 Subsidies referred to in Article I.1(g) shall not be prohibited where the fishery in question is within the EEZ of a developing country Member, provided that the agreement pursuant to which the rights have been acquired is made public, and contains provisions designed to prevent overfishing in the area covered by the agreement based on internationally-recognized best practices for fisheries management and conservation as reflected in the relevant provisions of international instruments aimed at ensuring the sustainable use and conservation of marine species, such as, inter alia, the Agreement for the Implementation of the Provisions of the United Nations Convention on the Law of the Sea of 10 December 1982 Relating to the Conservation and Management of Straddling Fish Stocks and Highly Migratory Fish Stocks ("Fish Stocks Agreement"), the Code of Conduct on Responsible Fisheries of the Food and Agriculture Organization ("Code of Conduct"), the Agreement to Promote Compliance with International Conservation and Management Measures by Fishing Vessels on the High Seas ("Compliance Agreement"), and technical guidelines and plans of action (including criteria and precautionary reference points) for the implementation of these instruments, or other related or successor instruments. These provisions shall include requirements and support for science-based stock assessment before fishing is undertaken pursuant to the agreement and for regular assessments thereafter, for management and control measures, for vessel registries, for reporting of effort, catches and discards to the national authorities of the host Member and to relevant international organizations, and for such other measures as may be appropriate.

---

[83] If the Member in question is not a member of the FAO, the peer review shall take place in another recognized and competent international organization.

III.4 Members shall give due regard to the needs of developing country Members in complying with the requirements of this Annex, including the conditions and criteria set forth in this Article and in Article V, and shall establish mechanisms for, and facilitate, the provision of technical assistance in this regard, bilaterally and/or through the appropriate international organizations.

Article IV
General Discipline on the Use of Subsidies

IV.1 No Member shall cause, through the use of any subsidy referred to in paragraphs 1 and 2 of Article 1, depletion of or harm to, or creation of overcapacity in respect of, (a) straddling or highly migratory fish stocks whose range extends into the EEZ of another Member; or (b) stocks in which another Member has identifiable fishing interests, including through user-specific quota allocations to individuals and groups under limited access privileges and other exclusive quota programmes. The existence of such situations shall be determined taking into account available pertinent information, including from other relevant international organizations. Such information shall include the status of the subsidizing Member's implementation of internationally-recognized best practices for fisheries management and conservation as reflected in the relevant provisions of international instruments aimed at the sustainable use and conservation of marine species, such as, inter alia, the Fish Stocks Agreement, the Code of Conduct, the Compliance Agreement, and technical guidelines and plans of action (including criteria and precautionary reference points) for the implementation of these instruments, or other related or successor instruments.

IV.2 Any subsidy referred to in this Annex shall be attributable to the Member conferring it, regardless of the flag(s) of the vessel(s) involved or the application of rules of origin to the fish involved.

Article V
Fisheries Management[84]

V.1 Any Member granting or maintaining any subsidy as referred to in Article II or Article III.2(b) shall operate a fisheries management system regulating marine wild capture fishing within its jurisdiction, designed to prevent overfishing. Such management system shall be based on internationally-recognized best practices for fisheries management and conservation as reflected in the relevant provisions of

---

[84] Developing country Members shall be free to implement and operate these management requirements on a regional rather than a national basis provided that all of the requirements are fulfilled in respect of and by each Member in the region.

international instruments aimed at ensuring the sustainable use and conservation of marine species, such as, inter alia, the Fish Stocks Agreement, the Code of Conduct, the Compliance Agreement, technical guidelines and plans of action (including criteria and precautionary reference points) for the implementation of these instruments, or other related or successor instruments. The system shall include regular science-based stock assessment, as well as capacity and effort management measures, including harvesting licences or fees; vessel registries; establishment and allocation of fishing rights, or allocation of exclusive quotas to vessels, individuals and/or groups, and related enforcement mechanisms; species-specific quotas, seasons and other stock management measures; vessel monitoring which could include electronic tracking and on-board observers; systems for reporting in a timely and reliable manner to the competent national authorities and relevant international organizations data on effort, catch and discards in sufficient detail to allow sound analysis; and research and other measures related to conservation and stock maintenance and replenishment. To this end, the Member shall adopt and implement pertinent domestic legislation and administrative or judicial enforcement mechanisms. It is desirable that such fisheries management systems be based on limited access privileges[85]. Information as to the nature and operation of these systems, including the results of the stock assessments performed, shall be notified to the relevant body of the FAO, where it shall be subject to peer review prior to the granting of the subsidy[86]. References for such legislation and mechanism, including for any modifications thereto, shall be notified to the Committee on Subsidies and Countervailing Measures ("the Committee") pursuant to the provisions of Article VI.4.

V.2 Each Member shall maintain an enquiry point to answer all reasonable enquiries from other Members and from interested parties in other Members concerning its fisheries management system, including measures in place to address fishing capacity and fishing effort, and the biological status of the fisheries in question. Each Member shall notify to the Committee contact information for this enquiry point.

---

[85] Limited access privileges could include, as appropriate to a given fishery, community-based rights systems, spatial or territorial rights systems, or individual quota systems, including individual transferable quotas.

[86] If the Member in question is not a member of the FAO, the notification for peer review shall be to another relevant international organization. The specific information to be notified shall be determined by the relevant body of the FAO or such other organization.

Article VI
Notifications and Surveillance

VI.1 Each Member shall notify to the Committee in advance of its implementation any measure for which that Member invokes the provisions of Article II or Article III.2; except that any subsidy for natural disaster relief[87] shall be notified to the Committee without delay[88]. In addition to the information notified pursuant to Article 25, any such notification shall contain sufficiently precise information to enable other Members to evaluate whether or not the conditions and criteria in the applicable provisions of Article II or Article III.2 are met.

VI.2 Each Member that is party to an agreement pursuant to which fishing rights are acquired by a Member government ('payer Member') from another Member government to fisheries within the jurisdiction of such other Member shall publish that agreement, and shall notify to the Committee the publication references for it.

VI.3 The terms on which a payer Member transfers fishing rights it has obtained pursuant to an agreement as referred to in paragraph 2 shall be notified to the Committee by the payer Member in respect of each such agreement.

VI.4 Each Member shall include in its notifications to the Committee the references for its applicable domestic legislation and for its notifications made to other organizations, as well as for the documents related to the reviews conducted by those organizations, as referred to in Article V.1.

VI.5 Other Members shall have the right to request information about the notified subsidies, including about individual cases of subsidization, about notified agreements pursuant to which fishing rights are acquired, and about the stock assessments and management systems notified to other organizations pursuant to Article V.1. Each Member so requested shall provide such information in accordance with the provisions of Article 25.9.

VI.6 Any Member shall be free to bring to the attention of the Committee information from pertinent outside sources (including intergovernmental organizations with fisheries management-related activities, regional fisheries management organizations and similar sources) as to any apparent illegal, unreported and unregulated fishing activities.

VI.7 Measures notified pursuant to this Article shall be subject to review by the Committee as provided for in Article 26.

[87] As provided for in Article I.1 and footnote 77.
[88] For the purposes of this provision, "without delay" shall mean not later than the date of entry into force of the programme, or in the case of an ad hoc subsidy, the date of commitment of the subsidy.

Article VII
Transitional Provisions

VII.1 Any subsidy programme which has been established within the territory of any Member before the date of entry into force of the results of the DDA and which is inconsistent with Article 3.1(c) and Article I shall be notified to the Committee not later than 90 days, or in the case of a developing country Member 180 days, after the date of entry into force of the results of the DDA.

VII.2 Provided that a programme has been notified pursuant to paragraph 1, a Member shall have two years, or in the case of a developing country Member four years, from the date of entry into force of the results of the DDA to bring that programme into conformity with Article 3.1(c) and Article I, during which period the programme shall not be subject to those provisions.

VII.3 No Member shall extend the scope of any programme, nor shall a programme be renewed upon its expiry.

Article VIII
Dispute Settlement

VIII.1 Where a measure is the subject of dispute settlement claims pursuant to Article 3.1(c) and Article I, the relevant provisions of Article 4 and of this Article shall apply. Article 30 and the relevant provisions of this Article shall apply to disputes arising under other provisions of this Annex.

VIII.2 Where a subsidy that has not been notified as required by Article VI.1 is the subject of dispute settlement pursuant to the DSU and Article 4, such subsidy shall be presumed to be prohibited pursuant to Article 3.1(c) and Article I. It shall be for the subsidizing Member to demonstrate that the subsidy in question is not prohibited.

VIII.3 Where a further transfer of access rights as referred to in Article I.1(g) is the subject of a dispute arising under this Annex, and the terms of that transfer have not been notified as required by Article VI.3, the transfer shall be presumed to give rise to a subsidy. It shall be for the payer Member to demonstrate that no such subsidy has arisen.

VIII.4 Where a dispute arising under this Annex raises scientific or technical questions related to fisheries, the panel should seek advice from fisheries experts chosen by the panel in consultation with the parties. To this end, the panel may, when it deems it appropriate, establish an advisory technical fisheries expert group, or consult recognized and competent international organizations, at the request of either party to the dispute or on its own initiative.

VIII.5 Nothing in this Annex shall impair the rights of Members to resort to the good offices or dispute settlement mechanisms of other international organizations or under other international agreements.

B: FINAL TEXT OF THE FAO-CITES MEMORANDUM OF UNDERSTANDING

As agreed by CITES and FAO in September-October, 2006.[*]

FAO AND CITES IN ORDER TO STRENGTHEN THE COOPERATION BETWEEN THEM HAVE DECIDED AS FOLLOWS:

1. The signatories will communicate and exchange information regularly and bring to each other's attention general information of common interest and areas of concern where there is a role for the other to play. The signatories will be invited as observers to meetings under their respective auspices where subjects that are of common interest will be discussed.
2. The signatories will cooperate as appropriate to facilitate capacity building in developing countries and countries with economies in transition on issues relating to commercially-exploited aquatic species listed on the CITES Appendices.
3. FAO will continue to provide advice to CITES on, and be involved in any future revision of, the CITES listing criteria.
4. The FAO will work together with CITES to ensure adequate consultations in the scientific and technical evaluation of proposals for including, transferring or deleting commercially-exploited aquatic species in the CITES Appendices based on the criteria agreed by the Parties to CITES, and both signatories will address technical and legal issues relating to the listing and implementation of such listings.
5. As is required by the Convention, the CITES Secretariat will continue to inform FAO of all relevant proposals for amendment of Appendices I and II. Such information shall be provided to FAO to allow FAO to carry out a scientific and technical review of such proposals in a manner it deems appropriate and for the resulting output to be transmitted to the CITES Secretariat. The CITES Secretariat shall communicate the views expressed and data provided from this review and its own findings and recommendations, taking due account of the FAO review, to the Parties to CITES.
6. In order to ensure maximum coordination of conservation measures, the CITES Secretariat will respect, to the greatest extent possible, the results of the FAO scientific and technical review of proposals to amend the Appendices, and technical and legal issues of common interest and the responses from all the relevant bodies associated with management of the species in question.
7. The Secretariats to CITES and FAO will periodically report on work completed under the MOU to the Conference of the Parties to CITES and the FAO Committee on Fisheries, respectively.

[*] Reproduced at CITES SC 53 Doc 10.1 and FAO *Fisheries Report No 807* (2006), Appendix F.

8. This MOU will take effect on the date of signature by both signatories. It will remain in force unless terminated by 90 days' written notice served by one upon the other, or replaced by another agreement. It may be amended by written mutual agreement.
9. Unless otherwise agreed, neither signatory will be legally or financially liable in any way for activities carried out jointly or independently under this MoU. Separate letters of agreement or other arrangements, with specific budgets and resource identification, will be concluded for individual activities involving the commitment of financial resources by either signatory.

Director General, FAO, 29 September 2006
Secretary-General, CITES Secretariat, 3 October 2006

Bibliography

BOOKS AND EDITED COLLECTIONS

Abbott, Kenneth and Duncan Snidal, 'International Action on Bribery and Corruption: Why the Dog Didn't Bark in the WTO' in Daniel Kennedy and James Southwick (eds.) *The Political Economy of International Trade Law: Essays in Honour of Robert E Hudec* (Cambridge University Press, 2002) 177

Allain, Marc, *Trading Away Our Oceans* (Greenpeace, 2007)

Allott, Philip, *Eunomia: New Order for a New World* (Oxford University Press, 1990)
  The Health of Nations: Society and Law Beyond the State (Cambridge University Press, 2002)

Alvarez, José, *International Organizations as Law-Makers* (Oxford University Press, 2005)

Appleton, Arthur, *Environmental Labelling Programmes: International Trade Law Implications* (Kluwer Law International, 1997)

Barnes, Richard, 'The LOSC: An Effective Framework for Domestic Fisheries Conservation?' in David Freestone, Richard Barnes and David Ong (eds.) *The Law of the Sea: Progress and Prospects* (Oxford University Press, 2006) 233
  Property Rights and Natural Resources (Hart Publishing, 2009)

Bernstein, Steven, *The Compromise of Liberal Environmentalism* (Columbia University Press, 2001)

Bhagwati, Jagdish, *In Defense of Globalization* (Oxford University Press, 2004)
  Termites in the Trading System: How Preferential Agreements Undermine Free Trade (Oxford University Press, 2008)

Birnie, Patricia and Alan Boyle, *International Law and the Environment* (2nd edn, Oxford University Press, 2002)

Boyle, Alan and Christine Chinkin, *The Making of International Law* (Oxford University Press, 2007)

Brooks, Richard, Ross Jones and Ross Virginia, *Law and Ecology: The Rise of the Ecosystem Regime* (Ashgate Publishing, 2002)

Brownlie, Ian, *Principles of Public International Law* (Oxford University Press, 6th edn, 2003)

Burke, William, *The New International Law of Fisheries: UNCLOS 1982 and Beyond* (Clarendon Press, 1994)

Burns, William C G, 'Potential Causes of Action for Climate Change Impacts under the United Nations Fish Stocks Agreement' in William C G Burns and Hari M Osofsky (eds.) *Adjudicating Climate Change: State, National and International Approaches* (Cambridge University Press, 2009) 314

Chayes, Abram and Antonia Handler Chayes, *The New Sovereignty: Compliance with International Regulatory Agreements* (Harvard University Press, 1995)

Chayes, Antonia Handler, Abram Chayes and Ronald B Mitchell, 'Active Compliance Management in Environmental Treaties' in Winfried Lang (ed.) *Sustainable Development and International Law* (Kluwer Academic Publishers Group, 1995) 75.

Chesterman, Simon, *Just War or Just Peace? Humanitarian Intervention and International Law* (Oxford University Press, 2002)

Chinkin, Christine and Ruth Mackenzie, 'Intergovernmental Organizations as "Friends of the Court"' in Laurence Boisson de Chazournes, Cesare Romano and Ruth Mackenzie (eds.) *International Organizations and International Dispute Settlement: Trends and Prospects* (Transnational Publishers, 2002) 135

Churchill, Robin and Vaughan Lowe, *The Law of the Sea* (3rd edn, Manchester University Press, 1999)

Cohen, Joshua and Charles Sabel, 'Sovereignty and Solidarity: EU and US' in Jonathan Zeitlin and David Trubek (eds.) *Governing Work and Welfare in a New Economy: European and American Experiments* (Oxford University Press, 2003) 345

Cordonier Segger, Marie-Claire and C G Weeramantry (eds.) *Sustainable Justice: Reconciling Economic, Social and Environmental Law* (Martinus Nijhoff: Brill Academic, 2005)

Courville, Sasha 'Understanding NGO-based Social and Environmental Regulatory Systems: Why We Need New Models of Accountability' in Michael W Dowdle (ed.) *Public Accountability: Designs, Dilemmas and Experiences* (Cambridge University Press, 2006) 271

Crawford, James, *The Creation of States in International Law* (2nd edn, Oxford University Press, 2006)

de Búrca, Gráinne and Joanne Scott, 'Introduction: New Governance, Law and Constitutionalism' in Gráinne de Búrca and Joanne Scott (eds.) *Law and New Governance in the EU and the US* (Hart Publishing, 2006) 1

Deere, Carolyn, *Net Gains: Linking Fisheries Management, International Trade and Sustainable Development* (IUCN, 2000)

Delbrück, Jost, '"Laws in the Public Interest" – Some Observations on the Foundations and Identification of *Erga Owner* Norms in International Law' in Volkmar Götz, Peter Selwer and Rüdiger Wolfrum (eds.) *Liber Amicorum Günther Jaenicke: Zum 85. Geburtstag* (Springer, 1998)

Dollar, David, 'Fostering Equity through International Institutions' in Roger B Porter, Pierre Sauve, Arvind Subramanian, Americo Beviglia Zampetti (eds.) *Efficiency, Equity, and Legitimacy: The Multilateral Trading System at the Millennium* (Brookings Institution Press, 2001) 212

Dommen, Caroline, 'Fish for Thought: Fisheries, International Trade and Sustainable Development' (1999) *Natural Resources, International Trade and Sustainable Development Series, No. 1* (ICTSD and IUCN).

Dryzek, John, 'Political and Ecological Communication' in John Dryzek and David Schlosberg (eds.) *Debating the Earth* (Oxford University Press, 1998; reprinted 1999) 584

Eckersley, Robyn, *Environmentalism and Political Theory: Toward an Ecocentric Approach* (State University of New York Press, 1992)

Ellen Hey, *Teaching International Law: State-Consent as Consent to a Process of Normative Development and Ensuing Problems* (Kluwer Law International, 2003)

Eritja, Mar Campins (ed.) *Sustainability Labelling and Certification* (Marcial Pons, 2004)

Esty, Daniel, *Greening the GATT: Trade, Environment, and the Future* (Institute for International Economics, 1994)

Farber, Daniel, *Eco-Pragmatism: Making Sensible Environmental Decisions in an Uncertain World* (University of Chicago Press, 1999)

Footer, Mary, *An Institutional and Normative Analysis of the World Trade Organization* (Martinus Nijhoff, 2006)

Franck, Thomas, *The Power of Legitimacy Among Nations* (Oxford University Press, 1990)

Franckx, Erik, 'The Protection of Biodiversity and Fisheries Management: Issues Raised by the Relationship Between CITES and LOSC' in David Freestone, Richard Barnes and David Ong (eds.) *The Law of the Sea: Progress and Prospects* (Oxford University Press, 2006), 210

Gillespie, Alexander, *Whaling Diplomacy* (Edward Elgar Publishing, 2005)

Grynberg, Roman and Natallie Rochester, 'Expert Opinion: Fixing Cotonou's Rules of Origin Regime' in Adil Najam, Mark Halle and Ricardo Meléndez-Ortiz (eds.) *Trade and Environment: A Resource Book* (IISD, ICTSD and the Regional and International Networking Group, 2007) 107

Guilfoyle, Douglas, *Shipping Interdiction and the Law of the Sea* (Cambridge University Press, 2009)

Hart, H. L. A., *The Concept of Law* (Oxford University Press, 1961)

Henriksen, Tore, Geir Hønneland and Are Sydnes, *Law and Politics in Ocean Governance: The UN Fish Stocks Agreement and Regional Fisheries Management Regimes* (Martinus Nijhoff Publishers, 2006)

Herr, Richard, 'The International Regulation of Patagonian Toothfish' in Olav Schram Stokke (ed.) *Governing High Seas Fisheries: The Interplay of Global and Regional Regimes* (Oxford University Press, 2001) 303

Higgins, Rosalyn, *Problems and Process: International Law and How We Use It* (Clarendon Press, 1994)

Hindley, Brian, 'What Subjects are Suitable for WTO Agreement?' in Daniel Kennedy and James Southwick (eds.) *The Political Economy of International Trade Law: Essays in Honour of Robert E. Hudec* (Cambridge University Press, 2002) 157

Hønneland, Geir, 'Recent Global Agreements on High Seas Fisheries: Potential Effects on Fishermen Compliance' in Olav Schram Stokke (ed.) *Governing*

High Seas Fisheries: The Interplay of Global and Regional Regimes (Oxford University Press, 2001) 121

Horn, Henrik and Joseph Weiler, 'European Communities – Trade Description of Sardines: Textualism and its Discontent' in Henrik Horn and Petros Mavroidis (eds.) *The WTO Case Law of 2002* (2003) 248

Howse, Robert, 'A New Device for Creating International Normativity: The WTO Technical Barriers to Trade Agreement and "International Standards"' in Christian Joerges and Ernst-Ulrich Petersmann (eds.) *Constitutionalism, Multilevel Trade Governance and Social Regulation* (Hart Publishing, 2006) 383

Jackson, John, *Sovereignty, the WTO, and Changing Fundamentals of International Law* (Cambridge University Press, 2006)

Jackson, John, William Davey and Alan Sykes, *Legal Problems of International Economic Relations* (4th edn, West Group, 2002)

Jennings, Simon, Michel Kaiser and John Reynolds, *Marine Fisheries Ecology* (Blackwell Publishing, 2001; 2006 edn)

Johnston, Douglas, *The International Law of Fisheries: A Framework for Policy-Oriented Inquiries* (New Haven Press; Martinus Nijhoff Publishers, 1985) (first published Yale University Press, 1965)

Karkkainen, Bradley, 'Information-forcing Regulation and Environmental Governance' in Gráinne de Búrca and Joanne Scott (eds.) *Law and New Governance in the EU and the US* (Hart Publishing, 2006) 293

Kaye, Stuart, *International Fisheries Management* (Kluwer Law International, 2001)

Keck, Margaret E and Kathryn Sikkink, *Activists Beyond Borders: Advocacy Networks in International Politics* (Cornell University Press, 1998)

Keohane, Robert, 'Global Governance and Democratic Accountability' in David Held and Mathias Koenig-Archibugi (eds.) *Taming Globalization* (Polity Press, 2003) 130

Keohane, Robert and Joseph Nye, 'The Club Model of Multilateral Cooperation and Problems of Democratic Legitimacy' in Roger Porter, Pierre Sauve, Arvind Subramanian and Americo Beviglia Zampetti (eds.) *Efficiency, Equity, and Legitimacy: The Multilateral Trading System at the Millennium* (Brookings Institution Press, 2001) 264

Keohane, Robert, Peter Haas and Marc Levy, 'The Effectiveness of International Environmental Institutions' in Peter Haas, Robert Keohane and Marc Levy (eds.) *Institutions for the Earth: Sources of Effective International Environmental Protection* (MIT Press, 1993) 3

Klein, Natalie, *Dispute Settlement in the UN Convention on the Law of the Sea* (Cambridge University Press, 2005)

Koskenniemi, Martti, *From Apology to Utopia: The Structure of International Legal Argument* (Cambridge University Press reissue, 2006)

The Gentle Civilizer of Nations: The Rise and Fall of International Law, 1870–1960 (Cambridge University Press, 2002)

Krasner, Stephen, 'Structural Causes and Regime Consequences: Regimes as Intervening Variables' in Stephen Krasner (ed.) *International Regimes* (Cornell University Press, 1983) 2

Kuijper, Pieter Jan, 'Some Institutional Issues Presently Before the WTO' in Daniel Kennedy and James Southwick (eds.) *The Political Economy of International Trade Law: Essays in Honour of Robert E Hudec* (Cambridge University Press, 2002) 81

La Fayette, Louise de, 'The Role of the United Nations in International Oceans Governance' in David Freestone, Richard Barnes and David Ong (eds.) *The Law of the Sea: Progress and Prospects* (Oxford University Press, 2006) 63

Langille, Brian, 'General Reflections on the Relationship of Trade and Labor (Or: Fair Trade Is Free Trade's Destiny)' in Jagdish Bhagwati and Robert E Hudec, *Fair Trade and Harmonization: Prerequisites for Free Trade?* (Volume 2: Legal Analysis) (MIT Press, 1996) 231

Latour, Bruno, *Reassembling the Social: An Introduction to Actor-Network-Theory* (Oxford University Press, 2005)

Lauterpacht, Hersh, *The Development of International Law by the International Court* (Stevens, 1958)

The Function of Law in the International Community (The Clarendon Press, 1933)

Levering, Ralph B and Miriam L Levering, *Citizen Action for Global Change: The Neptune Group and the Law of the Sea* (Syracuse University Press, 1999)

Lindblom, Anna-Karin *Non-Governmental Organisations in International Law* (Cambridge University Press, 2005)

Marks, Susan, *The Riddle of All Constitutions: International Law, Democracy, and the Critique of Ideology* (Oxford University Press, 2003)

Matsushita, Mitsuo, Thomas Schoenbaum and Petros Mavroidis, *The World Trade Organization: Law, Practice, and Policy* (Oxford University Press, 2nd edn, 2006)

Mavroidis, Petros, '*Amicus curiae* Briefs Before the WTO: Much Ado About Nothing' in Armin von Bogdandy, Petros Mavroidis and Yves Meny (eds.) *European Integration and International Co-ordination, Studies in Transnational Economic Law in Honour of Claus-Dieter Ehlermann* (Kluwer, 2002) 317

The General Agreement on Tariffs and Trade: a Commentary (Oxford University Press, 2005)

McDorman, Ted, 'Fisheries Conservation and Management and International Trade Law' in Ellen Hey (ed.) *Developments in International Fisheries Law* (Kluwer Law International, 1999) 501

Morgenthau, Hans, *Politics Among Nations* (2nd edn, Knopp, 1954)

Motaal, Doaa Abdel, 'The Trade and Environment Policy Formulation Process' in Adil Najam, Mark Halle and Ricardo Meléndez-Ortiz (eds.) *Trade and Environment: A Resource Book* (IISD, ICTSD and the Regional and International Networking Group, 2007) 17

Nanz, Patrizia, 'Democratic Legitimacy and Constitutionalisation of Transnational Trade Governance: A View from Political Theory' in Christian Joerges and Ernst-Ulrich Petersmann (eds.) *Constitutionalism, Multilevel Trade Governance and Social Regulation* (Hart Publishing, 2006) 60

Nussbaum, Martha, *Sex and Social Justice* (Oxford University Press, 1999)

Women and Human Development: The Capabilities Approach (Cambridge University Press, 2000)
Oberthür, Sebastian and Thomas Gehring (eds.), *Institutional Interaction in Global Environmental Governance: Synergy and Conflict Among International and EU Policies* (MIT Press, 2006)
Ostrom, Elinor, *Governing the Commons: The Evolution of Institutions for Collective Action* (Cambridge University Press, 1990)
Ostry, Sylvia, 'World Trade Organization: Institutional Design for Better Governance' in Roger B Porter, Pierre Sauve, Arvind Subramanian and Americo Beviglia Zampetti (eds.) *Efficiency, Equity, and Legitimacy: The Multilateral Trading System at the Millennium* (Brookings Institution Press, 2001) 361
Pauwelyn, Joost, *Conflict of Norms in Public International Law: How WTO Law Relates to Other Rules of International Law* (Cambridge University Press, 2003)
Perez, Oren, *Ecological Sensitivity and Global Legal Pluralism: Rethinking the Trade and Environment Conflict* (Hart Publishing, 2004)
Peterson, M J, 'International Fisheries Management' in Peter Haas, Robert Keohane and Marc Levy (eds.) *Institutions for the Earth: Sources of Effective International Environmental Protection* (MIT Press, 1993) 249
Qureshi, Asif, *Interpreting WTO Agreements: Problems and Perspectives* (Cambridge University Press, 2006)
Rayfuse, Rosemary, 'The Interrelationship between the Global Instruments of International Fisheries Law' in Ellen Hey (ed.) *Developments in International Fisheries Law* (Kluwer Law International, 1999) 107
Sand, Inger-Johanne, 'Polycontextuality as an Alternative to Constitutionalism' in Christian Joerges, Inger-Johanne Sand and Gunther Teubner (eds.) *Transnational Governance and Constitutionalism* (Hart Publishing, 2004) 41
Sands, Philippe, *Principles of International Environmental Law* (2nd edition, Cambridge University Press, 2003)
Sarooshi, Dan, *International Organizations and their Exercise of Sovereign Powers* (Oxford University Press, 2005)
Schermers, Henry and Niels Blokker, *International Institutional Law* (4th edn, Martinus Nijhoff Publishers, 2003)
Scott, Joanne, *The WTO Agreement on Sanitary and Phytosanitary Measures: A Commentary* (Oxford University Press, 2007)
Selznick, Philip, 'Self-Regulation and the Theory of Institutions' in Gunther Teubner, Lindsay Farmer and Declan Murphy (eds.) *Environmental Law and Ecological Responsibility: The Concept and Practice of Ecological Self-Organization* (John Wiley & Sons, 1994) 395
Sen, Amartya, *Commodities and Capabilities* (Oxford University Press, 1985)
  Development as Freedom (Oxford University Press, 1999)
Shaffer, Gregory, *Defending Interests: Public – Private Partnerships in WTO Litigation* (Brookings Institution Press, 2003)
Shaw, Malcolm, *International Law* (4th edn, Cambridge University Press, reprinted 2001)

Shepel, Harm, *The Constitution of Private Governance: Product Standards in the Regulation of Integrating Markets* (Hart Publishing, 2005)
Singer, Peter, *Animal Liberation: A New Ethics for our Treatment of Animals* (Cape, 1976)
Slaughter, Anne-Marie, *A New World Order* (Princeton University Press, 2005)
Snyder, Francis, 'Thinking about "Interests": Legislative Process in the European Community' in June Starr and Jane Collier (eds.) *History and Power in the Study of Law: New Directions in Legal Anthropology* (Cornell University Press, 1989) 168
Stiglitz, Joseph, *Globalization and its Discontents* (Norton, 2002)
Stokke, Olav Schram and Clare Coffey, 'Institutional Interplay and Responsible Fisheries: Combating Subsidies, Developing Precaution' in Sebastian Oberthür and Thomas Gehring, *Institutional Interaction in Global Environmental Governance* (MIT Press, 2006) 127
Stokke, Olav Schram (ed.), *Governing High Seas Fisheries: The Interplay of Global and Regional Regimes* (Oxford University Press, 2001)
Stone, Christopher, 'Should Trees Have Standing? Toward Legal Rights for Natural Objects' in Christopher Stone, *Should Trees Have Standing? And Other Essays on Law, Morals, and the Environment* (Oceana Publications, 1996) 1 (first published *Southern California Law Review* 1972)
  'Should We Establish A Guardian For Future Generations?' in Christopher Stone, *Should Trees Have Standing? And Other Essays on Law, Morals, and the Environment* (Oceana Publications, 1996), 65
  The Gnat is Older than Man: Global Environment and Human Agenda (Princeton University Press, 1993)
  'Too Many Fishing Boats, Too Few Fish: Can Trade Laws Trim Subsidies and Restore the Balance in Global Fisheries?' in Keven P Gallagher and Jacob Werksman (eds.) *International Trade and Sustainable Development* (Earthscan Publications Ltd, 2002) 286
Toope, Stephen, 'Emerging Patterns of Governance and International Law' in Michael Byers (ed.) *The Role of Law in International Politics: Essays in International Relations and International Law* (Oxford University Press, 2000) 91
Torgerson, Douglas, 'Limits of the Administrative Mind: The Problem of Defining Environmental Problems' in John Dryzek and David Schlosberg (eds.) *Debating the Earth* (Oxford University Press, 1999) 110
Trubek, Louise, 'New Governance Practices in US Health Care' in Gráinne de Búrca and Joanne Scott (eds.) *Law and New Governance in the EU and the US* (Hart Publishing, 2006) 245
Victor, David, Kal Raustiala and Eugene Skolnikoff, 'Introduction and Overview' in David Victor, Kal Raustiala and Eugene Skolnikoff (eds.) *The Implementation and Effectiveness of International Environmental Commitments: Theory and Practice* (International Institute for Applied Systems Analysis, 1998) 1
Vicuña, Francisco Orrego, *The Changing International Law of High Seas Fisheries* (Cambridge University Press, 1999)
Warner, Robin, *Protecting the Oceans Beyond National Jurisdiction: Strengthening the International Law Framework* (Martinus Nijhoff Publishers, 2009)

Watts, Arthur (ed.) *The International Law Commission 1949–1998* (Oxford University Press, 1999)

Young, Margaret A., *Regime Interaction in International Law: Facing Fragmentation* (Cambridge University Press, forthcoming)

Young, Oran, *International Cooperation: Building Regimes for Natural Resources and the Environment* (Cornell University Press, 1989)

ARTICLES

Abbott, Frederick, 'Distributed Governance at the WTO-WIPO: An Evolving Model for Open-Architecture Integrated Governance' (2000) 3 *JIEL* 63

Abi-Saab, Georges, 'Whither the International Community?' (1997) 9 *EJIL* 248

Alston, Philip, 'Resisting the Merger and Acquisition of Human Rights by Trade Law: A Reply to Petersmann' (2002) 13 *EJIL* 815

Alvarez, José, 'The New Treaty Makers' (2002) 25 *Boston College International and Comparative Law Journal* 213

Anonymous, 'Rethinking Regulation: Negotiation as an Alternative to Traditional Rulemaking' (1981) 94 *Harvard Law Review* 1871

Aust, Anthony, 'The Theory and Practice of Informal International Instruments' (1986) 35 *ICLQ* 787

Bagwell, Kyle, Petros Mavroidis and Robert Staiger, 'It's a Question of Market Access' (2002) 96 *AJIL* 56

Bartels, Lorand, 'Applicable Law in WTO Dispute Settlement Proceedings' (2001) 35 *JWT* 499

Benvenisti, Eyal, 'Exit and Voice in the Age of Globalization' (1999) 98 *Michigan Law Review* 167

   'The Interplay between Actors as a Determinant of the Evolution of Administrative Law in International Institutions' (2006) 68 *Law and Contemporary Problems* 319

Bhagwati, Jagdish, 'Afterword: The Question of Linkage' (2002) 96 *AJIL* 126

Bigdeli, Sadeq Z, 'Will the "Friends of Climate" Emerge in the WTO? The Prospects of Applying the "Fisheries Subsidies" Model to Energy Subsidies' (2008) 1 *Carbon & Climate Law Review* 78

Boyle, Alan, 'The Southern Bluefin Tuna Arbitration' (2001) 50 *ICLQ* 447

Bronckers, Marco, 'More Power to the WTO?' (2001) 4 *JIEL* 41

Brunnée, Jutta and Stephen Toope, 'International Law and Constructivism: Elements of an Interactional Theory of International Law' (2000) *Columbia Journal of Transnational Law* 19

Buckingham, Donald and Peter Phillips, 'Hot Potato, Hot Potato: Regulating Products of Biotechnology by the International Community' (2001) 35 *JWT* 1

Buzbee, William, 'Recognizing the Regulatory Commons: A Theory of Regulatory Gaps' (2003) 89 *Iowa Law Review* 1

Carmody, Chios, 'WTO Obligations as Collective' (2006) 17 *EJIL* 419

Carr, Christopher and Harry Scheiber, 'Dealing with a Resource Crisis: Regulatory Regimes for Managing the World's Marine Fisheries' (2002) 21 *Stanford Environmental Law Journal* 45

Cavallero, Eric, 'Federative Global Democracy' (2009) 40 *Metaphilosophy* 42

Chang, Seung Wha, 'WTO Disciplines on Fisheries Subsidies: A Historic Step Towards Sustainability?' (2003) 6 *JIEL* 879

Charnovitz, Steve, '*Belgian Family Allowances* and the Challenge of Origin-Based Discrimination' (2005) 4 *World Trade Review* 7
  'Nongovernmental Organizations and International Law' (2006) 100 *AJIL* 348
  'The WTO's Environment Progress' (2007) 10 *JIEL* 685
  'Triangulating the World Trade Organization' (2002) 96 *AJIL* 28
  'WTO Cosmopolitics' (2002) 34 *NYU Journal of International Law and Politics* 299

Chaytor, Beatrice, 'An Examination of Trade and Environmental Policy Responses to the Challenge of Fisheries Subsidies' (1999) 8 *RECIEL* 275

Cho, Sungjoon, 'Linkage of Free Trade and Social Regulation: Moving beyond the Entropic Dilemma' (2005) 5 *Chicago Journal of International Law* 625

Churchill, Robin and Joanne Scott, 'The Mox Plant Litigation: The First Half-Life' (2004) 53 *ICLQ* 643

Churchill, Robin and Geir Ulfstein, 'Autonomous Institutional Arrangements in Multilateral Environmental Agreements: A Little-Noticed Phenomenon in International Law' (2000) 94 *AJIL* 623

Cohen, Joshua and Charles Sabel, 'Global Democracy?' (2005) 37 *NYU Journal of International Law and Politics* 763

Cooney, Rosie, 'CITES and the CBD: Tensions and Synergies' (2001) 10 *RECIEL* 259

Cooney, Rosie and Andrew Lang, 'Taking Uncertainty Seriously: Adaptive Governance and International Trade' (2007) 18 *EJIL* 523

Currie, Duncan E J and Kateryna Wowk, 'Climate Change and CO_2 in the Oceans and Global Oceans Governance' (2009) 4 *Carbon and Climate Law Review* 387

Curtin, Deirdre, 'Private Interest Representation or Civil Society Deliberation? A Contemporary Dilemma for European Union Governance' (2003) 12 *Social and Legal Studies* 55

Davey, William, 'The Case for a WTO Permanent Panel Body' (2003) 6 *JIEL* 177

Delbrück, Jost, 'The Earl A Snyder Lecture in International Law: Transnational Federalism: Problems and Prospects of Allocating Public Authority Beyond the State' (2004) 11 *Indiana Journal of Global Legal Studies* 31

Dillon, Sara, 'A Farewell to "Linkage": International Trade Law and Global Sustainability Indicators' (2002) 55 *Rutgers Law Review* 87

Dorf, Michael and Charles Sabel, 'A Constitution of Democratic Experimentalism' (1998) 98 *Columbia Law Review* 267

Dunoff, Jeffrey, 'Border Patrol at the World Trade Organization' (1998) 9 *Yearbook of International Environmental Law* 20
  'Does Globalization Advance Human Rights?' (1999) 25 *Brooklyn Journal of International Law* 125
  'Rethinking International Trade' (1998) 19 *University of Pennsylvania Journal of International Economic Law* 347

'The Death of the Trade Regime' (1999) 10 *EJIL* 733

Eckersley, Robyn, 'A Green Public Sphere in the WTO?: The Amicus Curiae Interventions in the Transatlantic Biotech Dispute' (2007) 13 *European Journal of International Relations* 329

Edwards, Holly, 'When Predators Become Prey: The Need for International Shark Conservation' (2007) 12 *Ocean and Coastal Law Journal* 305

Esty, Daniel 'Stepping Up to the Global Environmental Challenge' (1996–97) 8 *Fordham Environmental Law Journal* 103

Fischer-Lescano, Andreas and Gunther Teubner, 'Regime-collisions: The Vain Search for Legal Unity in the Fragmentation of Global Law' (2004) 25 *Michigan Journal of International Law* 999

Franckx, Erik, '*Pacta Tertiis* and the Agreement for the Implementation of the Straddling and Highly Migratory Fish Stocks Provisions of the UNCLOS' (2000) 8 *Tulane Journal of International and Comparative Law* 49

French, Duncan 'Treaty Interpretation and the Incorporation of Extraneous Legal Rules' (2006) 55 *ICLQ* 281

Gazzini, Tarcisio, 'Can Authoritative Interpretation under Article IX:2 of the Agreement Establishing the WTO Modify the Rights and Obligations of Members?' (2008) 57 *ICLQ* 169

Gillespie, Alexander, 'Forum Shopping in International Environmental Law: the IWC, CITES, and the Management of Cetaceans' (2002) 33 *Ocean Development and International Law* 17

Goote, Maas, 'Convention on Biological Diversity: The Jakarta Mandate on Marine Coastal Biological Diversity' (1997) 12 *International Journal of Marine and Coastal Law* 377

Goodin, Robert E, 'Enfranchising All Affected Interests, and Its Alternatives' (2007) 35 *Philosophy & Public Affairs* 1

Grynberg, Roman and Natallie Rochester, 'The Emerging Architecture of a World Trade Organization Fisheries Subsidies Agreement and the Interests of Developing Coastal States' (2005) 39 *JWT* 503

Guzman, Andrew, 'Global Governance and the WTO' (2004) 45 *Harvard International Law Journal* 303

Haas, Peter, 'Addressing the Global Governance Deficit' (2004) 4 *Global Environmental Politics* 1

'Introduction: Epistemic Communities and International Policy Coordination' (1992) 46 *International Organization* 1

Hallum, Victoria, 'International Tribunal for the Law of the Sea: The *Mox Nuclear Plant Case*' (2002) 11 *RECIEL* 372

Hardin, Garrett, 'The Tragedy of the Commons' (1968) *Science* 1243

Heerey, Justice Peter, 'Recent Australian Developments' (2004) 23 *Civil Justice Quarterly* 386

Henriksen, Tore, 'Revisiting the Freedom of Fishing and Legal Obligations on States Not Party to Regional Fisheries Management Organizations' (2009) 40 *Ocean Development and International Law* 80

Hershkoff, Helen and Benedict Kingsbury, 'Crisis, Community, and Courts in Network Governance: A Response to Liebman and Sabel's Approach to Reform of Public Education' (2003) 28 *NYU Review of Law & Social Change* 319

Higgins, Rosalyn, 'Time and the Law: International Perspectives on an Old Problem' (1997) 46 *ICLQ* 501

Hocking, Brian, 'Changing the Terms of Trade Policy Making: From the "Club" to the "Multistakeholder" Model' (2004) 3 *World Trade Review* 3

Hollis, Duncan, 'Why State Consent Still Matters – Non-State Actors, Treaties, and the Changing Sources of International Law' (2005) 23 *Berkeley Journal of International Law* 1

Horn, Henrik and Petros Mavroidis, 'Still Hazy after All These Years: The Interpretation of National Treatment in the GATT/WTO Case Law on Tax Discrimination' (2004) 15 *EJIL* 39

Howse, Robert, 'From Politics to Technocracy – and Back Again: The Fate of the Multilateral Trading Regime' (2002) 96 *AJIL* 94

 'Human Rights in the WTO: Whose Rights, What Humanity? Comment on Petersmann' (2002) 13 *EJIL* 651

 'Membership and its Privileges: The WTO, Civil Society, and the *Amicus* Brief Controversy' (2003) 9 *European Law Journal* 496

 'The Sardines Panel and AB Rulings – Some Preliminary Reactions' (2002) 29(3) *Legal Issues of Economic Integration* 247

Hsieh, Chih-hao *et al.*, 'Fishing Elevates Variability in the Abundance of Exploited Species' (2006) 443 *Nature* 859

Jackson, John, 'Sovereignty Modern: A New Approach to an Outdated Concept' (2003) 97 *AJIL* 782

Jenks, Wilfred, 'Conflict of Law-Making Treaties' (1953) 30 *BYIL* 401

Joyner, Christopher C, 'Biodiversity in the Marine Environment: Resource Implications for the Law of the Sea' (1995) 28 *Vanderbilt Journal of Transnational Law* 635

Karkkainen, Bradley, 'Post-Sovereign Environmental Governance' (2004) 4 *Global Environmental Politics* 72

Kelly, Claire, 'Power, Linkage and Accommodation: The WTO as an International Actor and Its Influence on Other Actors and Regimes' (2006) 24 *Berkeley Journal of International Law* 79

 'The Value Vacuum: Self-enforcing Regimes and the Dilution of the Normative Feedback Loop' (2001) 22 *Michigan Journal of International Law* 673

Kelsen, Hans, 'Sovereignty and International Law' (1960) 48 *Georgetown Law Journal* 627

Kennedy, David, 'The Mystery of Global Governance' (2008) 34 *Ohio Northern University Law Review* 827

Kingsbury, Benedict 'First Amendment Liberalism as Global Legal Architecture: Ascriptive Groups and the Problems of the Liberal NGO Model of International Civil Society' (2002) 2 *Chicago Journal of International Law* 183

 'Sovereignty and Inequality' (1998) 9 *EJIL* 599

Kingsbury, Benedict, Nico Krisch and Richard Stewart, 'The Emergence of Global Administrative Law' (2005) 68 *Law and Contemporary Problems* 15

Koh, Harold, 'Why Do Nations Obey International Law?' (1997) 106 *Yale Law Journal* 2599

Koskenniemi, Martti, 'Hierarchy in International Law: A Sketch' (1997) 8 *EJIL* 566

'International Law and Hegemony: A Reconfiguration' (2004) 17 *Cambridge Review of International Affairs* 197

'International Law: Constitutionalism, Managerialism and the Ethos of Legal Education' (2007) 1 *European Journal of Legal Studies* (online)

'The Fate of Public International Law: Between Technique and Politics' (2007) 70 *Modern Law Review* 1

Koskenniemi, Martti and Päivi Leino, 'Fragmentation of International Law? Postmodern Anxieties' (2002) 15 *Leiden Journal of International Law* 553

Krisch, Nico, 'The Pluralism of Global Administrative Law' (2006) 17 *EJIL* 247

Kysar, Douglas, 'Preferences for Processes: The Process/Product Distinction and the Regulation of Consumer Choice' (2004) 118 *Harvard Law Review* 525

Lamy, Pascal, 'The Place of the WTO and its Law in the International Legal Order' (2006) 17 *EJIL* 969

Lang, Andrew, 'Reflecting on "Linkage": Cognitive and Institutional Change in The International Trading System' (2007) 70 *Modern Law Review* 523

Leebron, David, 'Linkages' (2002) 96 *AJIL* 5

Levit, Janet Koven, 'A Bottom-Up Approach to International Lawmaking: The Tale of Three Trade Finance Instruments' (2005) 30 *Yale Journal of International Law* 125

Lindroos, Anja and Michael Mehling, 'Dispelling the Chimera of "Self-Contained Regimes": International Law and the WTO' (2006) 16 *EJIL* 857

Little, Laura and Marcos A Orellana, 'Can CITES Play a Role in Solving the Problems of IUU Fishing?: The Trouble with Patagonian Toothfish' (2004) 16 *Colorado Journal of International Environmental Law and Policy* 21

Lloyd, Peter, 'When Should New Areas of Rules be Added to the WTO?' (2005) 4 *World Trade Review* 275

Lowe, Vaughan, 'Book review: Sovereignty, the WTO and Changing Fundamentals of International Law' (2007) 101 *AJIL* 234

Martínez-Garmendia, Josué and James L Anderson, 'Conservation, Markets, and Fisheries Policy: The North Atlantic Bluefin Tuna and the Japanese Sashimi Market' (2005) *Agribusiness* 17

Marceau, Gabrielle, 'Conflict of Norms and Conflicts of Jurisdictions: The Relationship Between the WTO Agreement and MEAs and Other Treaties' (2001) 35 *JWT* 1081

Marceau, Gabrielle and Matthew Stilwell, 'Practical Suggestions for Amicus Curiae Briefs Before WTO Adjudicating Bodies' (2001) 4 *JIEL* 155

Matthews, Paul, 'Problems Related to the Convention on the International Trade in Endangered Species' (1996) 45 *ICLQ* 421

McLachlan, Campbell, 'The Principle of Systemic Integration and Article 31(3)(c) of the Vienna Convention' (2005) 54 *ICLQ* 279

McLaughlin, Richard, 'Settling Trade-Related Disputes Over the Protection of Marine Living Resources: UNCLOS or the WTO?' (1997) 10 *Georgetown International Environmental Law Review* 29

Motaal, Doaa Abdel, 'The "Multilateral Scientific Consensus" and the World Trade Organization' (2004) 38 *JWT* 855

Murphy, James B, 'Alternative Approaches to the CITES "Non-Detriment" Finding for Appendix II Species' (2006) 36 *Environmental Law* 531

Orellana, Marcos, 'The Swordfish Dispute between the EU and Chile at the ITLOS and the WTO' (2002) 71 *Nordic Journal of International Law* 55

Palmeter, David and Petros Mavroidis, 'The WTO Legal System: Sources of Law' (1998) 92 *AJIL* 398

Pauwelyn, Joost, 'A Typology of Multilateral Treaty Obligations: Are WTO Obligations Bilateral or Collective in Nature?' (2003) 14 *EJIL* 907

'How to Win a World Trade Organization Dispute Based on Non-World Trade Organization Law?' (2003) 37 *JWT* 997

'Recent Books on Trade and Environment: GATT Phantoms Still Haunt the WTO' (2004) 15 *EJIL* 575

Peraute, N C, M G Pajaro, J J Meeuwig and A C J Vincent, 'Biology of a Seahorse Species, Hippocampus Comes in the Central Philippines' (2002) 60 *Journal of Fish Biology* 821

Petersmann, Ernst-Ulrich, 'Addressing Institutional Challenges to the WTO in the New Millennium: A Longer-Term Perspective' (2005) 8 *JIEL* 647

'Taking Human Dignity, Poverty and Empowerment of Individuals More Seriously: Rejoinder to Alston' (2002) 13 *EJIL* 845

'Time for a United Nations "Global Compact" for Integrating Human Rights in the Law of Worldwide Organizations: Lessons from European Integration' (2002) 13 *EJIL* 621

Ratner, Stephen, 'Regulatory Takings in Institutional Context: Beyond the Fear of Fragmented International Law' (2008) 102 *AJIL* 475

Raustiala, Kal and David Victor, 'The Regime Complex for Plant Genetic Resources' (2004) 58 *International Organization* 277

Rayfuse, Rosemary, 'The Future of Compulsory Dispute Settlement Under The Law of the Sea Convention' (2005) 36 *Victoria University of Wellington Law Review* 683

'The United Nations Agreement on Straddling and Highly Migratory Fish Stocks as an Objective Regime: A Case of Wishful Thinking?' (1999) 20 *Australian Year Book of International Law* 253

Rayfuse, Rosemary and Robin Warner, 'Securing a Sustainable Future for the Oceans Beyond National Jurisdiction: The Legal Basis for an Integrated Cross-Sectoral Regime for High Seas Governance for the 21st Century' (2008) 23(3) *The International Journal of Marine and Coastal Law* 399

Rayfuse, Rosemary et al., 'Australia and Canada in Regional Fisheries Organizations: Implementing the United Nations Fish Stocks Agreement' (2003) 36 *The Dalhousie Law Journal* 47

Reisman, W Michael, 'International Lawmaking: A Process of Communication' (1981) 75 *American Society of International Law Proceedings* 101
Rittel, Horst and Melvin Webber, 'Dilemmas in a General Theory of Planning' (1973) 4 *Policy Sciences* 155
Roessler, Frieder, 'Comment on a WTO Permanent Panel Body – The Cobra Effects of the WTO Panel Selection Procedures' (2003) 6 *JIEL* 230
Rose, Gregory, 'Marine Biodiversity Protection through Fisheries Management – International Legal Developments' (1999) 8 *RECIEL* 284
Safrin, Sabrina, 'Treaties in Collision? The Biosafety Protocol and the World Trade Organization Agreements' (2002) 96 *AJIL* 606
Sand, Peter, 'Japan's "Research Whaling" in the Antarctic Southern Ocean and the North Pacific Ocean in the Face of the Endangered Species Convention (CITES)' (2008) 17(1) *RECIEL* 56
Sarfaty, Galit, 'Why Culture Matters in International Institutions: The Marginality of Human Rights at the World Bank' (2009) 103 *AJIL* 647
Schachter, Oscar, 'The Twilight Existence of Nonbinding International Agreements' (1977) 71 *AJIL* 296
Scheiber, Harry N, Kathryn J Mengerink and Yann-huei Song, 'Ocean Tuna Fisheries, East Asian Rivalries, and International Regulation: Japanese Policies and the Overcapacity/IUU Fishing Conundrum' (2007) 30 *University of Hawai'i Law Review* 97
Scheuerman, William, 'Democratic Experimentalism or Capitalist Synchronization? Critical Reflections on Directly-Deliberative Polyarchy' (2004) XVII *Canadian Journal of Law and Jurisprudence* 101
Scott, Joanne, 'International Trade and Environmental Governance: Relating Rules (and Standards) in the EU and the WTO' (2004) 15 *EJIL* 307
Scott, Joanne and Susan Sturm, 'Courts as Catalysts: Rethinking the Judicial Role in New Governance' (2007) 13 *Columbia Journal of European Law* 565
Shaffer, Gregory, 'The WTO Under Challenge: Democracy and the Law and Politics of the WTO's Treatment of Trade and Environment Matters' (2001) 25 *Harvard Environmental Law Review* 1
Shaffer, Gregory and Mark A Pollack, 'Hard vs Soft Law: Alternatives, Complements and Antagonists in International Governance' (2010) 94 *Minnesota Law Review* 706
Shell, G Richard, 'The Trade Stakeholders Model and Participation by Nonstate Parties in the World Trade Organization' (1996) 17 *University of Pennsylvania Journal of International Economic Law* 359
'Trade Legalism and International Relations Theory: An Analysis of the World Trade Organization' (1995) 44 *Duke Law Journal* 829
Simma, Bruno, 'Self-Contained Regimes' (1985) 16 *Netherlands Yearbook of International Law*, 111
Simma, Bruno and Andreas L Paulus, 'The "International Community": Facing the Challenge of Globalisation' (1998) 9 *EJIL* 266
Slaughter, Anne-Marie, 'International Law and International Relations Theory: A Dual Agenda' (1993) 87 *AJIL* 205

Somek, Alexander, 'Kelsen Lives' (2007) 18 *EJIL* 409
Stein, Eleanor, 'The Arc of Consequence: Time for a New Ecocriticism?' (2008) 17 *Social and Legal Studies* 123
Stein, Eric, 'International Integration and Democracy: No Love at First Sight' (2001) 95 *AJIL* 489
Stephens, Tim, 'The Limits of International Adjudication in International Environmental Law: Another Perspective on the Southern Bluefin Tuna Case' (2004) 19 *International Journal of Marine and Coastal Law* 177
Stiglitz, Joseph 'A New Agenda for Global Warming' (2006) 3:7 *The Economists' Voice* available at http://www.bepress.com/ev/vol3/iss7/art3
Sturm, Susan, 'The Architecture of Inclusion: Advancing Workplace Equity in Higher Education' (2006) 29 *Harvard Journal of Law & Gender* 247
Swanson, Timothy, 'The Evolving Trade Mechanisms in CITES' (1992) 1 *RECIEL* 57
Tarullo, Daniel, 'Logic, Myth, and the International Economic Order' (1985) 26 *Harvard International Law Journal* 533
Tietje, Christian, 'Global Governance and Inter-Agency Co-operation in International Economic Law' (2002) 26 *JWT* 501
Trachtman, Joel, 'Institutional Linkage: Transcending "Trade and ..."' (2002) 96 *AJIL* 77
  'Trade and ... Problems, Cost-Benefit Analysis and Subsidiarity' (1998) 9 *EJIL* 32
Umbricht, Georg, 'An "Amicus Curiae Brief" on Amicus Curiae Briefs at the WTO' (2001) 4 *JIEL* 773
Walker, Terence, 'Can Shark Resources be Harvested Sustainably?' (1998) 49 *Marine & Freshwater Research* 553
Weiler, Joseph, 'The Geology of International Law – Governance, Democracy and Legitimacy' (2004) 64 *ZaöRV* 547
Weiler, Joseph and Andreas Paulus, 'The Structure and Change in International Law or Is There a Hierarchy of Norms in International Law?' (1997) 8 *EJIL* 545
Widdows, Kelvin, 'What is an Agreement in International Law' (1979) 50 *BYIL* 117
Wiener, Jonathan, 'On the Political Economy of Global Environmental Regulation' (1999) 97 *Georgetown Law Journal* 749
Wiers, Jochem, 'French Ideas on Climate and Trade Policies' (2008) 1 *Carbon and Climate Law Review* 18
Witbooi, Emma, 'Governing Global Fisheries: Commons, Community Law and Third Country Coastal Waters' (2008) 17 *Social Legal Studies* 369
Wold, Chris, 'Multilateral Environment Agreements and the GATT: Conflict and Resolution?' (1996) 26 *Environmental Law* 841
  'Natural Resource Management and Conservation: Trade in Endangered Species' (2002) 13 *Yearbook of International Environmental Law* 389
Worm, Boris *et al.*, 'Impacts of Biodiversity Loss on Ocean Ecosystem Services' (2006) 314 *Science* 787
  et al., 'Rebuilding Global Fisheries' (2009) 325 *Science*
Yeater, Marceil and Juan Vasquez, 'Demystifing the Relationship Between CITES and the WTO' (2001) 10 *RECIEL* 271

Young, Margaret A., 'Fragmentation or Interaction: The WTO, Fisheries Subsidies, and International Law' (2009) 8 *World Trade Review* 477
  'Protecting Endangered Marine Species: Collaboration between the Food and Agriculture Organization and the CITES Regime' (2010) 11 *Melbourne Journal of International Law* 441
  'The WTO's Use of Relevant Rules of International Law: An Analysis of the Biotech Case' (2007) 56 *ICLQ* 907
  'WTO Undercurrents at the Court of Justice' (2005) 30 *European Law Review* 211
Young, Oran, 'The Politics of International Regime Formation: Managing Natural Resources and the Environment' (1989) 43 *International Organization* 349
Zahrnt, Valentin, 'Domestic Constituents and the Formulation of WTO Negotiating Positions: What the Delegates Say' (2008) 7 *World Trade Review* 393
Zonnekeyn, Geert A 'The Appellate Body's Communication on *Amicus Curiae* Briefs in the *Asbestos* Case: An Echternach Procession?' (2001) 35 *JWT* 553

POLICY PAPERS, NEWS SOURCES AND UNPUBLISHED WORKS

  'EU, Mauritania Renegotiate Fisheries Access Agreement' (2008) 8:5 *Bridges Trade BioRes (Trade and Biological Resources News Digest)* (20 March 2008)
  'Friends of Fish Denounce EU Aid Package', *Financial Times* (17 June 2006)
  'Shark Species Face Extinction Amid Overfishing and Appetite for Fins' *Guardian* (18 February 2008) 11
Benitah, Marc, 'Five Suggestions for Clarifying the Draft Text on Fisheries Subsidies (2008) 12:1 *Bridges Trade and Biological Resources News Digest* 21
  'Ongoing WTO Negotiations on Fisheries Subsidies' *ASIL Insight* (June 2004)
Bernasconi-Osterwalder, Nathalie, 'Interpreting WTO Law and the Relevance of Multilateral Environmental Agreements in *EC-Biotech*', Background note to presentation at BIICL WTO Conference, May 2007, available on the website of the Center for International Environmental Law (www.ciel.org)
Butler, Leah, 'Effects and Outcomes of Amicus Curiae Briefs at the WTO: An Assessment of NGO Experiences' (2006) (unpublished manuscript at http://ist-socrates.berkeley.edu/~es196/projects/2006final/butler.pdf)
Gaski, Andrea L, *An Examination of the International Trade in Bluefin Tuna with an Emphasis on the Japanese Market* (1993, TRAFFIC International)
Khor, Martin, 'Debate on Policy Space Dominates UNCTAD Review' TWN Info Service on WTO and Trade Issues (May06/10) (13 May 2006)
Knirsch, Juergen, Daniel Mittler, Martin Kaiser, Karen Sack, Christoph Thies, Larry Edwards, *Deadly Subsidies: How Government Funds are Killing Oceans and Forests and Why the CBD Rather Than the WTO Should Stop This Perverse Use of Public Money* (Greenpeace International (undated, circa 2006)
McKie, Robin, 'Push to Ban Trade in Endangered Bluefin Tuna' *The Observer* (14 February 2010)
Pfahl, Stefanie, 'Is the WTO the Only Way?' *Briefing Paper for Greenpeace International and Friends of the Earth*, (undated, circa 2005)

Schorr, David, *Healthy Fisheries, Sustainable Trade: Crafting New Rules on Fishing Subsidies in the World Trade Organization* (WWF Position Paper and Technical Resource, 2004)
  'Towards Rational Disciplines on Subsidies to the Fishery Sector: A Call for New International Rules and Mechanisms' in WWF, *The Footprint of Distant Water Fleets on World Fisheries* (1998)
Schorr, David and John Caddy, *Sustainability Criteria for Fisheries Subsidies: Options for the WTO and Beyond* (commissioned by UNEP and WWF) (2007)
Shaffer, Gregory and Victor Mosoti, 'EC Sardines: A New Model for Collaboration in Dispute Settlement?' (2002) 6:7 BRIDGES 15
Turner, Pamela S, '*Struggling to Save the Seahorse*' National Wildlife (6 January 2005)
WWF, *Hard Facts, Hidden Problems, A Review of Current Data on Fishing Subsidies* (2001) 18
WWF, TRAFFIC, IUCN, 'Regulating International Trade in Commercial Marine Fisheries Products' (February 2002). See also excerpts from *Bridges Digest* (*BRIDGES Weekly Trade News Digest*) (various)

PAPERS SUBMITTED TO OR PRODUCED BY INTERNATIONAL ORGANISATIONS

APEC, *Study into the Nature and Extent of Subsidies in the Fisheries Sector of APEC Member Economies* (2000)
CITES, Animal Committee, AC19 WG3 Doc. 1, Summary of the Report of the 19th meeting of the CITES Animal Committee (18–21 August 2003)
CBD, CoP5 Decision, 'Ecosystem Approach', Nairobi, 15–26 May 2000 (Decision V/6); CoP7 Decision, 'Ecosystem Approach', Kuala Lumpur, 9–20 February 2004 (Decision VII/11)
CITES CoP12 Doc. 16.2.1, 'Synergy and cooperation between CITES and FAO' (Japan) (30 August 2002)
CITES CoP12 Doc. 16.2.2, 'FAO collaboration with CITES through a Memorandum of Understanding' (United States) (01/10/02)
CITES CoP12 Doc. 58 (Criteria for amendment of Appendices I and II) (2002)
CITES CoP13 Doc. 12.3 ('Revision of Resolution Conf. 12.4 on Cooperation between CITES and the Commission for the Conservation of Antarctic Marine Living Resources regarding trade in toothfish [Australia]')
CITES CoP13 Doc. 57 (Criteria for amendment of Appendices I and II) (2004)
CITES CoP14 Doc.18.1 (Cooperation with the Food and Agriculture Organization of the United Nations)
CITES Doc. CoP14 Inf. 38, Information Submitted for fourteenth meeting of the Conference of the Parties, The Hague (Netherlands) 3–15 June 2007
CITES SC 54. Doc 19 'Introduction from the Sea' (15 August 2006) submitted to fifty-fourth meeting of the CITES Standing Committee (Switzerland) (2–6 October 2006)
CITES Doc SC49 Summary Report (Rev. 1), Forty-ninth meeting of the Standing Committee, Geneva, 22–25 April 2003
CITES Doc SC50-Summary Report, Fiftieth meeting of the Standing Committee (Geneva (Switzerland), 15–19 March 2004

CITES Doc SC53 Doc 10, Annex (Geneva (Switzerland), 27 June–1 July 2005)
CITES Doc SC53 Summary Report (Geneva (Switzerland), 27 June–1 July 2005)
CITES Doc SC54 Doc.8 (Geneva (Switzerland), 2–6 October 2006)
CITES Press Release, 'CITES Conference to Consider New Trade Rules for Marine, Timber and Other Wildlife Species', Geneva, May 2007 http://www.cites.org/eng/news/press/2007/0705_presskit.shtml
CITES SC49 Doc. 6.3, reproducing Notification No. 2003/030, Geneva, 6 May 2003
CITES SC49 Doc. Inf. 3 'Proposal from Japan', Informal Document submitted to forty-ninth meeting of the Standing Committee, Geneva, 22–25 April 2003
CITES SC51 Doc.8 (Bangkok, (Thailand) 1 October 2004)
European Commission, *Handbook for Trade Sustainability Impact Assessment* (2006)
FAO, *Fisheries and Aquaculture Report No. 902*, Report of the twenty-eighth session of the Committee on Fisheries, 2–6 March 2009
FAO, FAO Fisheries Circular No. 940, Rome, April 1999, Gail Lugten, 'A Review of Measures Taken by Regional Marine Fishery Bodies to Address Contemporary Fishery Issues'
FAO, *FAO Fisheries Report No. 293*, William Burke, '1982 Convention on the Law of the Sea Provisions on Conditions of Access to Fisheries Subject to National Jurisdiction', Report of the Export Consultation on the Conditions of Access to the Fish Resources of the Exclusive Economic Zone (1983) 23
FAO, *FAO Fisheries Report No. 613*, Report of the Technical Consultation on the Measurement of Fishing Capacity (FIPP/R615(En)) (FAO, 2000)
FAO, *FAO Fisheries Report No. 655*, Report of the 24th Session of the Committee on Fisheries, 26 February–2 March 2001
FAO, *FAO Fisheries Report No. 667*, Second Technical Consultation on the Suitability of the CITES Criteria for Listing Commercially-Exploited Aquatic Species, Namibia, 22–25 October 2001
FAO, *FAO Report No. 673*, Report of the eighth session of the Sub-Committee on Fish Trade, Bremen, Germany, 12–16 February 2002
FAO, *Fisheries Report No. 702*, Report of the twenty-fifth session of the Committee on Fisheries, 24–28 February 2003
FAO, *Fisheries Report No. 736*, Report of the ninth session of the Sub Committee on Fish Trade, Bremen, Germany, 10–14 February 2004
FAO, *FAO Fisheries Report No. 741*, Report of the Expert Consultation on Implementation Issues Associated with Listing Commercially-Exploited Aquatic Species on CITES Appendices, 25–28 May 2004
FAO, *Fisheries Report No. 746*, Report of the Expert Consultation on Legal Issues Related to CITES and Commercially-Exploited Aquatic Species FAO, 22–25 June 2004
FAO, *FAO Fisheries Report No. 780*, Report of the twenth-sixth session of the Committee on Fisheries, 7–11 March 2005
FAO, *Fisheries Report No. 807*, Report of the tenth session of the Sub-Committee on Fish Trade, 30 May–2 June 2006
FAO, *Fisheries Report No. 830*, Report of the twenty-seventh session of the Committee on Fisheries, 5–9 March 2007

FAO, *Fisheries Report No. 833*, Report of the second FAO Ad Hoc Expert Advisory Panel for the Assessment of Proposals to Amend Appendices I and II of CITES Concerning Commercially-Exploited Aquatic Species, 26–30 March 2007

FAO, *FAO Fisheries Technical Paper No. 389*, Stefania Vannuccini, 'Shark Utilization, Marketing and Trade' (1999) (www.fao.org/docrep/005/x3690e/x3690e00.htm)

FAO, *FAO Fisheries Technical Paper No. 433/1*, J M Ward, J E Kirkley, R Metzner and S Pascoe, 'Measuring and Assessing Capacity in Fisheries: 1. Basic Concepts and Management Options' (2004)

FAO, *FAO Fisheries Technical Paper No. 475*, 'Global Study of Shrimp Fisheries' (2008)

FAO, *FAO Fisheries Technical paper No. 495*, 'The State of World Highly Migratory, Straddling and Other High Seas Fishery Resources and Associated Species' (2006)

FAO, Guidelines for the Eco-labelling of Fish and Fish Products (2005)

FAO, Strategy for Improving Information on Status and Trends of Capture Fisheries (Strategy-STF) 2003, approved by consensus at COFI: see FAO, *Fisheries Report No. 702* (2003)

FAO, *The States of World Fisheries and Aquaculture* (2002)

FAO, *The State of World Fisheries and Aquaculture* (2004)

FAO, *The State of World Fisheries and Aquaculture* (2006)

FAO, *The State of World Fisheries and Aquaculture* (2008)

ILC, Report of the Study Group, Fragmentation of International Law: Difficulties arising from the Diversification and Expansion of International Law: Conclusions (A/CN.4/L.702) (18 July 2006)

ILC, Report of the Study Group, Fragmentation of International Law: Difficulties arising from the Diversification and Expansion of International Law: Analytical Study finalised by the Chairman (A/CN.4/L.682 and Corr.1) (13 April 2006)

IUCN, Workshop in High Seas Governance for the 21st Century, Co-Chairs Summary Report, December 2007 (2008)

Nuclear Supplier Group (NSG) Guidelines for Transfers of Nuclear-Related Dual-Use Equipment, Materials, Software and Related Technology (INFCIRC/254, Part 2)

OECD, *Subsidies: A Way Towards Sustainable Fisheries?* Policy Brief (December 2005)

UNCTAD Doc, Eleventh Session, Sao Paulo, 13–18 June 2004, TD/410 (24 June 2004) 'Sao Paulo Consensus'

UN Doc A/55/274, 'Report of the Work of the United Nations Open-ended Informal Consultative Process on Oceans and the Law of the Sea at its First Meeting' (31 July 2000)

UN Doc A/57/80, 'Report of the Work of the United Nations Open-ended Informal Consultative Process on Oceans and the Law of the Sea at its Third Meeting' (2 July 2002)

UN Doc A/58/95, 'Report of the Work of the United Nations Open-ended Informal Consultative Process on Oceans and the Law of the Sea' (26 July 2003)
UN Doc A/59/122, 'Report of the work of the United Nations Open-ended Informal Consultative Process on Oceans and the Law of the Sea at its fifth meeting' (1 July 2004)
UN Doc A/60/99, 'Report of the Work of the United Nations Open-ended Informal Consultative Process on Oceans and the Law of the Sea at its Sixth Meeting' (7 July 2005)
UN Doc A/61/156, 'Report of the Work of the United Nations Open-ended Informal Consultative Process on Oceans and the Law of the Sea at its Seventh Meeting' (17 July 2006)
UN Doc A/62/169, 'Report of the Work of the United Nations Open-ended Informal Consultative Process on Oceans and the Law of the Sea at its Eighth Meeting' (30 July 2007)
UN Doc A/63/174, 'Report of the Work of the United Nations Open-ended Informal Consultative Process on Oceans and the Law of the Sea at its Ninth Meeting' (25 July 2008)
UN Doc A/64/131, 'Report of the Work of the United Nations Open-ended Informal Consultative Process on Oceans and the Law of the Sea at its Tenth Meeting' (13 July 2009)
UN Doc UNEP/CBD/ExCOP/1/2, Final Report of the 6th Ordinary Meeting of the Open-Ended Ad Hoc Working Group on Biosafety, Cartagena, Colombia (15 February 1999)
United Nations Department of Public Information, 'Special Session of the General Assembly to Review and Appraise the Implementation of Agenda 21' (23–27 June 1997) www.un.org/ecosocdev/geninfo/sustdev/fishery.htm
WHO, Report of the Committee of Experts on Tobacco Industry Documents, Tobacco Company Strategies to Undermine Tobacco Control Activities at the World Health Organization (July 2000)
World Bank and the FAO, *The Sunken Billions: The Economic Justifications for Fisheries Reform* (2008)
WTO, Consultative Board, *The Future of the WTO: Addressing Institutional Challenges in the New Millennium* (WTO, 2004) (Report to the Director-General of the WTO by Peter Sutherland, Jagdish Bhagwati, Kwesi Botchwey, Niall FitzGerald, Koichi Hamada, John Jackson, Celso Lafer and Thierry de Montbrial)
WTO Doc. G/SPS/R/47 'Information Management System' – Note by the Secretariat, 'Workshop on Transparency 15–16 October 2007' (8 January 2008)
WTO Doc. G/TBT/1/Rev. 8, 'Decision of the Committee on Principles for the Development of International Standards, Guides and Recommendations with relation to Articles 2, 5 and Annex 3 of the Agreement' (23 May 2002)
WTO Doc. G/TBT/M/43, 'Committee on Technical Barriers to Trade – Minutes of the Meeting of 9 November 2007 – Note by the Secretariat' (21 January 2008)
WTO Doc. TN/DS/20 'Report by the Chairman, Ambassador Ronald Saborío Soto, to the Trade Negotiations Committee' (26 July 2007)

338   BIBLIOGRAPHY

WTO Doc. TN/DS/W/38 'Communication from the European Communities Dispute Settlement Body Special Session' (23 January 2003)

WTO Doc. TN/MA/W/35, Negotiating Group on Market Access, 'Draft Elements of Modalities for Negotiations on Non-Agricultural Products' (16 May 2003)

WTO Doc. TN/RL/1 'Negotiating Group on Rules – Meeting Held on 11 March 2002 – Report by the Chairman to the Trade Negotiating Committee' (11 April 2002)

WTO Doc. TN/RL/GEN/36 'Fisheries Subsidies to Management Services' (New Zealand) (22 March 2005)

WTO Doc. TN/RL/GEN/41 'Communication from United States to the Rules Group' (13 May 2005)

WTO Doc. TN/RL/GEN/57/Rev.2 'WTO Fisheries Subsidies Disciplines: Architecture on Fisheries Subsidies Disciplines' (Antigua and Barbuda, Barbados, Dominican Republic, Fiji, Grenada, Guyana, Jamaica, Papua New Guinea, St Kitts and Nevis, St Lucia, Solomon Islands and Trinidad and Tobago) (13 September 2005)

WTO Doc. TN/RL/GEN/79 'Negotiating Group on Rules – Further Contribution to the Discussion on the Framework for Disciplines on Fisheries Subsidies – Paper from Brazil' (16 November 2005); see also TN/RL/GEN/79R3 (2 June 2006)

WTO Doc. TN/RL/GEN/92 'Negotiating Group on Rules – Fisheries Subsidies: Small-Scale Fisheries – Paper from the Republic of Korea' (18 November 2005)

WTO Doc. TN/RL/GEN/114 'Fisheries Subsidies: Framework for Disciplines' (Japan, Korea, Separate Customs Territory of Taiwan, Penghu, Kinmen and Matsu) (21 April 2006); see also WTO Doc. TN/RL/GEN/114R1 (2 June 2006)

WTO Doc. TN/RL/GEN/134 'Negotiating Group on Rules – Fisheries Subsidies – Submission of the European Communities' (24 April 2006)

WTO Doc. TN/RL/GEN/138 'Negotiating Group on Rules – Fisheries Subsidies: Special and Differential Treatment – Paper from Argentina' (1 June 2006); Rev1 (26 January 2007)

WTO Doc. TN/RL/GEN/145 'Negotiating Group on Rules – Fisheries Subsidies: Proposed New Disciplines – Proposal from the United States' (22 March 2007)

WTO Doc. TN/RL/M/1 'Note by the Secretariat, Summary Report of the Meeting Held on 11 March 2002' (8 April 2002)

WTO Doc. TN/RL/M/2 'Note by the Secretariat, Summary Report of the Meeting held on 6 and 8 May 2002' (11 June 2002)

WTO Doc. TN/RL/M/7 'Note by the Secretariat, Summary Report of the Meeting held on 19–21 March 2003' (11 April 2003)

WTO Doc. TN/RL/M/8 'Summary Report of the Meeting held on 5–7 May 2003' (8 June 2003)

WTO Doc. TN/RL/M/10 'Note by the Secretariat, Summary Report of the Meeting Held on 18-19 June 1003' (17 June 2003)
WTO Doc. TN/RL/M/17 'Note by the Secretariat, Summary Report of the Meeting held on 7 & 8 June 2004' (31 August 2004)
WTO Doc. TN/RL/M/18 'Note by the Secretariat, Summary Report of the Meeting held on 28 September 2004' (15 October 2004)
WTO Doc. TN/RL/W/3 'Submission from Australia, Chile, Ecuador, Iceland, New Zealand, Peru, Philippines and the United States: The Doha Mandate to Address Fisheries Subsidies: Issues' (24 April 2002)
WTO Doc. TN/RL/W/11 'Negotiating Group on Rules – Japan's Basic Position on the Fisheries Subsidies Issue' (2 July 2007)
WTO Doc. TN/RL/W12 'Submission from New Zealand: Fisheries Subsidies: Limitations of Existing Subsidy Disciplines' (4 July 2002)
WTO Doc. TN/RL/W/17 'Negotiating Group on Rules – Korea's Views on the Doha Development Agenda Discussions on Fisheries Subsidies' (2 October 2002)
WTO Doc. TN/RL/W/21 'Negotiating Group on Rules – Adverse Trade and Conservation Effects of Fisheries Subsidies – Communication from the United States' (15 October 2002)
WTO Doc. TN/RL/W/52 'Negotiating Group on Rules – Japan's Contribution to Discussion on Fisheries Subsidies Issue' (6 February 2003)
WTO Doc. TN/RL/W/58 'Negotiating Group on Rules – Subsidies in the Fisheries Sector: Possible Categorizations – Submission from Argentina, Chile, Iceland, New Zealand, Norway, and Peru' (10 February 2003)
WTO Doc. TN/RL/W/77 'Communication from the United States, Possible Approaches to Improved Disciplines on Fisheries Subsidies' (19 March 2003)
WTO Doc. TN/RL/W/69 'Negotiating Group on Rules – Korea's Views on the Suggested Categorization of Fishery Subsidies' (18 March 2003)
WTO Doc. TN/RL/W/97 (Negotiating Group on Rules – Korea's Comments on the Submission from the United States (TN/RL/W/77) – Submission from Korea' (5 May 2003)
WTO Doc. TN/RL/W/115 'Possible Approaches to Improved Disciplines on Fisheries Subsidies – Communication from Chile' (10 June 2003)
WTO Doc. TN/RL/W/136 'Fisheries Subsidies' (Antigua and Barbuda, Belize, Fiji Islands, Guyana, the Maldives, Papua New Guinea, Solomon Islands, St Kitts and Nevis) (14 July 2003)
WTO Doc. TN/RL/W/154 'Communication from New Zealand, Fisheries Subsidies: Overcapacity and Overexploitation' (26 April 2004)
WTO Doc. TN/RL/W/159 'Fisheries Subsidies: Proposed Structure of the Discussion' (Japan) (7 June 2004)
WTO Doc. TN/RL/W/160 'Negotiating Group on Rules – Questions and Comments from Korea on New Zealand's Communication on Fisheries Subsidies (TN/RL/W/154)' (Korea) (8 June 2004)
WTO Doc. TN/RL/W/164 'Paper by Japan' (27 September 2004)

WTO Doc. TN/RL/W/166 'Communication from Argentina, Chile, Ecuador, New Zealand, Philippines, Peru, Fisheries Subsidies' (2 November 2004)

WTO Doc. TN/RL/W/169 'Negotiating Group on Rules – Additional Views on the Structure of the Fisheries Subsidies Negotiation – Communication from the United States' (13 December 2004), TN/RL/W/196 (22 November 2005)

WTO Doc. TN/RL/W/172 'Contribution to the Discussion of the Framework for the Disciplines of the Fisheries Subsidies' (Japan, Korea, the Separate Customs Territory of Taiwan, Penghu, Kinmen and Matsu) (22 February 2005)

WTO Doc. TN/RL/W/197 'Note by the Secretariat, Definitions Related to Artisanal, Small-scale and Subsistence Fishing' (24 November 2005)

WTO Doc. TN/RL/W/195 'Note from the Chairman' (22 November 2005)

WTO Doc. TN/RL/W/207 'Negotiating Group on Rules – Promoting Development and Sustainability in Fisheries Subsidies Disciplines: an Informal Dialogue on Select Technical Issues – Communication from New Zealand' (2 April 2007)

WTO Doc. TN/RL/W/213 'Draft Consolidated Chair Texts of the AD and SCM Agreements' (30 November 2007)

WTO Doc. TN/RL/W/232 'Working Document from the Chairman' (Annex C – Fisheries Subsidies) (28 May 2008)

WTO Doc. TN/RL/W/236 'New Draft Consolidated Chair Texts of the AD and SCM Agreements' (19 December 2008)

WTO Doc. TN/TE/INF/9/Rev.1 'Statement by the Secretariat of the Convention on Biological Diversity (CBD) at the Committee on Trade and Environment Special Session of 12–13 October 2004 – Paragraph 31 (i)' (13 October 2004)

WTO Doc. TN/TE/R/4 'Committee on Trade and Environment – Special Session – Summary Report on the MEA Information Session on Paragraph 31(ii) of the Doha Declaration – 12 November 2002 – Note by the Secretariat' (21 January 2003)

WTO Doc. TN/TE/R/5 'Committee on Trade and Environment – Special Session – Summary Report on the Fifth Meeting of the Committee on Trade and Environment in Special Session – 12–13 February 2003 – Note by the Secretariat' (14 April 2003)

WTO Doc. TN/TE/S/2/Rev.2 'Note by the Secretariat: Existing Forms of Cooperation and Information Exchange between UNEP/MEAs and the WTO' (16 January 2007)

WTO Doc. TN/TE/W/28 'Committee on Trade and Environment – Special Session – Paragraph 31(i) of the Doha Declaration – Specific Trade Obligations set out in MEAs – Implementation of CITES in Hong Kong, China (30 April 2003)

WTO Doc. TN/TE/W/39 'The Relationship between WTO Rules and MEAS in the Context of the Global Governance System – Submission by the EC' (24 March 2004)

WTO Doc. TN/TE/W/40 'Committee on Trade and Environment – Special Session – Sub-Paragraph 31(i) of the Doha Declaration – Submission by the United States' (21 June 2004)

WTO Doc. TN/TE/W/45 'Committee on Trade and Environment – Special Session – Paragraph 31(i) of the Doha Declaration – Australia's Experience – Submission by Australia' (12 October 2004)

WTO Doc. TN/TE/W/53 'Committee on Trade and Environment – Special Session – Putting MEA/WTO Governance into Practice: The EC's Experience in the Negotiation and Implementation of MEAs' (4 July 2003)

WTO Doc. TN/TE/W/58 'Committee on Trade and Environment – Special Session – The Relationship between Existing WTO Rules and Specific Trade Obligations (STOs) Set Out in MEAs – A Swiss Perspective on National Experiences and Criteria used in the Negotiation and Implementation of MEAs Switzerland' (6 July 2005)

WTO Doc. TN/TE/W/68 'Committee on Trade and Environment – Special Session – Proposal for a Decision of the Ministerial Conference on Trade and Environment – Submission by the EC' (30 June 2006)

WTO Doc. TN/TE/W/72/Rev.1 'Committee on Trade and Environment – Special Session – Proposal for an Outcome on Trade and Environment Concerning Paragraph 31(i) of the Doha Ministerial Declaration – Submission from Australia and Argentina – Revision' (7 May 2007)

WTO Doc. WT/CTE/GEN/5 'Regular Session of the CTE of 8 October 2002 – Statement from the Secretariat of the CITES' (20 November 2002)

WTO Doc. WT/CTE/GEN/6 'Economic Incentives and Trade Policy – Communication from the Secretariat of the CITES' (20 November 2002)

WTO Doc. WT/CTE/GEN/7 'Decision on Economic Incentives and Trade Policy – Communication from the Secretariat of the CITES' (11 February 2003)

WTO Doc. WT/CTE/GEN/10 'Environment-related issues in the Negotiations on WTO Rules – Statement by Mr. Jan Woznowski at the Regular Session of the Committee on Trade and Environment of 29–30 April 2003' (11 April 2003)

WTO Doc. WT/CTE/INF/6/Rev.4 'Committee on Trade and Environment – International Intergovernmental Organizations – Observer Status in the Committee on Trade and Environment – Revision' (2 February 2007)

WTO Doc. WT/CTE/M/30 Note by the Secretariat (11 September 2002)

WTO Doc. WT/CTE/W/1 'Environmental Benefits of Removing Trade Restrictions and Distortions – Note by the Secretariat' (16 February 1995)

WTO Doc. WT/CTE/W/15 and Corr.1, Note by the Secretariat, 'Recent Developments in the FAO Code of Conduct for Responsible Fisheries' (1 December 1995)

WTO Doc. WT/CTE/W/62 'The 1994 Agreement Relating to the Implementation of Part XI of the 1982 UN Convention on the Law of the Sea – Communication from the UN Division for Ocean Affairs and the Law of the Sea' (16 September 1997)

WTO Doc. WT/CTE/W/126 'Committee on Trade and Environment – The FAO International Plan of Action for the Management of Fishing Capacity and Related Initiatives for Sustainable Fisheries' (Communication from the FAO) (12 October 1999)

342 BIBLIOGRAPHY

WTO Doc. WT/CTE/W/135 'CTE – Update of FAO Activities Related to Fisheries – Communication from the Food and Agriculture Organization (FAO)' (29 Feb 2000)

WTO Doc. WT/CTE/W/160/Rev.4 'Matrix on Trade Measures Pursuant to Selected Multilateral Environmental Agreements' (2005)

WTO Doc. WT/CTE/W/165 'The Relationship between the Convention on International Trade in Endangered Species of Wild Fauna and Flora and the WTO – Communication from the Secretariat of the CITES' (13 October 2000)

WTO Doc. WT/CTE/W/167/Add.1 'Committee on Trade and Environment – Environmental Benefits of Removing Trade Restrictions and Distortions: The Fisheries Sector – Item 6 of the Work Program by the Secretariat – Addendum' (19 June 2001)

WTO Doc. WT/CTE/W/189 'Committee on Trade and Environment – Update of FAO Activities Related to Fisheries – Report of the Expert Consultation on Economic Incentives and Responsible Fisheries – Communication from the FAO' (18 June 2001)

WTO Doc. WT/CTE/W/205 'UNEP Workshop on the Impact of Trade-Related Policies on Fisheries and Measures Required for their Sustainable Management' (8 May 2002)

WTO Doc. WT/CTE/W/235 'Decisions of the Seventh Meeting of the Conference of the Parties to the Convention on Biological Diversity and the First Meeting of the Parties to the Cartagena Protocol on Biosafety of Relevance to the WTO – Note by the Executive Secretary of the CBD' (8 June 2004)

WTO Doc. WT/CTE/W/216 'Committee on Trade and Environment – Technical Assistance and Capacity Building Activities in 2002 – Note by the Secretariat – Paragraph 33' (30 September 2002)

WTO Doc. WT/DSB/19 'Indicative List of Governmental and Non-Governmental Panelists' 29 March 2000

WTO Doc. WT/DS58/9 'United States – Import Prohibition of Certain Shrimp and Shrimp Products – Constitution of the Panel Established at the Requests of Malaysia, Thailand, Pakistan and India – Note by the Secretariat' (17 April 1997)

WTO Doc. WT/GC/M/60 'Minutes of the General Council Meeting of 22 November 2000' (23 January 2001)

WTO, *World Trade Report* (WTO, 2006)

SELECTED WEBSITES

www.acwl.ch
www.ccamlr.org
www.ciel.org
www.cites.org
www.defenders.org/index.php
www.fao.org
www.fishsubsidy.org

SELECTED WEBSITES 343

www.globalsubsidies.org
www.hsus.org
www.iarc.fr
www.icc-cpi.int
www.icsid.worldbank.org
www.ictsd.org
www.ifaw.org
www.isealalliance.org
www.iucn.org
www.iwmc.org/iwmcinfo/statement.htm
www.msc.org
www.oceana.org
www.oceansatlas.org/www.un-oceans.org/Index.htm
www.trade-environment.org/page/southernagenda/description.htm
www.un.org/Depts/los/doalos_activities/about_doalos.htm
www.unep.org
www.wto.org

Index

Aarhus Convention, 231
abbreviations list, xxix–xxxii
accountability, regime interaction and
    legal framework for, 290
    legitimacy of, 278–84
Agenda 21
    drafting of, 63–4
    shrimp import ban, WTO adjudication
        impact on, 199, 201
    submissions by, 209
Agreement on Agriculture, 73n.288, 73
Agreement on Import Licensing
    Procedures, 72
Agreement on Rules of Origin, 72
Agreement on Subsidies and
    Countervailing Measures (SCM
    Agreement), 6, 73–4
    'Chair's text' on WTO agreement with,
        91–3, 95, 114, 118n.178, 121–3,
        291, 294–5
    fisheries subsidies restrictions and, 91–3,
        97, 129–33, 247
    conditionality peer review, 91n.29,
        91n.30, 92n.33, 121–3
    dispute resolution, 125–6, 132
    notification requirements, 115, 291
Agreement on Technical Barriers to Trade
    (TBT), 71
    conditionality review, 123, 132
    EC – Biotech case and, 210–11
    legitimacy issues and exogenous sources
        usage, 76, 237, 268
    mutual agreement principles and
        legitimacy of, 270
    regime interaction and, 76, 292
    standards classifications, 116–17
Agreement on the Application of Sanitary
    and Phytosanitary Measures (SPS
    Agreement), 72
    conditionality review, 123, 132
    dispute resolution, 126
    EC – Biotech case and, 210–11, 212,
        289–92
    legitimacy issues with
        exogenous sources usage, 237, 268
    mutual agreement principles
        and, 270
    risk of managerialism and, 277
    regime interaction and, 76, 292
    standards classifications of, 116–17
    VCLT Article 31(1) principles and, 234
Agreement on Trade-Related Aspects of
    Intellectual Property (TRIPS),
    70n.270, 100
amber-box test for fisheries subsidies, 91–3,
    115–19, 120
amicus curiae briefs
    as adjudication tool, 228
    legal framework for regime interaction
        and, 297
    legitimacy issues and exogenous sources
        usage, 238
    from NGOs, 220–4
Animal and Plants Committees, CITES. *See
    under* Convention on the
    International Trade in
    Endangered Species of Wild Flora
    and Fauna
Appellate Body (WTO)
    legitimacy guidelines of, 283
    shrimp import ban and sea turtle
        conservation and role of
        GATT Article XX *chapeau*,
            interpretation of, 200–1
        GATT Article XX(g) interpretation,
            198–200
        parallel membership of relevant rules,
            need for, 201–4
        regime interaction dispute settlement
            procedures and, 206–9

INDEX   345

applicable law principle, shrimp import ban, WTO adjudication, regime interaction and role of, 195–7
arbitration tribunals
  Convention for the Conservation of Southern Bluefin Tuna and, 50
  fragmentation of international fishing regulations and, 8
  UNCLOS and EEZ regimes and, 44–5
Argentina
  Argentina – Textiles and Apparel case and, 218
  EC – Biotech case and, 202–4, 210
  fishing subsidies exemptions, 118
  MOU between CITES Secretariat and FAO and, 164
Argentina – Textiles and Apparel case, IGO consultations on, 218
Asia-Pacific Fishery Commission, 39
Australia
  MOU between CITES Secretariat and FAO policy coordination with, 175
  national policy coordination with global initiatives in, 251
Australia – Salmon case, *amicus curiae* briefs filed in, 222
authority-sharing proposals, fisheries subsidies dispute resolution and, 127

Bagwell, Kyle, 99
basking shark, CITES Appendices listing for, 135n.2, 140
benchmarking procedures, fisheries subsidies conditionality and, 119–24
biases of regimes
  fisheries subsidies restrictions and, allegations of, 98
  legitimacy issues and use of exogenous sources and, 235–9, 245
  trade restrictions on endangered marine species and allegations of, 143–4
binding resolutions, dispute settlements and importance of, 229
Biosafety Protocol
  EC – Biotech case and, 202–3, 209–10, 212, 285
  establishment of, 149
  legal framework for regime interaction and, 296–7
  parallel membership issues and, 264
Blokker, Niels, 177
bluefin tuna, endangered species classification, 137, 141n.38
Brazil, fisheries subsidies restrictions exemptions, 118

notification requirements, 115
Brazil – Retreaded Tyres case, 211, 222, 228
Bridges (ICTSD) publication, 257
'buybacks' of decommissioned fishing vessels, 93n.40, 93
by-catch phenomenon, shrimp fisheries and, 190
Cairns group (WTO), dispute resolution and, 126
Canada
  EC – Biotech case and, 202–4, 210
  EC – Hormones case and, 213–14
  fisheries subsidies in, 88
Canada – Herring and Salmon case, 199
capacity building
  MOU between CITES Secretariat and FAO and, 179, 185, 188
  regime interaction and, 228
  resource allocation and, 256–8
Cartland, Michael, 207n.126
case-study research methodology, on international fishing laws, 16–19
CBD. *See* Convention on Biological Diversity
CCAMLR (Commission for the Conservation of Antarctic Marine Living Resources), 40, 41n.63, 177n.248, 177–9
chapeau of GATT Article XX, 75–7
  WTO fisheries import ban adjudication and, 193–4, 200–1
Charnovitz, Steve, 99–101
Chayes, Abram, 185
Chayes, Antonia Handler, 185
Chile, fisheries subsidies restrictions, notification requirements, 114
China
  Endangered Species Advisory Committee, 175
  shark fin demand in, 135–6
CITES. *See* Convention on the International Trade in Endangered Species of Wild Flora and Fauna
CMS. *See* Convention on the Conservation of Migratory Species of Wild Animals
coastal states jurisdiction, exclusive economic zone regime and, 34n.13, 34–8
Code of Conduct for Responsible Fishing, FAO, 12, 48–9, 52–3
  conditionality issues, 119
  fisheries subsidies restraints and, 89–104, 293–4
  inter-regime acceptance of, 112
  marine by-catch and shrimp fisheries bans and, 191

346   INDEX

Code of Conduct for Responsible (cont.)
   sea turtle status determination and,
      205, 212
   trade restrictions on endangered marine
      species, 138, 149, 153
Codex Alimentarius Commission
   EC - Hormones case and, 213-14
   standards classification, 116-17
COFI (Committee on Fisheries) (FAO), 47,
     161, 162-7, 172-4
Cohen, Joshua, 290
'Coherence in Global Economic Policy
     Making', 78
Commission for the Conservation of
     Antarctic Marine Living Resources
     (CCAMLR), 40, 41n.63, 177n.248,
     177-9
Committee on Fisheries (COFI) (FAO), 47,
     161, 162-7, 172-4
Committee on Trade and Environment
     (CTE) (WTO), 29, 80-1
   fisheries subsidies and, 94-6
   inter-regime information sharing
   national policy coordination with global
     initiatives in, 251
   non-WTO participants in, 109-10
   observership practices, 177
   shrimp import ban, WTO adjudication
     of, 201
'common interest' issues
   MOU between FAO and CITES Secretariat
     and, 177, 188
   parallel membership and, 265-6
'communicative systems' model,
     conflicting norms of fishing laws
     and, 15
competence challenges
   as impediment to regime interactions,
     258-9
   regime legitimacy and, 245
   WTO fisheries subsidies restrictions
     and, 97
Compliance Agreement (FAO)
   adoption of, 47
   fisheries subsidies conditionality and,
     119, 293-4
   trade restrictions on endangered fish
     species and, forum shopping
     and, 138
compliance problems
   fisheries subsidies restrictions and, 91-3
   MOU between CITES Secretariat and FAO
     and, 184-6
   resource allocation and, 257
compulsory dispute settlement
   fragmentation of the law of the sea and
     weakness of, 51
   UNCLOS and EEZ regimes and, 44
   WTO provisions for, 70-5
compulsory jurisdiction, UNCLOS and EEZ
     regimes and, 45
Conference of the Parties (COP)
   CITES provisions for, 57, 59
   CITES species listing criteria and, 58-60,
     140, 179-81
   Convention on Biological Diversity
     and, 64
   MOU between FAO and CITES Secretariat
     and, 157, 159
   trade restrictions on endangered fish
     species and
   inter-regime interaction and, 153-4
   marine species listings and, 140
   opposition to, 142
conflicting norms. *See also* normative
     principles
   fragmentation of international fishing
     regulations and, 10-16
   ILC Study Group on fragmentation of
     international law and role of,
     11-14, 261
   institutional analysis of, 13
   International Law Commission and,
     152n.101
   legal framework for regime interaction
     and, 302-4
   MOU between CITES Secretariat and FAO
     and, 186
consent, legitimacy of regime interaction
     and, 271, 276
constitutionality, fragmentation of
     international law and, 10
consumer activists, as regime interaction
     stakeholders, 28
Convention for the Conservation of
     Southern Bluefin Tuna, 50
Convention on Biological Diversity (CBD), 29
   binding decisions of, 229
   Biosafety Protocol, 149
   EC - Biotech case and, 202-3, 210
   fisheries subsidies and, 98
   historical overview, 32
   international environmental law
     and, 63-5
   law of the sea and, 65-7
   shrimp import ban, WTO adjudication
     of, impact on, 199
   trade restrictions on endangered fish
     species and, forum shopping and,
     138-54
   VCLT Article 31(1) principles and, 234
Convention on the Conservation of
     Migratory Species of Wild Animals
     (CMS), 189

fisheries subsidies restrictions and, 247
historical overview, 32
international environmental law and,
    61-3
law of the sea and, 67
mutual agreement principles and, 269
parallel memberships and effectiveness
    of, 265
regime exclusivity and, 244
sea turtles conservation and, 191, 204
shrimp import ban, WTO adjudication
    and regime interaction involving,
    196-7, 199, 200, 201
trade restrictions on endangered fish
    species and
    forum shopping and, 138-54
    parallel membership impact on, 263
Convention on the International Trade in
    Endangered Species of Wild Flora
    and Fauna (CITES), 6-8
    Animal and Plants Committees, 60
    conch species review, 140
    expertise limitations of, 143
    FAO involvement in, 181-4
    FAO, incursion into role of, 18
    FAO, MOU with. *See under* Memorandum
        of Understanding
    fisheries sustainability provisions, 85,
        134-5
    fragmentation of international law
        and, 12
    historical overview of, 32
    institutional analysis of, 16-19
    international environmental law and,
        29, 56-61
    law of the sea and, 65-7
    market failures and, 99n.73
    regulatory mechanisms of, 58-9
    species Appendices. *See under*
        endangered species classification
    states as stakeholders in, 28
    trade restrictions on fish species and,
        6-8, 18, 30
        bias allegations by states against,
            143-4
        coherence in management of, 144-6
        expertise questioned by other regimes,
            142-3
        institutional bias and insensitivity
            concerns, 148-51
        marine species Appendices listings,
            139-41
        opposition to role of, 141-6
        policy adaptation by, 147-8
        regime interaction and, 153-4
        regime overlap vs. normative conflict
            in, 151-3

responses to opposition to, 146-54
textual mandate, 146-54
US policy coordination with, 251
Conventions, table of, xxi-xxviii
Cooney, Rosie, 150n.91
COP. *See* Conference of the Parties
Cozendey, Carlos, 207n.126
Crawford, James, 23
criminal law, international, fragmentation
    and, 8-16
CTE. *See* Committee on Trade and
    Environment

Decision on Trade and Environment, 236
Declaration on Coherence in Global
    Economic Policy Making
    (WTO), 269
declarations, table of, xxi-xxviii
Defenders of Wildlife, 166n.185, 166
Delbrück, Jost, 207n.126
Derrida, Jacques, 233n.289
developing countries
    as stakeholders in sustainable fishing,
        24n.104, 24-5
    trade restrictions on endangered species
        and, 150n.91, 150
discursive networks, conflicting norms of
    fishing laws and, 15
Dispute Settlement Body (DSB) (WTO),
    adjudication of US shrimp import
    ban, establishment of, 193
dispute settlement. *See also* compulsory
    dispute settlements *See under*
    specific disputes, e.g. fisheries
    import bans
    fisheries subsidies restrictions, 124-9,
        132
    fragmentation of the law of the sea and
        weakness of, 51
    legal framework for regime interaction
        and, 295-8
    MOU between FAO and CITES Secretatiat
        and, 184-6
    parallel membership and, 269
    shrimp import ban, WTO adjudication
        of, 70-5, 85, 195-224
    UNCLOS and EEZ regimes and, 44
Division for Ocean Affairs and the Law of
    the Sea (DOALOS), UN, 29,
    38n.40, 38
Doha Round negotiations (WTO), 29, 74-5
    adjudicator selection criteria and, 226
    fisheries subsidies restrictions and, 86,
        94-6, 98
    institutional arrangements and, 271
    legitimacy issues and
        duties to take account of others, 284-5

348   INDEX

Doha Round negotiations (cont.)
 exogenous sources usage, 236
 national policy coordination with, 246n.14, 249, 250
 non-WTO participants in, 109
 parallel membership issues and, 229–39, 263, 268
 regime interaction reform and, 81n.338, 81
dolphin conservation
 Irrawaddy dolphin, endangered species classification, 140
 tuna harvesting and, 7, 193
domestic law
 fragmentation of international law and, 10
 legal framework for regime interaction and, 300
 WTO fisheries subsidies and, 99–101
Dorf, Michael, 290
DSB (Dispute Settlement Body) (WTO), adjudication of US shrimp import ban, establishment of, 193
DSU. *See* Understanding on Rules and Procedures Governing the Settlement of Disputes
dumping practices, WTO provisions on, 73
Dunoff, Jeffrey, 225
duties to take account of others, regime interaction legitimacy and, 284–7

Earth Summit (United Nations Conference on Environment and Development or UNCED), 63
EC. *See* European Commission
EC – Asbestos case
 amicus curiae briefs in, 221
 IGO consultations on, 219
EC – Biotech case, 11n.42, 31
 amicus curiae briefs in, 223
 IGO consultations in, 217–18
 institutional arrangements in, 270
 legal framework for regime interaction in, 296–7
 legitimacy issues in
  duty to take account of others and, 285
  risk of managerialism and, 277
  use of exogenous sources, 235–9
 parallel membership issues and, 229–32, 264
 party submissions in, 209–10
 scientific experts' consultations in, 212
 VCLT Article 31(1) analysis, 234
 WTO adjudication of, 202–4
EC – Chicken Cuts case, 215
EC – Geographical Indications case, IGO consultations on, 217

EC – Hormones case, scientific experts' consultations in, 213–14
EC – Sardines case, 71
 amicus curiae briefs filed in, 222–3
 IGO consultations on, 218
EC – Tariff Preferences case, 128
 Enabling Clause principle in, 205
 IGO consultations on, 219–20
eco-labelling
 non-governmental organisation involvement in, 7n.25
 stakeholders in, 25
economic development
 fisheries subsidies and overfishing, 87–9
 institutional bias concerning, trade restrictions on endangered marine species and, 149
ecosystem management paradigm, regime interaction on, 261, 274
ecotourism, endangered marine species and, 137n.17, 137, 140, 150
EEC (European Economic Community), fisheries import ban adjudication and, 193
EEZs. *See* exclusive economic zones
Enabling Clause principle
 disputed norms and, 219–20
 fisheries subsidies dispute resolution and, 128
 shrimp import ban, WTO adjudication based on, 205
endangered species classification
 CITES species Appendices, 58n.184, 58n.187, 58–60, 261
  accountability issues with, 278
  evaluation of, 181–4
  listing criteria for, 58–60, 179–81
  'look-alike' clauses in, 145, 148, 153
  regime interaction concerning, 292
  'split-listing' in, 146, 148, 153
  trade restrictions on endangered species and, 139–41, 247
  transfer of species between listings, 144
 in CMS, 62–3
 marine species, 135–7
  look-alike clause concerning, 145, 148, 153
  split-listing of, 146, 148, 153
entrenched regime interaction
 fisheries subsidies and, 113–24, 132
 MOU CITES Secretariat and FAO and, 176–84
environmental damage, fisheries subsidies and, 94–6, 99, 104
environmental law, international. *See* international environmental law

INDEX 349

epistemic networks, information sharing and institutional learning and, 255
equivocally overfished standard, fisheries subsidies dispute resolution and, 127
ethnographic research, vocabulary of regimes and, 21
EU. *See* European Union
European Commission (EC). *See also* entries beginning *EC* - for specific case law
  adjudicator selection criteria reforms proposed by, 226
  endangered species classifications and, 141
  national policy coordination with global initiatives in, 251
  standards classifications, 117
  trade and environmental policy overlap and, 284
European Court of Justice, 124, 231
European Economic Community (EEC), fisheries import ban adjudication and, 193
European eel, endangered species classification, 140
European Union (EU)
  accountability and openness recommendations for, 290
  exclusive economic zones and fishing agreements of, 37
  fisheries subsidies
    exemptions, 118
    notification requirements, 115
    statistics, 88n.9
  regional fishing agreements and, 69
  sustainability impact assessments by, 251
exclusive economic zones (EEZs)
  Fish Stocks Agreement and, 41-6
  fisheries subsidies restrictions and, 91, 119
  state institutions and regime of, 33
  trade restrictions on endangered marine species
    allowable catch limits in, 145
    CITES mandate concerning, 142
    domestic consumption exemptions, 147
    forum shopping in, 138-54
  UNCLOS and, 34-8
'exhaustible natural resources' principle, US shrimp import ban, complaints about, 198-200

exogenous regimes
  amicus curiae briefs filed by, 238
  biases of, 235-9, 245
  fisheries import bans and, 235-9
  intergovernmental organisations and, 237-9
  legal framework for regime interaction and, 268
  legitimacy issues with usage of, 76, 236, 237
  mutual agreement principles and legitimacy of, 236, 270
  VCLT provisions and use of, 238
  WTO adjudication and use of, 235-9
expertise of regimes. *See also* scientific experts' consultations
  accountability of, 279
  EC - Hormones case and, 213-14
  fisheries subsidies restrictions, competence challenges to, 97
  shrimp import ban, WTO adjudication and role of, 211-15
  trade restrictions on fish species and challenges to
    CITES restrictions, 142-3
    FAO restrictions, 7, 30
exports of fish and fishery products
  conditionality exemptions, 121-3
  global trends in, 24n.104
  shark exports, 135-6
  shrimp fisheries, production and export statistics, 190n.4
express powers principle, regime interaction legitimacy and, 272

fairness principles, WTO fisheries subsidies and, 99-101
FAO. *See* Food and Agriculture Organization
FAO State of the World Fisheries Report (SOFIA), 47
fish stocks
  international growth of fishing industry and, 5n.16
  predicted collapse of, 4
  UNCLOS 'high seas' provisions and regional regimes, 38-46
Fish Stocks Agreement
  FAO assistance in, 47
  fisheries subsidies restrictions and, 90, 129-33, 293-4
    conditionality issues, 119
    dispute resolution, 128
  historical overview, 32-3
  institutional bias and insensitivities in, 149
  institutional framework of, 29

Fish Stocks Agreement (cont.)
  marine by-catch and shrimp fisheries bans and, 190
  regime interaction and, 51
  regional fish management organisations and, 41-6
  trade restrictions on endangered marine species
    forum shopping and, 138-54
    regime overlap vs. normative conflict in, 151-3
  WTO interaction with, 77
Fisheries Global Information System (FAO), 47
fisheries subsidies
  adjudicator selection criteria for disputes involving, 226
  conditionality through benchmarking and peer review, 119-24
  dispute resolution concerning, 124-9, 132
  Doha negotiations and, 86, 94-6
  entrenched regime interactions, 113-24
  EU. *See under* European Union
  FAO mandate. *See under* Food and Agriculture Organization
  Fish Stocks Agreement and, 90, 129-33, 293-4
    conditionality issues, 119
    dispute resolution, 128
  fishery sustainability and, 6
  forum shopping and. *See under* forum shopping
  fragmented policy and law-making involving, 246-7
  IGOs and. *See under* intergovernmental organisations
  inter-regime learning about, 105-13
  legal framework for regime interaction and, 289-92
  NGOs and. *See under* non-governmental organisations
  opposition to WTO restrictions, 96-9, 258, 267
  overfishing and, 87-9
  parallel membership and restrictions on, 263, 268-9
  regime interaction on. *See under* regime interaction
  SCM Agreement restrictions on. *See under* Agreement on Subsidies and Countervailing Measures
  SPS Agreement and. *See under* Agreement on the Application of Sanitary and Phytosanitary Measures
  standards classifications and, 115-19
  state institutions' reactions to, 89, 107, 289-92
  trade distortions and, 94-6
  trade policy coherence and effectiveness concerning, 98-9
  UNEP and restrictions on
    inter-regime learning and, 105, 131
    non-UN participants, 111-13
  in United States. *See under* United States
  voluntary approach to, 90-1
  WTO draft consolidated text for, 307-15
  WTO provisions concerning, 29, 73-4, 85-133
    benchmarking/peer review-based conditionality, 119-24
    bias alleged in, 98
    dispute resolution, 124-9, 247
    expertise and judicial competence issues, 97
    forum shopping to regime interaction transition, 103-4
    inter-regime learning and, 105-13
    ongoing indeterminacy in, 101-3
    opposition to, 96-9, 258, 267
    policy adaptation, 99-101
    standards classifications and, 115-19
    strengths and limitations of, 129-33, 246-7
    support for, 99-104
    textual basis for, 101
  WWF and. *See under* World Wide Fund for Nature
fisheries sustainability
  international trade law and, 5-8
  regime interaction and legislation promoting, 85
  trade restrictions on certain species, 6-8
fishing capacity
  defined, 87
  fisheries subsidies as enhancement of, 88
  IPOA-Capacity and, 90-1
fishing industry
  exports. *See* exports of fish and fishery products
  imports. *See* imports of fish and fishery products
  international growth of, 5
  as regime interaction stakeholder, 28
  shark fins, trading of, 135-6
fishing licence retirements, 93
flags of states
  FAO Compliance Agreement and, 47
  Fish Stocks Agreement, 42-6, 43n.76
  fisheries subsidies and, 114
  illegal fishing classifications and, 116
  regional regime jurisdiction and, 40

trade restrictions on endangered fish species and
  forum shopping and, 138
  on-board fish processing and, 142
Food and Agriculture Organization (FAO) (UN)
  Asia-Pacific Fishery Commission and, 39
  CITES incursion into role of, 18
  CITES Secretariat, MOU with. *See under* Memorandum of Understanding
  Code of Conduct for Responsible Fishing. *See* Code of Conduct for Responsible Fishing, FAO
  COFI, 47, 161, 162-7, 172-4
  Compliance Agreement. *See* Compliance Agreement
  constitutive instruments of, 12
  fish and marine food sources promotion by, 3
  fish stock monitoring by, 4, 135n.4, 136nn.5-6
  Fisheries and Aquaculture Department, 55
  fisheries management regime of, 46-9, 134-5
  fisheries subsidies mandate, 6, 96-7, 102, 129-33, 246-7, 258, 267, 289-92
    conditionality peer review, 120-1, 132
    inter-regime learning and, 105
    notification requirements, 115
  fisheries sustainability and, 85
  historical overview of fishing codes in, 32-3
  IATTC, interaction with, 66
  institutional analysis of, 16-19, 29
  IUU fishing standard and, 115, 293-4
  JECFA, 214
  legitimacy issues for, 272
  multilateral fishing subsidies regulations oversight by, 18
  mutual agreement principles and, 269
  national policy coordination with initiatives of, 249
  normative and institutional fisheries conservation initiatives, 33
  parallel membership and effectiveness of, 265
  participating states as stakeholders in, 28
  regime exclusivity and, 244-9
  regime interaction and, 55, 66, 239-40
  standards classifications, 115, 117-18, 119
  trade restrictions on endangered fish species and
    concerns over CITES mandate, 142-6, 147-8
    expertise of, 7, 30
    fisheries management regime, 134-5
    forum shopping and, 138-54

regime interaction on, 153-4
regime overlap vs. normative conflict in, 151-3
forum exclusivity, as impediment to regime interaction, 258-61. *See also* regime exclusivity
forum shopping
  fisheries subsidies and, 129-33
  ongoing indeterminacy and, 101-3
  policy adaptation and, 99-101
  regime interaction, shift to, 103-4
  regimes of relevance and. *See* regimes of relevance and forum shopping
  shift to regime interaction from, 103-4, 243-9
  trade restrictions on endangered marine species and, 138-54
  WTO adjudication of shrimp import ban and, 190-5
fragmentation of international law, 8-16
  fisheries subsidies negotiations as example of, 246-7
  ILC Fragmentation Study. *See under* International Law Commission
  regime interaction and, 243-66
  weakness of compulsory dispute agreements and, 51
Framework Convention on Tobacco Control, 283
framing practices, shrimp import ban adjudication and, 227-9
Franckx, Eric, 151-3
'Friends of Fish,' 95, 105

General Agreement on Tariffs and Trade (GATT)
  Argentina – Textiles and Apparel case and, 218
  Article XI, adjudication of US shrimp import ban and, 193
  Article XX, 75-7
    chapeau interpretations of shrimp import ban, 200-1
    fisheries import ban adjudication and, 193-4, 198-200
    parallel membership and, 265
  Article XXIII, shrimp import ban, WTO adjudication and regime interaction, 196
  dumping practices provisions, 73
  IGO consultations with, 216
  national fisheries policies and, 70
  national treatment provision, 102
  origins of, 245
  regime interdependencies, 75-9
  trade liberalisation and, 3
  Uruguay Round. *See* Uruguay Round

352   INDEX

General Agreement on Tariffs (cont.)
    WTO interaction with, 77
General Agreement on Trade in Services (GATS), 70n.270
General Council
    non-WTO participants in, 109, 112n.138
    WTO interaction and, 79-81, 80n.323
genetically modified organisms, 209-10
    IGO consultations on, 217-18
    parallel membership issues and, 264
    standards classifications, 202-4
Germany, MOU between FAO and CITES Secretariat and, evaluation of CITES species listings, 183
global administrative law, regime legitimacy and accountability and, 281
Global Environment Facility, 257
global policy-making
    international law and, 31
    regime exclusivity and, 244
    regional fishing management organisations and, 35
governance models
    future use in non-fisheries context, 299-302
    ILC Fragmentation Study recommendations and, 298-9
    information-sharing and institutional learning and, 253-6
    inter-regime learning, 131
    legal framework for regime interaction and, situating of participants and, 304-6
    legitimacy issues and, 248
        express and implied powers, 272
        law-making activities and, 289-92
    national policy coordination and regime interaction, 249-53
    regime interaction and, 266
    resource allocation and, 256-8
great white shark, endangered species classification, 140
green-box fisheries subsidies, 91n.28, 91-3, 115-19
Greenpeace
    monitoring activities of, 257
    on WTO fisheries subsidies, 98
Guzman, Andrew, 290

Haas, Peter, 40, 255
high seas regime
    regime interdependency and, 65
    regional fishing management organisations and, 38-46
Hong Kong
    Ministerial Declaration, 94
    MOU between CITES Secretariat and FAO, policy coordination with, 174
horizontal coordination, resource allocation and, 256
hormone-treated beef, scientific experts' consultations concerning, 213-14
HSI. *See* Humane Society International
human rights law, fragmentation and, 8-16, 26n.115
Humane Society International (HSI), 211
    amicus curiae briefs filed by, 222
    as regime interaction stakeholder, 28, 166

IARC (International Agency for Research on Cancer), 213n.167, 213
IATTC (Inter-American Tropical Tuna Commission), FAO interaction with, 66
ICC (International Criminal Court), 8n.30
ICCAT (International Commission for the Conservation of Atlantic Tunas), 40, 66
ICJ. *See* International Court of Justice
ICTSD. *See* International Centre for Trade and Sustainable Development
IGOs. *See* intergovernmental organisations
ILC. *See* International Law Commission
illegal, unreported and unregulated (IUU) fishing
    defined, 41n.62
    FAO definition, 115
    inter-regime learning and, 106
    IPOA-IUU fishing subsidies restrictions, 90-1
    legal framework for regime interaction concerning, 293-4
    Port State Measures Agreement, 43
    regional fish management organisations and, 41
IMF. *See* International Monetary Fund
implied powers principle, regime interaction legitimacy and, 272
imports of fish and fishery products
    ban on shrimp, WTO adjudication of, 18, 189-240
        adjudicator selection, 226-7
        Agenda 21. *See under* Agenda 21
        amicus curiae briefs from NGOs and others, 220-4
        Appellate Body. *See under* Appellate Body
        applicable law, 195-7
        complaints against US ban, 193-5
        conclusions concerning, 239-40
        DSU. *See under* Understanding on Rules and Procedures Governing the Settlement of Disputes

INDEX 353

exogenous sources, legitimacy and guidelines for, 235-9
forum shopping, relevant regimes, 190-5
framing by parties, 227-9
GATT Article XX interpretations, 198-201
historical evolution of, 191-3
IGO secretariats' consultations, 215-20
panellists, AB members, and Secretariat, role in dispute settlement, 206-9
parallel membership
  relevant rules, 201-4
  treaties and organisations, 229-39
parties' submissions, 209-11
problems and challenges in, 224-39
regime interaction and. *See under* regime interaction
relevant facts, 204-6
scientific experts' consultations, 211-15
sea turtle conservation and, 7, 18, 30-1, 189, 196
shrimp fisheries and marine by-catch, 190
treaty interpretation, 197-204
UNCLOS. *See under* UN Convention on the Law of the Sea
United States and. *See under* United States
VCLT provisions
  Article 31(1) provisions, 232-5
  Article 31(3)(c) provisions and, 229-32
fisheries import bans generally, WTO adjudication of. *See under* World Trade Organization
global trends in, 24n.104
India – Quantitative Restrictions case, 215
India, US shrimp import ban, complaints about, 193-5, 199
Indonesia, IPOA-Capacity plans submitted by, 90
information-sharing
  fisheries subsidies restrictions, inter-regime learning and, 105-8
  institutional learning through, 253-6
  MOU between FAO and CITES Secretariat and, 176-9
  needs-driving concept of, 285n.70
  regime interaction and, 253
institutional analysis
  case-study research methodology and, 16-19
  of CITES, 57, 148-51

historical overview, 32-3
international institutional law, MOU between CITES Secretariat and FAO and, 169-71
institutional arrangements, legal framework for regime interaction and, 268, 270-1
intellectual property
  TRIPS Agreement on, 70n.270, 100
  WIPO, 283
inter-agency collaboration, national policy coordination and, 250
Inter-American Convention for the Protection and Conservation of Sea Turtles, 197, 205
Inter-American Tropical Tuna Commission (IATTC), FAO interaction with, 66
inter-regime learning. *See also* regime interaction
  fisheries subsidies and, 99-101, 105-13, 131-3, 243
  information sharing and, 253-6
  promotion and obstruction of, 31
  resource allocation and, 256-8
  trade restrictions on endangered fish species and, 153-4
inter se modification, MOU between CITES Secretariat and FAO and, 171, 260
inter-state relations
  regime interaction and, 252-3
  WTO fisheries subsidies, policy adaptation, 99-101
intergovernmental organisations (IGOs)
  CITES interactions with, 57, 59, 153-4
  EC – Biotech case and, 31, 217-18, 223, 296-7
  EC – Hormones case and, 214
  fisheries subsidies and
    dispute resolution involving, 124-9
    inter-regime learning and, 105-8, 131-3
    notification requirements, 115
    General Council and, 79-81
    information-sharing and institutional learning and, 254-6
    institutional arrangements with, 16-19, 270
  law-making activities
    conflicting norms and, 15
    regime interaction legal framework and, 289-92
  legal framework for regime interaction and, 31, 296-7, 304-6
  legitimacy issues for
    accountability issues, 278-84
    duty to take account of others, 284-7
    exogenous sources usage and, 237-9

354   INDEX

intergovernmental organisations (cont.)
    express and implied powers and, 272
    risk of managerialism and, 276-8
    sovereignty and, 271
    MOU between FAO and CITES Secretariat
        information-sharing and observership
            practices, 177
        legal status, 155
        rules of procedure and, 172-4
    Multilateral Environmental Agreements
        compared with, 157n.132
    mutual agreement and, 269
    national policy coordination with, 249
    non-fisheries governance models and,
        299-302
    parallel membership and, 264
    participating states as stakeholders
        in, 28
    regime interaction and, 55, 289-92
    secretariats. *See* secretariats of IGOs
    shrimp import ban, WTO adjudication
        and role of
        legacy of, 240
        scientific experts' consultations and,
            211-15
        secretariats' consultations, 215-20
    UN consultative process and, 53
    WTO Rules Group exclusion of, 86,
        108-10, 131, 254, 261, 285, 289
International Agency for Research on
    Cancer (IARC), 213n.167, 213
International Centre for Trade and
    Sustainable Development (ICTSD)
    capacity-building initiatives and, 228
    as regime interaction stakeholder, 28
    resource allocation by, 257
    UNEP and, 112
International Coffee Organization, 219
International Commission for the
    Conservation of Atlantic Tunas
    (ICCAT), 40, 66
International Court of Justice (ICJ)
    amicus curiae briefs filed with, 224
    conflicting norms and rulings by, 10
    express and implied powers in rulings
        of, 272
    fisheries subsidies and, 98
    MOU between CITES Secretariat and FAO
        and, 170
    UNCLOS compulsory dispute settlement
        resolution and, 37-8
    UNCLOS legal framework and fisheries
        rulings of, 34
    Use of Nuclear Weapons case, 259
International Criminal Court (ICC), 8n.30
international criminal law, fragmentation
    and, 8-16

international environmental law
    CITES regime and, 56-61
    Convention on Biological Diversity and,
        63-5
    Convention on Migratory Species and,
        61-3
    Conventions and institutions of, 56-68
    fragmentation of, 8-16
    historical overview, 32
    independent development of, 3
    regime interaction and, 30-1, 65-7
    relevance to fisheries of, 29
    shrimp import ban, WTO adjudication
        based on, 205
    UN Environment Programme and, 67-8
International Fund for Animal Welfare
    (IFAW), 173, 174n.233
international institutional law, MOU
    between CITES Secretariat and
    FAO and, 169-71
international investment, fragmentation
    and, 8-16
international law
    fisheries trade and utilisation impact
        of, 29
    forum shopping vs. regime interaction
        and, 243-9
    fragmentation of. *See* fragmentation of
        international law
    global policy issues and, 31
    historical overview, 32-3
    ILC Fragmentation Study
        recommendations and, 298-9
    impediments to regime interaction
        in, 259
    legal framework for regime interaction
        and
        appropriate practices, 288-98
        risk of managerialism and, 302-4
        situating of participants in, 304-6
    parallel membership and, 264
    shrimp import ban, WTO adjudication
        and regime interaction involving,
        treaty interpretation and, 197-204
    sub-systems of, 19n.82
International Law Commission (ILC)
    Analytical Study, 19n.83
    conflicting norms and, 152n.101
    fisheries subsidies and, 98
    Fragmentation Study Group analyses, 261
    EC - Biotech case analysis, 230-2
    fragmentation of international law,
        11-14
    parallel membership analysis, 265-6
    recommendations of, 298-9
    VCLT Article 31(1) analysis, 233
    overlapping mandates and, 152

INDEX 355

International Maritime Organization, 52
International Monetary Fund (IMF)
   consultations with, 215-16
   GATT interaction with, 77
   WTO legitimacy and, 272
international oceans trust proposal, 46n.97
international organisations,
      law-making power, conflicting
      norms and, 15
International Plans of Action (IPOAs)
   FAO Code of Conduct and, 48-9, 129-33
   fisheries subsidies conditionality and, 119
   illegal, unregulated, and unreported
      fishing (IPOA-IUU), 90-1, 106
   IPOA-Sharks
      implementation difficulties for, 185
      trade restrictions on endangered
      marine species, 140
   Management of Fishing Capacity (IPOA-
      Capacity), 90-1, 112
international relations theory
   political realism in, 20n.86
   vocabulary of regimes and, 19-22
International Social and Environmental
      Accreditation (ISEAL) Alliance, 283
international trade law. *See also* trade
      restrictions on fish species
   CITES provisions and, 57
   fisheries sustainability and, 5-8
   fragmentation of, 8-16
   General Council and, 79-81
   independent development of, 3
   law of the sea and, 68-81
   market failures and, 99n.73
   regime interdependencies, 75-9
   regional agreements and, 69-70
   sustainable fisheries and, 29
   trade restrictions on marine species,
      134-88
   World Trade Organization agreements
      and, 70-5
   WTO coherence and effectiveness
      concerning, 98-9
International Trade Organization, 245,
   246n.10
International Tribunal of the Law of the Sea
   (ITLOS), 37-8
   compulsory dispute settlement
      provisions of, 45n.93, 70
   fisheries subsidies restrictions and,
      dispute resolution and, 126, 247
International Union for the Conservation
      of Nature and Natural Resources
      (IUCN)
   Red List, 136
   sea turtles status determination,
      204, 212

International Whaling Commission, CITES
      provisions and, 66
International Whaling Convention
   CITES provisions and, 66
   trade restrictions on marine species
      and, 142
International Wildlife Management
      Consortium (IWMC),
      174, 277
intersubjectivity, parallel membership
      and, 232-5
IPOAs. *See* International Plans of Action
Irrawaddy dolphin, endangered species
      classification, 140
ISEAL (International Social and
      Environmental Accreditation)
      Alliance, 283
ITLOS. *See* International Tribunal of the Law
      of the Sea
IUCN. *See* International Union for the
      Conservation of Nature and
      Natural Resources
IUU. *See* illegal, unreported and
      unregulated (IUU) fishing
IWMC (International Wildlife
      Management Consortium), 174,
      277

Japan
   acceptance of WTO fish subsidy
      restrictions, 101, 104n.93
   bluefin tuna imports, 137
   fisheries subsidies in, 88, 92, 96, 97
   WTO ongoing indeterminacy
      concerning, 102
   MOU between FAO and CITES Secretariat
      and, 160, 187
   non-WTO fisheries management
      advocacy, 106
   trade restrictions on endangered marine
      species and, opposition to CITES
      mandate, 141, 263
Jenks, Wilfred, 255
Joint FAO/WHO Expert Committee
      on Food Additives
      (JECFA), 214

Kelsen, Hans, 23
Keohane, Robert, 40
Kingsbury, Benedict, 271
Korea
   acceptance of WTO fish subsidy
      restrictions, 101, 104n.93
   fisheries subsidies in, 88, 96, 97
   non-WTO fisheries management
      advocacy, 106
Koskenniemi, Martti, 11, 302

Lamy, Pascal, 110, 226
Langille, Brian, 101
law-making initiatives, legal framework for regime interaction and, 289-92
law of the sea
  Conventions and institutions of, 29, 33-55
  EEZ regime and UNCLOS Agreement, 34-8
  'enclosure of the oceans' principle and, 36n.27, 82
  FAO fisheries management regime and, 46-9
  fisheries subsidies dispute resolution and, 126
  fragmentation of, 8-16
  independent development of, 3
  regime interdependency and, 50-3, 65-7
  regional regimes and, 38-46
  UN consultative process and, 53-5
  UN Environment Programme and, 67-8
  UNCLOS framework for, 34-46
  World Trade Organization and, 70-5
Leebron, David, 246n.14
legal framework for regime interaction, 22-4, 243-4, 267-87
  accountability issues, 278-84
  appropriate practices for regimes, 288-98
  consent and sovereignty, 271, 276
  dispute settlement initiatives and, 295-8
  duty to include other regimes, 284-7
  express and implied powers, 271, 276
  future regime interaction initiatives, 299-302
  ILC Fragmentation Study Group recommendations, 298-9
  implementation of existing commitments, 292-5
  information-sharing and institutional learning and, 253-6
  institutional arrangements, 270-1
  law-making activities and, 289-92
  legitimacy issues, 271-8
  managerialism risk and, 302-4
  mutual agreement and, 269-70
  parallel membership and, 268-9
  risk of managerialism and, 276-8
  state institutions and. *See under* state institutions
legal personality doctrine, MOU between FAO and CITES Secretariat, 155-7
legitimacy issues
  accountability of regimes and, 278-84
  consent and sovereignty principles, 271, 276
  Doha Round negotiations (WTO)
  duties to take account of others, 284-5
  exogenous sources usage, 236
  EC – Biotech case. *See under EC – Biotech* case
  exogenous sources usage and, 76, 235-9, 245, 268, 270
  express and implied powers principle, 272
  for FAO, 272, 277, 292
  global administrative law and, 281
  governance models and, 248
    express and implied powers, 272
    law-making activities and, 289-92
  for IGOs. *See under* intergovernmental organisations
  in international law making, 245
  legal framework for regime interaction and, 271-8
  MOU between FAO and CITES Secretariat, 167, 181-4, 277, 292
  for NGOs. *See under* non-governmental organisations
  normative principles, 279-84
  in regime interaction. *See under* regime interaction
  risk of managerialism and, 276-8
  SPS Agreement and. *See under* Agreement on the Application of Sanitary and Phytosanitary Measures
  for United Nations, 272-3, 276, 282
  US – Shrimp case. *See under US – Shrimp* case
  for WTO. *See under* World Trade Organization
Levy, Marc, 40
lex posterior derogat legi priori, fragmentation of international law and, 11, 298
lex specialis derogat legi generali, fragmentation of international law and, 11, 298
lex superior, fragmentation of international law and, 11, 298
'like' product claims, IGO consultations on, 219
Lloyd, Peter, 99
lobbyists
  fisheries subsidies and role of, 89
  as regime interaction stakeholder, 28
'look-alike' clauses, trade restrictions on endangered marine species and, 145, 148, 153
Lotus case, 23n.98

Malaysia, US shrimp import ban, complaints about, 193-5, 197, 199, 202
managerialism, risk of
  in *EC – Biotech* case, 277

for IGOs, 276-8
legal framework for regime interaction and, 276-8, 302-4
for NGOs, 277
in SPS Agreement, 277
marine species. *See also* specific species, e.g. sharks
  CITES Appendices listing, 57
    criteria for, 58-60, 179-81, 247
  endangered species classifications, 135-7
    look-alike clause concerning, 145, 148, 153
    split-listing of, 146, 148, 153
  migratory species, global nature of, 151n.96
  shrimp fisheries and marine by-catch, 190
  trade restrictions on endangered species. *See under* trade restrictions on fish species
Maritime Stewardship Council (MSC), 7n.25, 283
Marrakesh Agreement, 70, 74n.295, 79n.322
  fisheries import ban adjudication and, 198
  fisheries subsidies and, 101
Mauritania, EU agreement with, 37n.34
Mavroidis, Petros, 99
MEAs. *See* Multilateral Environmental Agreements
Memorandum of Understanding (MOU) between CITES Secretariat and FAO, 18, 30, 135, 154-76
  accountability issues in, 278
  alternative views of, 162
  capacity-building and, 179
  challenges to, 258
  CITES listing criteria issues, 179-81
  dispute resolution and, 184-6
  draft development process, 162-7
  duty to take account of others and, 285
  entrenched regime interaction through, 176-84
  evaluation of species listings, 181-4
  evolution of, 158-74
  final text of, 316-17
  finalisation process for, 167-9
  IGOs and. *See under* intergovernmental organisations
  information-sharing and institutional learning and, 253-6
  institutional arrangements in, 270
  international institutional law and, 169-71

  lack of transparency and openness concerning, 262
  legacy of, 186-8
  legal status, 155
  legitimacy of, 277, 292
  limitations on CITES' role, proposals for, 158-61
  national policy coordination and, 174-6, 249, 251
  regime exclusivity and, 247
  reporting activities and, 184
  reporting and resource allocation, 184
  resource allocation in, 184, 256-8
  rules of procedure uses and abuses, 172-4
  species listings legitimacy and, 167, 181-4
  substantive and procedural constraints, 169-74
  trade restrictions on endangered marine species and, 135, 154-76
  treaty amendment and verification and, 171-2
fragmentation of international law and, 9
Japan-US compromise text, 162-7
Mexico
  fisheries import ban adjudication and, 193
  MOU between FAO and CITES Secretariat and, 168
mobile global capital, regional regime jurisdiction and, 41
monitoring activities
  lack of, in regional regimes, 40
  MOU between CITES Secretariat and FAO and, 185
Multilateral Environmental Agreements (MEAs)
  development of, 3
  fisheries subsidies and, 94n.45, 97, 98, 102
  information-sharing and, 285
  intergovernmental organisations compared with, 157n.132
  international environmental law and, 56
  international trade laws and, 77
  national policy coordination with, 250, 252n.32
  parallel membership and, 268
  shrimp import ban, WTO adjudication and regime interaction, 195, 205
  WTO provisions and, 78-9
multilateral fishing subsidies regulations, WTO negotiation of, 17

358   INDEX

mutual agreement principle
  legal framework for regime interaction and, 268, 269-70
  legitimacy issues and use of exogenous sources and, 236

NAFO (Northwest Atlantic Fisheries Organization), regional agreements of, 39
Namibia, IPOA-Capacity plans submitted by, 90
national interest principle, fisheries subsidies restrictions and, 107
National Marine Fisheries Service (NMRF) (US), 192
national policy coordination
  MOU between CITES Secretariat and FAO and, 174-6, 249
  regime interaction and, 174-6, 249-53
'necessity' principle, express and implied powers and, 273-4
needs-driven information-sharing concept, 285n.70
Netherlands, fisheries import ban adjudication and, 193
'new governance', conflicting norms and, 15-16
New Haven Policy School, research methodology of, 17n.78
New Zealand, fisheries subsidies restrictions and, 105
Newfoundland cod fishery, overfishing and, 4n.6, 36n.27
NGOs. See non-governmental organisations
NMRF (National Marine Fisheries Service) (US), 192
'non-detriment' certificates, CITES trade restrictions on endangered marine species and, 139, 143
non-governmental organisations (NGOs)
  amicus curiae briefs from, 220-4
  CITES provisions and, 57, 59
  EC – Biotech case and, 220-4, 296-7
  eco-labelling and, 7n.25
  fisheries subsidies and, 98
    forum shopping transition to regime interaction, 103-4
    UN Environment Programme involvement, 111-13
  General Council and, 79-81
  information-sharing and institutional learning and, 253, 255
  institutional arrangements and work of, 271
  law-making activities
    conflicting norms of fishing laws and, 15

  regime interaction and, 291
  legal framework for regime interaction and, situating of participants and, 304-6
  legitimacy issues for
    accountability issues, 278-84
    exogenous sources usage, 237-9
    express and implied powers, 272-3, 276
    risk of managerialism and, 277
  MOU between FAO and CITES Secretariat, rules of procedure and, 172-4
  multilateral fishing subsidies regulations and, 18
  negotiations between CITES Secretariat and FAO and, 18
  resource allocation by, 258
  as stakeholders in regime interaction, 28
  UNCLOS framework for law of the sea and, 34
  WTO Rules Group exclusion of, 86, 108-10, 131, 254, 261, 285, 289
non-state actors, as stakeholders in sustainable fishing, 25
normative principles. See also conflicting norms
  evolutionary interpretations of, 297
  fisheries subsidies dispute resolution and, 127, 128
  IGO consultations on, 219-20
  ILC Fragmentation Study recommendations and, 298-9
  information-sharing and institutional learning and, 256
  regime legitimacy and accountability and, 279-84
  trade restrictions on fish species, 151-3
  vocabulary of regimes and, 21
Northwest Atlantic Fisheries Organization (NAFO), regional agreements of, 39
Norway, MOU between FAO and CITES Secretariat and, 162
notification requirements, fisheries subsidies restrictions, 114-15
nuclear weapons Conventions, fragmentation and, 8

observership
  MOU between FAO and CITES Secretariat and, 176-9, 188
  regime interaction and, 253
Oceana, as regime interaction stakeholder, 28
Oceans and Coastal Areas Network (UN-Oceans)
  law of the sea and, 54-5
  resource allocation and, 256

INDEX 359

OECD. *See* Organisation for Economic Cooperation and Development
on-board fish processing, trade restrictions on endangered fish species and, 142
ongoing determinacy, in WTO fisheries subsidies provisions, 101-3
openness
  lack of, as regime impediment, 261-2
  legal framework for regime interaction and, 290
'optimum objective', exclusive economic zone regime and, 36
'optimum utilization' of fish stocks, regional regimes and, 39
'ordinary meaning' of treaty terms
  IGO consultations on, 217-18, 233n.289, 233n.291
  parallel membership and, 232-5
  regime interaction and, 31
organisational structure, parallel membership and, 229-39
Organisation for Economic Cooperation and Development (OECD), 259, 282
  Export Credit Arrangement, 122-3, 293
  fisheries subsidies restrictions and, 88n.7, 105, 291
  conditionality exemptions, 121-3
Oslo and Paris Convention (OSPAR), 231
overfishing
  fisheries subsidies and, 87-9
  subsidies for ending of, 6
overlapping regimes. *See also* regime exclusivity; regime interaction
  parallel membership and, 262-6
  shrimp import ban consultations and, 215-20
  trade restrictions on fish species, 151-3

pacta sunt servanda principle, regime interaction and, 21
pacta tertius rule, Fish Stocks Agreement, 43
Pakistan, US shrimp import ban, complaints about, 193-5, 199
parallel membership
  legal framework for regime interaction and, 268-9, 289
  non-fisheries governance models and, 299-302
  regime interaction promotion through, 262-6
  shrimp import ban, WTO adjudication of, relevant rules, need for, 201-4
  of treaties and organisations, 229-39
  VCLT Article 31(1) and, 232-5
  VCLT Article 31(3)(c) and, 229-32, 269

Paris Convention for the Protection of Industrial Property, 217
parties' submissions, shrimp import ban, WTO adjudication and role of, 209-11
Patagonian toothfish, information-sharing and observership practices involving, 177-9
peer review
  fisheries subsidies conditionality and, 119-24
  regime legitimacy and accountability and, 279
peremptory norms
  fragmentation of international law and, 13
  ILC Fragmentation Study recommendations and, 299
Perez, Oran, 103-4, 225
Permanent Group of Experts (PGEs), fisheries subsidies restrictions and, dispute resolution, 124-9, 132
'plain meaning' doctrine, 233n.289
policy adaptation
  CITES species listing criteria, 181
  fisheries subsidies and, 99-101
  MOU between CITES Secretariat and FAO, 174-6, 181
  trade restrictions on endangered marine species, 147-8
policy space issue, regime interaction and, 252
Port State Measures Agreement
  dispute settlement procedures, 45
  'illegal, unreported and unregulated' (IUU) fishing and, 43
  marine species protection and, 138n.18
  national policy coordination and, 249
  regime interdependency and, 51
primacy of international law principle, regime interaction and, 23
public choice research
  fisheries subsidies and role of, 89
  stakeholders in sustainable fishing and, 27

red box of fisheries subsidies, 91-3, 115-19, 293-4
regime exclusivity
  forum shopping and, 244-9
  impediments to regime interaction and, 258-66
  self-contained regimes, challenges to, 245-6
regime interaction. *See also* inter-regime learning; overlapping regimes
  adjudicator selection and, 238

regime interaction (cont.)
  appropriate practice of, 288-98
  case-study research on, 17-19
  CITES activities and, 56-61
  Convention on Biological Diversity and,
    63-5
  Convention on Migratory Species and,
    61-3
  dispute settlement and, 295-8
  fisheries subsidies and, 30, 285
    dispute resolution, 127-9
    entrenchment of, 113-24
    forum shopping transition to, 103-4
    importance of, 129-33
    restrictions, 89-104
  fisheries sustainability laws, 85
  fragmentation of international law,
    243-66
  framing by parties in, 227-9
  future applications of, 299-302
  governance models and, 266
  impediments to, 258-66
    forum exclusivity, 258-61
    lack of transparency and openness,
      261-2
    parallel membership, 262-6, 268-9
  implementation of existing
    commitments, 292-5
  information-sharing and institutional
    learning and, 253-6
  international adjudication and
    interdependency, 30-1
  law-making and, 289-92
  law of the sea and, 65-7
  legal framework for. *See* legal framework
    for regime interaction
  legitimacy issues in, 248, 271-8
    consent and sovereignty, 271, 276
    exogenous sources usage and, 235-9
    express and implied powers, 272
    risk of managerialism and, 276-8
  MOU between FAO and CITES Secretariat
    and, 155, 176-84, 187
  multiple bases for, 267-71
  policy coordination and, 249-53
  promotion of, 249-58
  resource allocation and, 256-8
  risk of managerialism and, 276-8, 302-4
  shrimp import ban, WTO adjudication
    and role of, 195-224
    applicable law issues, 195-7
    dispute settlement methods, 206-24
    panellists, AB members, and
      Secretariat, role of, 206-9
    scientific consultations as promotion
      of, 211-15
  stakeholders in, 24-9

  state consent to, 31
  trade restrictions on marine species and,
    134-88
  UN consultative process and, 53-5
  vocabulary of, 19-22
regimes of relevance and forum shopping
  fisheries subsidies restrictions and,
    89-104
  trade restrictions on fish species and,
    138-54
  WTO adjudication of shrimp import ban
    and, 190-5
regional fishing management
    organisations (RFMOs)
  Fish Stocks Agreement and, 41-6
  fisheries subsidies and, 97, 102, 118,
    129-33, 258
  illegal fishing classifications, 116
  international trade law and, 69-70
  MOU between FAO and CITES Secretariat
    and, 159-60, 177, 183
  trade restrictions on endangered marine
    species, forum shopping and,
    138-54
  UNCLOS 'high seas' provisions and,
    38-46
  UNCLOS law of the sea framework
    and, 35
regulatory commons concept, legal
    framework for regime interaction
    and, 300
relevant facts principle, US shrimp
    import ban, WTO adjudication of,
    204-6
reporting procedures, MOU between FAO
    and CITES Secretariat, 184
Republic of Korea. *See* Korea
Resolution on Assistance to Developing
    Countries, 189, 199
resource allocation
  MOU between FAO and
    CITES Secretariat, 184
  regime interaction and, 256-8
RFMOs. *See* regional fishing management
    organisations
Rio Declaration
  Agenda 21 and, 64
  shrimp import ban, WTO adjudication
    of, impact on, 201
  UNCED or Earth Summit, 63
Rules Group, WTO, 94-7, 105-13
  CTE relations with, 111
  draft text for fisheries subsidies
    restrictions, 307-15
  fisheries subsidies reform mandate of,
    86, 94-6, 97, 101, 106, 113, 114,
    122, 293

inaccessibility during negotiations of, 86, 109–10, 113, 131, 254, 261, 285, 289
information dissemination by, 111
non-WTO members, 108–10
regime collaboration with, 105–10, 117–18, 121–3, 127, 132, 263
state institutions' scepticism concerning, 98, 263
UNEP and, 111–13
rules of procedure
MOU between CITES and FAO and abuses of, 172–4, 188
regime legitimacy and accountability and, 280
shrimp import ban, WTO adjudication of, parallel membership of relevant rules, need for, 201–4
table of, xxi–xxviii

Sabel, Charles, 290
Schermers, Henry, 177
Scheuerman, William, 304
scientific experts' consultations. *See also* expertise of regimes
CITES species listing criteria and, 179–81
EC – *Hormones* case and, 213–14
information-sharing and institutional learning and, 255
institutional bias and insensitivities, 148–51
shrimp import ban, WTO adjudication and role of, 211–15
trade restrictions on endangered marine species and reliance on, 143–4, 148–51
SCM Agreement. *See* Agreement on Subsidies and Countervailing Measures
Scott, Joanne, 123
seahorses, endangered species classification, 136, 140
sea turtle conservation
Appellate Body (WTO) on. *See under* Appellate Body
endangered listing, multilateral agreement concerning, 191
regime interaction on status of, 204
scientific experts' consultations concerning, 211–15
shrimp import ban in US and, 7, 18, 30–1, 189, 190, 191–3
turtle excluder devices, 192, 194, 196, 200
secretariats of IGOs
consultations with, WTO fisheries import ban adjudication, 215–20

regime interaction-based dispute settlement, 206–9
as stakeholders, 28, 215–20
self-contained regimes, challenges to, 245–6
self-restraint principles, WTO fisheries subsidies and, 99–101
sharks
basking shark, CITES Appendices listing for, 140
China, shark fin demand in, 135–6
CITES species listing and, 140
endangered classification for, 135–6, 140
great whites, 140
international trade increase and depletion of, 5, 135, 136
IPOA-Sharks
implementation difficulties for, 185
trade restrictions on endangered marine species, 140
whale shark
CITES Appendices listing for, 135n.2, 140
ecotourism and endangerment of, 137, 150
shrimp fisheries
import ban, WTO adjudication of. *See under* imports of fish and fishery products
marine by-catch and, 190
production and export statistics, 190n.4
SOCA (UN Subcommittee on Oceans and Coastal Areas), 53
SOFIA *(FAO State of the World Fisheries Report)*, 47
South Korea. *See* Korea
Southern Bluefin Tuna agreement, UNCLOS and EEZ regimes and, 44–5
sovereignty issues
fragmentation of international law and, 10
legal framework for regime interaction and, 23
legitimacy of regime interaction and role of, 271, 276
trade restrictions on marine species and, 134–88
special interest groups, stakeholders in sustainable fishing and, 27
specialty principle, as impediment to regime interaction, 259
species Appendices, CITES. *See under* endangered species classification
species-specific fish management, regimes for, 40 *See also* endangered species classification

362   INDEX

split-listing of endangered marine species, 146, 148, 153
SPS Agreement. *See* Agreement on the Application of Sanitary and Phytosanitary Measures
Staiger, Robert, 99
stakeholders in regime interaction, 31
  accountability of, 279
  identification of, 24-9
  information-sharing and institutional learning and, 253
  inter-regime learning and, 111-13
  law-making activities, involvement in, 289-92
  national policy coordination and, 250-1
  resource allocation by, 257
  secretariats of IGOs, 28, 215-20
standard-setting bodies, as regime interaction stakeholder, 28
state institutions
  CITES species listing criteria and, 58-60, 179-81
  exclusive economic zone regime and, 34-8
  fisheries subsidies and, 89, 107
  fragmentation of, 8
  inter-regime learning and, 107
  international trade disputes and, 3, 69
  legal framework for regime interaction and
    future non-fisheries law applications, 301
    law-making activities, 289-92
    situating of participants and, 304-6
  parallel membership and, 262-6
  regime exclusivity preferences of, 244-9
  regime interaction and, 31
  regional fish stock regimes and, 38-46
  sovereignty in international law and, 23
  as stakeholders in sustainable fishing, 22-4
  trade restrictions on endangered marine species, 138, 187
  UNCLOS regime and, 33
  WTO Rules Group and, 263
straddling fish stocks, 39
Subcommittee on Oceans and Coastal Areas (SOCA), UN, 53
successor undertaking, OECD Export Credit Arrangement, 122
surveillance regimes, fisheries subsidies restrictions and lack of, 92
sustainable fishing
  international trade law and, 5-8
  stakeholders in, 24-9
  subsidies for, 6
  trade restrictions on endangered marine species and, 150

Switzerland, national policy coordination with global initiatives in, 251
systemic integration principle, 11n.49
systems theory, conflicting norms of fishing laws and, 15

Taiwan
  acceptance of WTO fish subsidy restrictions, 101
  fisheries subsidies in, 88, 96, 269
  non-WTO fisheries management advocacy, 106
  statistics on fisheries subsidies, 88n.10
  tariff measures, fisheries subsidies and, 99, 101
TBT. *See* Agreement on Technical Barriers to Trade
technical guidelines, fisheries subsidies conditionality and, 119
TEDs (turtle excluder devices), 192, 194, 196, 200
textual mandate
  CITES trade restrictions on fish species and, 146-54
  legal framework for regime interaction and, 247
  WTO fisheries subsidies and, 101
Thai Cigarettes case, 216
Thailand
  cigarette import ban in, 216
  US shrimp import ban, complaints about, 193-5, 199
third-party states
  dispute settlements of UNCLOS and EEZ regimes and, 45
  regional fish management organisations circumvention by, 41
  shrimp import ban, WTO adjudication and submissions by, 200n.68, 209-11
Tietje, Christian, 285
TPRM (Trade Policy Review Mechanism), fisheries subsidies, conditionality peer review, 121
trade distorting subsidies, fisheries subsidies restrictions and, 94-6, 97
trade law, international. *See* international trade law
Trade Policy Review Mechanism (TPRM), fisheries subsidies, conditionality peer review, 121
trade regime, conflicting norms and, 14-15
trade restrictions on fish species, 6-8
  CITES and. *See under* Convention on the International Trade in

INDEX   363

Endangered Species of Wild Flora and Fauna
CMS and. *See under* Convention on the Conservation of Migratory Species of Wild Animals
COP and. *See under* Conference of the Parties
developing countries and, 150
EEZs and. *See under* exclusive economic zones
expertise of regimes and. *See under* expertise of regimes
Fish Stocks Agreement and. *See under* Fish Stocks Agreement
flags of states and. *See under* flags of states
forum shopping and, 138–54
marine species, endangered, 134–88
   CITES expertise questioned, 142–3
   CITES marine species Appendices listing, 139–41
   coherence in management of, 144–6
   forum shopping and, 138–54
   opposition to CITES oversight, 141–6
   regime interaction on, 153–4
regime interaction on, 30, 153–4
regime legitimacy and accountability and, 279
regime overlap concerning, 151–3
UNCLOS. *See under* UN Convention on the Law of the Sea
'tragedy of the commons', exclusive economic zone regime and, 36
transparency, lack of, as regime impediment, 261–2
treaty interpretation. *See also* Vienna Convention on the Law of Treaties
legitimacy issues in, 245
MOU between CITES Secretariat and FAO and, 171–2
mutual agreement principles and, 269
ordinary meaning. *See* 'ordinary meaning' of treaty terms
parallel membership and, 229–39, 265–6, 269
'plain meaning' doctrine, 233n.289
shrimp import ban, WTO adjudication and regime interaction involving, 197–204
TRIPS (Agreement on Trade-Related Aspects of Intellectual Property), 100
turtle excluder devices (TEDs), 192, 194, 196, 200

ultra vires principle, MOU between CITES Secretariat and FAO and, 169, 188

UN Charter, fragmentation of international law and principles of, 12
UN Conference on Environment and Development (UNCED, or Earth Summit), 63
UN Conference on Trade and Development (UNCTAD)
   disputed norms consultations with, 219
   mutual agreement principles and, 269
UN Consultative Process
   law of the sea and, 53-5
   MOU between FAO and CITES Secretariat, 155
   resource allocation and, 256
UN Convention on the Law of the Sea (UNCLOS), 3
   express and implied powers and, 274
   FAO-CITES Secretariat MOU and, role in, 159
   fisheries subsidies restrictions and, 6, 90, 97, 129-33
      dispute resolution and, 124n.222, 126, 247
   historical overview, 32-3
   institutional framework of, 29
   law of the sea and regime of, 34-46
   mandatory dispute settlement system, 37-8
   regime interdependency and, 50-3
   regional regimes and, 38-46
   shrimp import ban, WTO adjudication of impact on, 199, 239-40
   scientific experts' consultations and, 212
   submissions by, 209
   state institutions and creation of, 33
   trade restrictions on endangered marine species
      CITES affirmation of jurisdiction, 146-7
      coherence in management of, 144
      forum shopping and, 138-54
      policy adaptation and, 147-8
      regime overlap vs. normative conflict in, 151-3
      scientific bias allegations against, 143
   WTO interaction with, 77
UN Division for Ocean Affairs and the Law of the Sea (DOALOS), 29, 38
UN Environment Programme (UNEP)
   CITES interaction with, 57
   Convention on Migratory Species and, 62
   fisheries subsidies restrictions and inter-regime learning and, 105, 131
   non-UN participants in, 111-13
   governance role of, 257

364   INDEX

UN Environment Programme (cont.)
  information-sharing and institutional learning and, 253
  institutional arrangements in work of, 271
  law of the sea and, 67–8
  national policy coordination encouraged by, 250
  participating states as stakeholders in, 28
UN FAO. *See* Food and Agriculture Organization
UN, legitimacy issues for regime interaction and role of, 272–3, 276, 282
UN-Oceans. *See* Oceans and Coastal Areas Network
UN Reparations case
  duty to take account of others in, 286
  express and implied powers in, 272
UN Subcommittee on Oceans and Coastal Areas (SOCA), 53
Understanding on Rules and Procedures Governing the Settlement of Disputes (DSU)
  adjudicator selection criteria and, 226
  amicus curiae briefs and, 220–4
  US shrimp import ban, complaints about, 194–5
    applicable law principle, 195–7
    information-sharing and institutional learning in, 255
    regime interaction-based dispute settlement and, 206–9
    relevant facts principle, 204–6
    scientific experts' consultations, 211–15
unilateral trade regulations, fishing sustainability and, 7–8
United States
  EC – Biotech case and, 202–4
  EC – Hormones case and, 213–14
  fisheries subsidies in, 88, 90n.22
    notification requirements, 115
    statistics on, 88n.10
  IPOA-Capacity plans submitted by, 90
  MOU between CITES Secretariat and FAO, policy coordination with, 175
  national policy coordination with global initiatives in, 250, 290
  shrimp import ban in
    DSU and. *See under* Understanding on Rules and Procedures Governing the Settlement of Disputes
    historical evolution of, 191–3
    marine by-catch and, 190–5
    sea turtle conservation and, 7, 18, 30–1, 189

  US – Shrimp case. *See US – Shrimp* case
  WTO adjudication of, 189–240
Uruguay Round
  Committee on Trade and Environment established in, 80–1
  international trade agreements and, 70
  shrimp import ban, WTO adjudication of, impact on, 201
US – Shrimp case, 18, 30–1, 71
  absence of *amicus curiae* briefs in, 221–4
  adjudicator selection criteria and, 227
  IGO secretariat consultations in, 215–20
  information-sharing and institutional learning in, 255
  institutional arrangements in, 270
  legitimacy issues in
    duty to take account of others, 285–6
    use of exogenous sources, 235–9
  parallel membership of relevant rules, need for, 201–4
  regime interaction dispute settlement methods in, 206–24
  scientific experts' consultations and, 211–15
  treaty interpretations and, 198–200
  WTO adjudication of, regime interaction and, 195–224
    party submissions, 209–11
    relevant facts principle, 204–6
US – Tuna (Canada) case, 199
Use of Nuclear Weapons case, 259, 273

Valles Games, Guillermo, 95
Vienna Convention on the Law of Treaties (VCLT)
  Article 30, trade restrictions on endangered marine species, 152
  Article 31(1)
    EC – Biotech case and, 211
    parallel membership issues and, 232–5
  Article 31(3)(c)
    EC – Biotech case and, 202, 277
    fragmentation of international law and, 11
    parallel membership issues and, 229–32, 269
    US shrimp import ban, complaints about, 199, 202–3
  evolutionary interpretations of, 297
  inter se modification and, 50n.133, 171
  legitimacy issues and exogenous sources usage, 238
  parallel membership issues and, 266
  shrimp import ban, WTO adjudication, interpretations from, 197–8

voluntary conservation measures, trade restrictions on endangered marine species, 138

Weeramantry, C G (Judge), 259, 273
whale shark
  CITES Appendices listing for, 135n.2, 140
  ecotourism and endangerment of, 137, 150
White Paper on European Governance, 251
WHO. *See* World Health Organization
WIPO (World Intellectual Property Organization), 283
World Bank, WTO legitimacy and, 272
World Commission on Environment and Development, 199
World Environmental Court, proposal for, 14, 247
World Health Organization (WHO)
  competence challenges to, 259, 273
  Framework Convention on Tobacco Control, 283
  JECFA, 214
  Thai cigarette import ban and, 216
World Intellectual Property Organization (WIPO), 283
World Summit on Sustainable Development, fisheries subsidies restrictions and, 90, 97
World Trade Organization (WTO)
  Appellate Body. *See* Appellate Body
  Cairns group, dispute resolution and, 126
  CTE. *See* Committee on Trade and Environment
  Declaration on Coherence in Global Economic Policy Making, 269
  Director-General duties, 226
  dispute settlement, 70-5, 85, 269
  Doha Round. *See* Doha Round negotiations
  dolphin-safe labelling issue and, 7
  DSB, adjudication of US shrimp import ban, establishment of, 52-3
  EC – Biotech case and, 202-4, 296-7
  FAO Code of Conduct and, 52-3
  FAO fisheries subsidies mandate and, 96-7
  fisheries import bans, adjudication of, 18, 189-240
    adjudicator selection, 226-7
    amicus curiae briefs from NGOs and others, 220-4
    applicable law, 195-7
    complaints against US ban, 193-5
    conclusions concerning, 239-40
    exogenous sources, legitimacy and guidelines for, 235-9
    forum shopping, relevant regimes, 190-5
    framing by parties, 227-9
    GATT Article XX chapeau interpretation, 200-1
    GATT Article XX interpretation, 198-200
    IGO secretariats' consultations, 215-20
    marine by-catch and shrimp fisheries, 190
    panellists, AB members, and Secretariat, role in dispute settlement, 206-9
    parallel membership of relevant rules, 201-4
    parallel membership of treaties and organisations, 229-39
    parties' submissions, 209-11
    problems and challenges in, 224-39
    regime interaction in dispute settlement, 195-224
    relevant facts, 204-6
    scientific experts' consultations, 211-15
    shrimp. *See under* imports of fish and fishery products
    treaty interpretation, 197-204
    US shrimp products ban, impact of, 191-3
    VCLT Article 31(1) provisions, 232-5
    VCLT Article 31(3)(c) provisions and, 229-32
  fisheries subsidies provisions. *See under* fisheries subsidies
  historical overview, 32
  institutional analysis of, 16-19
  international environmental law and, 149
  international trade disputes and, 69, 70-5
  lack of transparency and openness in, 261
  legal framework for regime interaction and role of, law-making activities and, 289-92
  legitimacy issues for, 9, 248, 260
    accountability issues, 278-84
    adjudication, 235-9
    duties to take account of others, 284-5
    express and implied powers and, 272
    sovereignty and, 271
  MEAs and, 78-9
  multilateral fishing regulations negotiations, 17

World Trade Organization (cont.)
  national policy coordination with negotiations of, 249
  non-agricultural market access negotiations, 75n.297
  non-fisheries governance models and, 299-302
  observership practices, 177
  OECD Export Credit Arrangement and, 122-3
  parallel membership and role of, 262-6, 268
  participating states as stakeholders in, 28
  policy influences of, 246
  regime exclusivity and, 244-9
  regime interaction with, 75-9
    forum shopping transition to, 103-4, 154
    resource allocation and, 256
  Rules Group. *See* Rules Group, WTO
  SCM Agreement and, 6, 91-3, 121-3, 291, 294-5
  sea turtle conservation and shrimp import ban, 18, 30-1, 189-240
  as self-contained regime, 245-6
  shrimp import ban. *See under* imports of fish and fishery products
  stakeholder consultations with, 251
  trade liberalisation and, 3
  Trade Negotiations Committee, 74
  trade policy coherence and effectiveness and, 98-9
  trade restrictions on endangered fish species and, concerns over CITES mandate, 147
World Wide Fund for Nature (WWF)
  fisheries subsidies and, 131
    dispute resolution recommendations, 126-7
    forum shopping transition to regime interaction, 103-4
    multilateral fishing subsidies regulations and, 18
  resource allocation by, 257
  as stakeholder in regime interaction, 28
  UNEP and, 112
WTO. *See* World Trade Organization
WWF. *See* World Wide Fund for Nature

CAMBRIDGE STUDIES IN INTERNATIONAL AND COMPARATIVE LAW

Books in the series

Trading Fish, Saving Fish: The Interaction between Regimes in International Law
Margaret A. Young

The Individual in the International Legal System: State-Centrism, History and Change in International Law
Kate Parlett

The Participation of States in International Organisations: The Role of Human Rights and Democracy
Alison Duxbury

'Armed Attack' and Article 51 of the UN Charter: Evolutions in Customary Law and Practice
Tom Ruys

Science and Risk Regulation in International Law: The Role of Science, Uncertainty and Values
Jacqueline Peel

Theatre of the Rule of Law: The Theory, History and Practice of Transnational Legal Intervention
Stephen Humphreys

The Public International Law Theory of Hans Kelsen: Believing in Universal Law
Jochen von Bernstorff

Vicarious Liability in Tort: A Comparative Perspective
Paula Giliker

Legal Personality in International Law
Roland Portmann

Legitimacy and Legality in International Law: An Interactional Account
Jutta Brunnée and Stephen J. Toope

The Concept of Non-International Armed Conflict in International Humanitarian Law
Anthony Cullen

The Challenge of Child Labour in International Law
Franziska Humbert

Shipping Interdiction and the Law of the Sea
Douglas Guilfoyle

International Courts and Environmental Protection
Tim Stephens

Legal Principles in WTO Disputes
Andrew D. Mitchell

War Crimes in Internal Armed Conflicts
Eve La Haye

Humanitarian Occupation
Gregory H. Fox

The International Law of Environmental Impact Assessment: Process, Substance and Integration
Neil Craik

The Law and Practice of International Territorial Administration: Versailles, Iraq and Beyond
Carsten Stahn

Cultural Products and the World Trade Organization
Tania Voon

United Nations Sanctions and the Rule of Law
Jeremy Farrall

National Law in WTO Law: Effectiveness and Good Governance in the World Trading System
Sharif Bhuiyan

The Threat of Force in International Law
Nikolas Stürchler

Indigenous Rights and United Nations Standards
Alexandra Xanthaki

International Refugee Law and Socio-Economic Rights
Michelle Foster

The Protection of Cultural Property in Armed Conflict
Roger O'Keefe

Interpretation and Revision of International Boundary Decisions
Kaiyan Homi Kaikobad

Multinationals and Corporate Social Responsibility: Limitations and Opportunities in International Law
Jennifer A. Zerk

Judiciaries within Europe: A Comparative Review
John Bell

Law in Times of Crisis: Emergency Powers in Theory and Practice
Oren Gross and Fionnuala Ní Aoláin

Vessel-Source Marine Pollution: The Law and Politics of International Regulation
Alan Tan

Enforcing Obligations Erga Omnes *in International Law*
Christian J. Tams

Non-Governmental Organisations in International Law
Anna-Karin Lindblom

Democracy, Minorities and International Law
Steven Wheatley

Prosecuting International Crimes: Selectivity and the International Law Regime
Robert Cryer

Compensation for Personal Injury in English, German and Italian Law: A Comparative Outline
Basil Markesinis, Michael Coester, Guido Alpa, Augustus Ullstein

Dispute Settlement in the UN Convention on the Law of the Sea
Natalie Klein

The International Protection of Internally Displaced Persons
Catherine Phuong

Imperialism, Sovereignty and the Making of International Law
Antony Anghie

Necessity, Proportionality and the Use of Force by States
Judith Gardam

International Legal Argument in the Permanent Court of International Justice: The Rise of the International Judiciary
Ole Spiermann

Great Powers and Outlaw States: Unequal Sovereigns in the International Legal Order
Gerry Simpson

Local Remedies in International Law
C. F. Amerasinghe

Reading Humanitarian Intervention: Human Rights and the Use of Force in International Law
Anne Orford

Conflict of Norms in Public International Law: How WTO Law Relates to Other Rules of Law
Joost Pauwelyn

Transboundary Damage in International Law
Hanqin Xue

European Criminal Procedures
Edited by Mireille Delmas-Marty and John Spencer

The Accountability of Armed Opposition Groups in International Law
Liesbeth Zegveld

Sharing Transboundary Resources: International Law and Optimal Resource Use
Eyal Benvenisti

International Human Rights and Humanitarian Law
René Provost

Remedies Against International Organisations
Karel Wellens

Diversity and Self-Determination in International Law
Karen Knop

The Law of Internal Armed Conflict
Lindsay Moir

International Commercial Arbitration and African States: Practice, Participation and Institutional Development
Amazu A. Asouzu

The Enforceability of Promises in European Contract Law
James Gordley

International Law in Antiquity
David J. Bederman

Money Laundering: A New International Law Enforcement Model
Guy Stessens

Good Faith in European Contract Law
Reinhard Zimmermann and Simon Whittaker

On Civil Procedure
J. A. Jolowicz

Trusts: A Comparative Study
Maurizio Lupoi

The Right to Property in Commonwealth Constitutions
Tom Allen

International Organizations Before National Courts
August Reinisch

The Changing International Law of High Seas Fisheries
Francisco Orrego Vicuña

Trade and the Environment: A Comparative Study of EC and US Law
Damien Geradin

Unjust Enrichment: A Study of Private Law and Public Values
Hanoch Dagan

Religious Liberty and International Law in Europe
Malcolm D. Evans

Ethics and Authority in International Law
Alfred P. Rubin

Sovereignty Over Natural Resources: Balancing Rights and Duties
Nico Schrijver

The Polar Regions and the Development of International Law
Donald R. Rothwell

Fragmentation and the International Relations of Micro-States: Self-determination and Statehood
Jorri Duursma

Principles of the Institutional Law of International Organizations
C. F. Amerasinghe